Direct Energy Conversion

by Andrea M. Mitofsky

Copyright Information

© by Andrea M. Mitofsky 2018.

Direct Energy Conversion by Andrea M. Mitofsky is licensed under the Creative Commons Attribution-NonCommercial 4.0 International License. To view a copy of this license, visit http://creativecommons.org/licenses/by-nc/4.0/.

This is version 1.0.0 of the text.

The copyright license does not apply to the following figures: Fig. 2.10, Fig. 6.6, Fig. 6.14, Fig. 6.15, Fig. 6.17, Fig. 7.1, Fig. 7.3, Fig. 7.7, Fig. 7.8, Fig. 8.4, Fig. 8.5, Fig. 8.6, Fig. 9.8, and Fig. 9.10. These figures are used, with permission from other authors. The caption for each figure cites the source.

This textbook can be cited as:
A. M. Mitofsky, "Direct Energy Conversion", 2018.

This textbook can be downloaded from:
https://www.trine.edu/books/directenergy.aspx/

Andrea Mitofsky can be contacted at mitofskya@gmail.com

Contents

Contents i

1 Introduction 1
1.1 What is Direct Energy Conversion? 1
1.2 Preview of Topics 2
1.3 Conservation of Energy 9
1.4 Measures of Power and Energy 10
1.5 Properties of Materials 12
    1.5.1 Macroscopic Properties 12
    1.5.2 Microscopic Properties 13
1.6 Electromagnetic Waves 15
    1.6.1 Maxwell's Equations 15
    1.6.2 Electromagnetic Waves in Free Space 17
    1.6.3 Electromagnetic Waves in Materials 18
1.7 Problems 21

I Survey of Energy Conversion Devices 23

2 Capacitors and Piezoelectric Devices 23
2.1 Introduction 23
2.2 Capacitors 24
    2.2.1 Material Polarization 24
    2.2.2 Energy Storage in Capacitors 24
    2.2.3 Permittivity and Related Measures 26
    2.2.4 Capacitor Properties 28
2.3 Piezoelectric Devices 31
    2.3.1 Piezoelectric Strain Constant 32
    2.3.2 Piezoelectricity in Crystalline Materials 33
    2.3.3 Piezoelectricity in Amorphous and Polycrystalline Materials and Ferroelectricity 41
    2.3.4 Materials Used to Make Piezoelectric Devices .. 43
    2.3.5 Applications of Piezoelectricity 44
2.4 Problems 47

3 Pyroelectrics and Electro-Optics 53
3.1 Introduction 53
3.2 Pyroelectricity 53
    3.2.1 Pyroelectricity in Crystalline Materials 53

　　　　3.2.2　Pyroelectricity in Amorphous and Polycrystalline Materials and Ferroelectricity 55
　　　　3.2.3　Materials and Applications of Pyroelectric Devices . 55
　　3.3　Electro-Optics 56
　　　　3.3.1　Electro-Optic Coefficients 56
　　　　3.3.2　Electro-Optic Effect in Crystalline Materials 59
　　　　3.3.3　Electro-Optic Effect in Amorphous and Polycrystalline Materials 60
　　　　3.3.4　Applications of Electro-Optics 60
　　3.4　Notation Quagmire 61
　　3.5　Problems 64

4　Antennas　67
　　4.1　Introduction 67
　　4.2　Electromagnetic Radiation 69
　　　　4.2.1　Superposition 69
　　　　4.2.2　Reciprocity 69
　　　　4.2.3　Near Field and Far Field 72
　　　　4.2.4　Environmental Effects on Antennas 72
　　4.3　Antenna Components and Definitions 73
　　4.4　Antenna Characteristics 75
　　　　4.4.1　Frequency and Bandwidth 75
　　　　4.4.2　Impedance 77
　　　　4.4.3　Directivity 78
　　　　4.4.4　Electromagnetic Polarization 82
　　　　4.4.5　Other Antenna Considerations 85
　　4.5　Problems 86

5　Hall Effect　91
　　5.1　Introduction 91
　　5.2　Physics of the Hall Effect 91
　　5.3　Magnetohydrodynamics 96
　　5.4　Quantum Hall Effect 97
　　5.5　Applications of Hall Effect Devices 98
　　5.6　Problems 100

6　Photovoltaics　101
　　6.1　Introduction 101
　　6.2　The Wave and Particle Natures of Light 101
　　6.3　Semiconductors and Energy Level Diagrams 104
　　　　6.3.1　Semiconductor Definitions 104

		6.3.2	Energy Levels in Isolated Atoms and in Semiconductors	106
		6.3.3	Definitions of Conductors, Dielectrics, and Semiconductors	114
		6.3.4	Why Are Solar Cells and Photodetectors Made from Semiconductors?	115
		6.3.5	Electron Energy Distribution	117
	6.4	Crystallography Revisited		119
		6.4.1	Real Space and Reciprocal Space	119
		6.4.2	E versus k Diagrams	120
	6.5	Pn Junctions		122
	6.6	Solar Cells		127
		6.6.1	Solar Cell Efficiency	127
		6.6.2	Solar Cell Technologies	129
		6.6.3	Solar Cell Systems	130
	6.7	Photodetectors		132
		6.7.1	Types of Photodetectors	132
		6.7.2	Measures of Photodetectors	134
	6.8	Problems		136

7 Lamps, LEDs, and Lasers — 139

7.1	Introduction		139
7.2	Absorption, Spontaneous Emission, Stimulated Emission		139
	7.2.1	Absorption	139
	7.2.2	Spontaneous Emission	140
	7.2.3	Stimulated Emission	142
	7.2.4	Rate Equations and Einstein Coefficients	143
7.3	Devices Involving Spontaneous Emission		147
	7.3.1	Incandescent Lamps	147
	7.3.2	Gas Discharge Lamps	148
	7.3.3	LEDs	150
7.4	Devices Involving Stimulated Emission		152
	7.4.1	Introduction	152
	7.4.2	Laser Components	153
	7.4.3	Laser Efficiency	157
	7.4.4	Laser Bandwidth	159
	7.4.5	Laser Types	161
	7.4.6	Optical Amplifiers	165
7.5	Relationship Between Devices		165
7.6	Problems		169

8 Thermoelectrics — 173
- 8.1 Introduction — 173
- 8.2 Thermodynamic Properties — 173
- 8.3 Bulk Modulus and Related Measures — 176
- 8.4 Ideal Gas Law — 178
- 8.5 First Law of Thermodynamics — 179
- 8.6 Thermoelectric Effects — 180
  - 8.6.1 Three Related Effects — 180
  - 8.6.2 Electrical Conductivity — 183
  - 8.6.3 Thermal Conductivity — 184
  - 8.6.4 Figure of Merit — 186
- 8.7 Thermoelectric Efficiency — 188
  - 8.7.1 Carnot Efficiency — 188
  - 8.7.2 Other Factors That Affect Efficiency — 191
- 8.8 Applications of Thermoelectrics — 192
- 8.9 Problems — 195

9 Batteries and Fuel Cells — 201
- 9.1 Introduction — 201
- 9.2 Measures of the Ability of Charges to Flow — 202
  - 9.2.1 Electrical Conductivity, Fermi Energy Level, and Energy Gap Revisited — 203
  - 9.2.2 Mulliken Electronegativity — 204
  - 9.2.3 Chemical Potential and Electronegativity — 205
  - 9.2.4 Chemical Hardness — 207
  - 9.2.5 Redox Potential — 207
  - 9.2.6 pH — 208
- 9.3 Charge Flow in Batteries and Fuel Cells — 211
  - 9.3.1 Battery Components — 211
  - 9.3.2 Charge Flow in a Discharging Battery — 212
  - 9.3.3 Charge Flow in a Charging Battery — 213
  - 9.3.4 Charge Flow in Fuel Cells — 214
- 9.4 Measures of Batteries and Fuel Cells — 216
  - 9.4.1 Cell Voltage, Specific Energy, and Related Measures — 216
  - 9.4.2 Practical Voltage and Efficiency — 219
- 9.5 Battery Types — 223
  - 9.5.1 Battery Variety — 223
  - 9.5.2 Lead Acid — 225
  - 9.5.3 Alkaline — 226
  - 9.5.4 Nickel Metal Hydride — 227
  - 9.5.5 Lithium — 228

CONTENTS

9.6 Fuel Cells 229
    9.6.1 Components of Fuel Cells and Fuel Cell Systems . . . 229
    9.6.2 Types and Examples 231
    9.6.3 Practical Considerations of Fuel Cells 232
9.7 Problems 234

10 Miscellaneous Energy Conversion Devices   237
10.1 Introduction 237
10.2 Thermionic Devices 237
10.3 Radiation Detectors 237
10.4 Biological Energy Conversion 239
10.5 Resistive Sensors 239
10.6 Electrofluidics 241

II Theoretical Ideas   245

11 Calculus of Variations   245
11.1 Introduction 245
11.2 Lagrangian and Hamiltonian 245
11.3 Principle of Least Action 247
11.4 Derivation of the Euler-Lagrange Equation 249
11.5 Mass Spring Example 251
11.6 Capacitor Inductor Example 258
11.7 Schrödinger's Equation 262
11.8 Problems 263

12 Relating Energy Conversion Processes   269
12.1 Introduction 269
12.2 Electrical Energy Conversion 270
12.3 Mechanical Energy Conversion 276
12.4 Thermodynamic Energy Conversion 282
12.5 Chemical Energy Conversion 286
12.6 Problems 288

13 Thomas Fermi Analysis   291
13.1 Introduction 291
13.2 Preliminary Ideas 293
    13.2.1 Derivatives and Integrals of Vectors in Spherical Coordinates 293
    13.2.2 Notation 294
    13.2.3 Reciprocal Space Concepts 296

- 13.3 Derivation of the Lagrangian 296
- 13.4 Deriving the Thomas Fermi Equation 305
- 13.5 From Thomas Fermi Theory to Density Functional Theory . 308
- 13.6 Problems. 309

14 Lie Analysis 311
- 14.1 Introduction 311
  - 14.1.1 Assumptions and Notation 312
- 14.2 Types of Symmetries 313
  - 14.2.1 Discrete versus Continuous 313
  - 14.2.2 Regular versus Dynamical 314
  - 14.2.3 Geometrical versus Nongeometrical 314
- 14.3 Continuous Symmetries and Infinitesimal Generators 315
  - 14.3.1 Definition of Infinitesimal Generator 315
  - 14.3.2 Infinitesimal Generators of the Wave Equation 316
  - 14.3.3 Concepts of Group Theory 320
- 14.4 Derivation of the Infinitesimal Generators 322
  - 14.4.1 Procedure to Find Infinitesimal Generators 322
  - 14.4.2 Thomas Fermi Equation Example 323
  - 14.4.3 Line Equation Example 326
- 14.5 Invariants 330
  - 14.5.1 Importance of Invariants 330
  - 14.5.2 Noether's Theorem 330
  - 14.5.3 Derivation of Noether's Theorem 331
  - 14.5.4 Line Equation Invariants Example 333
  - 14.5.5 Pendulum Equation Invariants Example 334
- 14.6 Summary 336
- 14.7 Problems. 337

Appendices 341
- A. Variable List 341
- B. Select Units of Measure 349
- C. Overloaded Terminology 351
- D. Specific Energies 353

References 355

Index 370

About the Book 374

About the Author 374

Acknowledgements

I would like to thank many people who helped me complete this text. Each chapter was reviewed by two technical reviewers, all of whom have a Ph.D. degree in a relevant field. I appreciate the feedback I got from these reviewers. More specifically, I would like to thank my colleagues at Trine University who reviewed chapters including Brett Batson, Jamie Canino, Steve Carr, Maria Gerschutz, Ira Jones, Vicki Moravec, Chet Pinkham, Sameer Sharma, VK Sharma, Kendall Teichert, Deb Van Rie, and Kevin Woolverton. I would also like to thank my colleagues from other institutions who reviewed chapters including William D. Becher, Wei-Choon Ng and James Tian. I especially want to thank Eric Johnson for his helpful comments and critiques on the entire text, and I would like to thank Sue Radtke for proofreading the entire text. I appreciate the helpful feedback I received from students who took my Direct Energy Conversion course too.

1 Introduction

1.1 What is Direct Energy Conversion?

Energy conversion devices convert between electrical, magnetic, kinetic, potential, optical, chemical, nuclear, and other forms of energy. Energy conversion processes occur naturally. For example, energy is converted from optical electromagnetic radiation to heat when sunlight warms a house, and energy is converted from potential energy to kinetic energy when a leaf falls from a tree. Alternatively, energy conversion devices are designed and manufactured by a wide range of scientists and engineers. These energy conversion devices range from tiny integrated circuit components such as thermocouples which are used to sense temperature by converting microwatts of power from thermal energy to electricity to enormous coal power plants which convert gigawatts of energy stored in the chemical bonds of coal into electricity.

A *direct* energy conversion device converts one form of energy to another through a single process. For example, a solar cell is a direct energy conversion device that converts optical electromagnetic radiation to electricity. While some of the sunlight that falls on a solar cell may heat it up instead, that effect is not fundamental to the solar cell operation. Alternatively, *indirect* energy conversion devices involve a series of direct energy conversion processes. For example, some solar power plants involve converting optical electromagnetic radiation to electricity by heating a fluid so that it evaporates. The evaporation and expansion of the gas spin a rotor of a turbine. The energy from the mechanical motion of the rotor is converted to a time varying magnetic field which is then converted to an alternating electrical current in the coils of the generator.

This text focuses on direct energy conversion devices which convert between electrical energy and another form. Because of the wide variety of devices that fit in this category, energy conversion is a topic important to all types of electrical engineers. Some electrical engineers specialize in building instrumentation systems. Many sensors used by these engineers are direct energy conversion devices, including strain gauges used to measure pressure, Hall effect sensors which measure magnetic field, and piezoelectric sensors used to detect mechanical vibrations. The electrical energy produced in a sensor may be so small that amplification is required. Other electrical engineers specialize in the production and distribution of electrical power. Batteries and solar cells are direct energy conversion devices used to store and generate electricity. They are particularly useful in remote locations or in hand held gadgets where there is no easy way to connect to the power grid. Relatedly, direct energy conversion devices such as thermoelectric devices and fuel cells are used to power satellites, rovers, and other aerospace systems. Many electrical engineers work in the automotive industry. Direct energy conversion devices found in cars include batteries, optical cameras, Hall effect sensors in tachometer used to measure rotation speed, and pressure sensors.

Direct energy conversion is a fascinating topic because it does not fit neatly into a single discipline. Energy conversion is fundamental to the fields of electrical engineering, but it is also fundamental to mechanical engineering, physics, chemistry, and other branches of science and engineering. For example, springs are energy storage devices often studied by mechanical engineers, capacitors are energy storage devices often studied by electrical engineers, and batteries are energy storage devices often studied by chemists. Relatedly, energy storage and energy conversion devices, such as springs, capacitors, and batteries, are not esoteric. They are commonplace, cheap, and widely available. While they are found in everyday objects, they are active subjects of contemporary research too. For example, laptop computers are limited by the lifetime of batteries, and cell phone reception is often limited by the quality of an antenna. Batteries, antennas, and other direct energy conversion devices are studied by both consumer companies trying to build better products and academic researchers trying to understand fundamental physics.

1.2 Preview of Topics

This book is intended to both illustrate individual energy conversion technologies and illuminate the relationship between them. For this reason, it is organized in two parts. The first part is a survey of energy conversion processes. The second part introduces calculus of variations and uses it as a framework to relate energy conversion processes.

Due to the wide variety, it is not possible to discuss all energy conversion devices, even all direct devices, in detail. However, by studying the example direct energy conversion processes, we can gain an understanding of indirect processes and other applications. The devices discussed in this book involve energy conversion between electrical form and another form. Additionally, devices involving magnets and coils will not be discussed. Many useful devices, including motors, generators, wind turbines, and geothermal power plants, convert energy electromagnetically using magnets and coils. Approximately 90% of power supplied to the electrical grid in the United States comes from generators that use magnets and coils [1]. Also, about 2/3 to 3/4 of energy used by manufacturing facilities goes towards motors

[2, ch. 1]. However, plenty of good resources discussing these topics exist. Furthermore, this book emphasizes device that operate near room temperature and at relatively low power (<1 kW). Many interesting devices, such as nuclear power plants, operate at high temperatures. One reason not to discuss more powerful devices is that the vast majority of large electrical generators in use today involve turbines with coils and magnets. Another reason is that these devices are often limited by material considerations. Finding materials to construct high temperature devices is a challenging problem, but it is not the purpose of this book. Additionally, only technologies commercially available on the market today are discussed in this book. Also, many quality texts exist on the topics of renewable and alternative energy sources. For this reason, this book will not focus on renewable or alternative energy technologies. Topics like wind turbines, which involve electromechanical energy conversion with magnets and coils, are not discussed. Solar cells, piezoelectric devices, and other direct energy conversion devices are discussed and can be considered both direct energy conversion devices and renewable energy devices.

While a few books on direct energy conversion exist, there are few things which set this book apart. First, many of the books on direct energy conversion, including [3] and [4], are written at the graduate level while this book is aimed at a more general audience. This book is used for the course Direct Energy Conversion taught at Trine University, which is a junior undergraduate level course for electrical engineers. This book is not intended only for electrical engineering students. It is also aimed at researchers who are interested in how energy conversion is studied by scientists and engineers in other disciplines. The idea of energy conversion is fundamental to physics, chemistry, mechanical engineering, and multiple other disciplines. This book discusses fundamental physics behind energy conversion processes, introduces terminology used, and relates concepts of material science used for building devices. The chapters were written so that someone who is not an antenna designer, for example, can read the relevant chapter as an introduction and gain insights into some of the terminology and key concepts used by electromagnetics researchers. Second, a number of good books on the topic, including [3] and [5] were written decades ago. The concepts of these books remain relevant, and these books often predicted which technologies would be of interest. However, there is a need for a book which discusses the most accessible and commonplace direct energy conversion technologies in use today. Additionally, many of these classic texts are out of print, and contemporary texts are needed.

The reader is assumed to be familiar with introductory chemistry and physics. Background in electrical circuits and materials may also be helpful.

Math through Calculus I is used in the first part of the book, and math through Calculus III (including partial derivatives) is used in the second part. Many topics in this text are discussed qualitatively. No attempt is made to be mathematically rigorous, and proofs are not given. The physics of devices is emphasized over excessive mathematics. Additionally, all physical systems will be discussed semiclassically, which means that explanations will involve electrons and electromagnetic fields, but the wave-particle duality of these quantities will not be discussed. While quantum mechanical, quantum field theoretical, and other more precise theories exist to describe many physical situations, semiclassical discussions will be used to make this book more easily accessible to readers without a background in quantum mechanics.

Chapters 2 - 10 comprise the first part of this book. As mentioned above, they survey various direct energy conversion processes which convert to or from electricity and which do not involve magnets and coils. Table 1.1 lists many of the processes studied along with where in the text they are discussed, and Table 1.2 lists some of the devices detailed. This text is not intended to be encyclopedic or complete. Instead, it is intended to highlight the physics behind some of the most widely available and accessible energy conversion devices which convert to or from electrical energy. One way to understand energy conversion devices used to convert to or from electricity is to classify them as most similar to capacitors, inductors, resistors, or diodes. While not all devices fit neatly in these categories, many do. The second column of Table 1.2 lists the category for various devices. Similarly, energy conversion processes may be capacitive, inductive, resistive, or diode-like.

Capacitive energy conversion processes are discussed in Chapters 2 and 3. Capacitors, piezoelectric devices, pyroelectric devices, and electro-optic devices are discussed. A *piezoelectric device* is a device which converts mechanical energy directly to electricity or converts electricity directly to mechanical energy [6] [3]. A material polarization and voltage develop when the piezoelectric device is compressed. A *pyroelectric device* converts a temperature differential into electricity [6]. The change in temperature induces a material polarization and a voltage in the material. *Electro-optic devices* convert an optical electromagnetic field to energy of a material polarization. In these devices, an external optical field typically from a laser induces a material polarization and a voltage across the material. Chapters 4 and 5 discuss inductive energy conversion devices including antennas, Hall effect devices, and magnetohydrodynamic devices. An *antenna* converts electrical energy to an electromagnetic field or vice versa. A *Hall effect device* converts a magnetic field to or from electricity. A *magnetohy-*

1 INTRODUCTION

Process	Forms of Energy	Example Devices	Discussed in Section
Piezoelectricity	Electricity $\updownarrow$ Mechanical Energy	Piezoelectric Vibration Sensor, Electret Microphone	2.3
Pyroelectricity	Electricity $\updownarrow$ Heat	Pyroelectric Infrared Detector	3
Electro-optic Effect	Optical Electromagnetic Energy $\updownarrow$ Material Polarization	Controllable Optics, Liquid Crystal Displays	3.3
Electromagnetic Transmission and Reception	Electricity $\updownarrow$ Electromagnetic Energy	Antenna	4
Hall Effect	Electricity $\updownarrow$ Magnetic Energy	Hall Effect Device	5
Magnetohydrodynamic Effect	Electricity $\updownarrow$ Magnetic Energy	Magnetohydrodynamic Device	5.3
Absorption	Optical Electromagnetic Energy $\downarrow$ Electricity	Solar cell, Semiconductor Optical Photodetector	6

Table 1.1: Variety of energy conversion processes.

Process	Forms of Energy	Example Devices	Discussed in Section
Spontaneous Emission	Electricity ↓ Optical Electromagnetic Energy	Lamp, LED	7.3
Stimulated Emission	Electricity ↓ Optical Electromagnetic Energy	Laser, Optical Amplifier	7.4
Thermoelectric Effects (Incl. Seebeck, Peltier and Thomson)	Electricity ↕ Heat	Thermoelectric cooler, Peltier device, Thermocouple	8.8
(Battery or Fuel Cell) Discharging	Chemical Energy ↓ Electricity	Battery, Fuel Cell	9
(Battery or Fuel Cell) Charging	Electricity ↓ Chemical Energy	Battery, Fuel Cell	9
Thermionic Emission	Heat ↓ Electricity	Thermionic Device	10.2
Electrohydrodynamic Effect	Electricity ↕ Fluid flow	Microfluidic Pump, Valve	10.6

Table 1.1 continued: Variety of energy conversion processes.

1 INTRODUCTION

Device	Similar to Component	Forms of Energy	Discussed Section
Piezoelectric Device	Capacitor	Electricity $\updownarrow$ Mechanical Energy	2.3
Pyroelectric Device	Capacitor	Electricity $\updownarrow$ Heat	3
Electro-optic Device	Capacitor	Optical Energy $\updownarrow$ Material Polarization	3.3
Antenna	Inductor	Electricity $\updownarrow$ Electromagnetic	4
Hall Effect Device	Inductor	Electricity $\updownarrow$ Magnetic Energy	5
Magnetohydrodynamic Device	Inductor	Electricity $\updownarrow$ Magnetic Energy	5.3
Solar Cell	Diode	Optical Energy $\downarrow$ Electricity	6
LED, Laser	Diode	Electricity $\downarrow$ Optical Energy	7
Thermoelectric Device	Diode	Electricity $\updownarrow$ Heat	8.8
Geiger Counter	Diode	Radiation $\downarrow$ Electricity	10.3
Resistance Temp. Detector	Resistor	Heat $\downarrow$ Electricity	10.5
Potentiometer	Resistor	Electricity $\downarrow$ Heat	10.5
Strain Gauge	Resistor	Mechanical Energy $\downarrow$ Electricity	10.5

Table 1.2: Variety of energy conversion devices.

drodynamic device converts kinetic energy of a conducting material in the presence of a magnetic field into electricity.

Optical devices are discussed in Chapters 6 and 7. These chapters discuss devices made from diode-like pn junctions such as solar cells, LEDs, and semiconductor lasers as well as other types of devices such as incandescent lamps and gas lasers. *Thermoelectric devices* convert a temperature differential into electricity [3, p. 146]. They are also made from junctions of materials in which heat and charges flow at different rates, and they are discussed in Chapter 8. Batteries and fuel cells are discussed in Chapter 9. A *battery* is a device which stores energy as a chemical potential. Batteries range in size from tiny hearing aid button sized batteries which store tens of milliamp-hours of charge to large car batteries which can store 10,000 times as much energy. A *fuel cell* is a device which converts chemical energy to electrical energy through the oxidation of a fuel [3]. During battery operation, the electrodes are consumed, and during fuel cell operation, the fuel and oxidizer are consumed instead. A variety of resistor-like energy conversion devices, among other devices, are discussed briefly in Chapter 10.

Chapters 11 - 14 comprise the second part of this book. These chapters are more theoretical, and they establish a mathematical framework for understanding energy conversion. This mathematics allows relationships to be studied between energy conversion devices built by electrical engineers, mechanical engineers, chemists, and scientists of other disciplines. Chapters 11 and 12 introduce the idea of calculus of variations and apply it to a wide variety of energy conversion processes. Chapter 13 applies the idea of calculus of variations to energy conversion within an individual atom. Chapter 14 shows how a study of the symmetries of the equations produced from calculus of variations can provide further insights into energy conversion processes.

1.3 Conservation of Energy

Energy conservation is one of the most fundamental ideas in all of science and engineering. Energy can be converted from one form to another. For example, kinetic energy of a moving ball can be converted to heat by friction, or it can be converted to potential energy if the ball rolls up a hill. However, energy cannot be created or destroyed. The idea of energy conservation will be considered an axiom, and it will not be questioned throughout this book. Sometimes people use somewhat loose language when describing energy conversion. For example, one might say that energy is lost to friction when a moving block slides along a table or when electricity flows through a resistor. In both cases, the energy is not lost but is instead converted to heat. Thermoelectric devices and pyroelectric devices can convert a temperature differential back to electricity. Someone might say that energy is generated by a coal power plant. What this phrase means is that chemical energy stored in the coal is converted to electrical energy. When a battery is charged, electrical energy is converted back to chemical energy. This imprecise language will occasionally be used in the text, but in all cases, energy conservation is assumed. While it might seem like an abstract theoretical law, energy conservation is used regularly by circuit designers, mechanical engineers modeling mechanisms, civil engineers designing pipe systems, and other types of engineers.

Efficiency of an energy conversion device, η_{eff}, is defined as the power output of the desired energy type over the power input.

$$\eta_{eff} = \frac{P_{out}}{P_{in}} \qquad (1.1)$$

Efficiency may be written as a fraction or a percent. For example, if we say that an energy conversion device is 75% efficient, we mean that 75% of the energy is converted from the first form to the second while the remaining energy either remains in the first form or is converted to other undesired forms of energy. Energy conversion devices are rarely 100% efficient, and some commercial energy conversion devices are only a few percent efficient. Multiple related measures of efficiency exist where the input and output powers are chosen slightly differently. To accurately compare efficiency measures of devices, consistent of input and output power must be used.

1.4 Measures of Power and Energy

This book brings together topics from a range of fields including chemistry, electrical engineering, and thermodynamics. Scientists in each branch of study use symbols to represent specific quantities, and the choice of variables by scientists in one field often contradict the choice by scientists in another field. In this text, different fonts are used to represent different symbols. For example, S represents entropy, $\$$ represents the Seebeck coefficient, and $\mathbb{S}$ represents action. A list of variables used in this text along with their units can be found in Appendix A. Use the tables in the appendix as tools.

Power P and energy E are fundamental measures. Power absorbed by a system is the derivative of the energy absorbed with respect to time.

$$P = \frac{dE}{dt} \tag{1.2}$$

In SI units, energy is measured in joules and power is measured in watts. While these are the most common measures, many other units are used. Every industry, from the petroleum industry to the food industry to the electrical power industry, seems to have its own favorite units. Tables 1.3 and 1.4 list energy and power conversion factors. Values in the tables are from references [7] and [8].

Conversions between joules and some units, including calories, ergs, kilowatt hours, and tons of TNT are exact definitions [7]. The calorie is approximately the energy needed to increase the temperature of one gram of water by a temperature of one degree Celsius, but it is defined to be 4.1868 J [7]. Note that there is both a calorie and food calorie (also called kilocalorie). The food calorie or kilocalorie is typically used when specifying the energy content of foods, and it is a thousand times as large as the (lowercase c) calorie. Other conversions listed in Table 1.3, including the conversion for energy in barrels of crude oil, are approximate average values instead of exact definitions [8]. Values in Table 1.3 are listed to the significant precision known or to four significant digits. Other inexact values throughout this text are also specified to four significant digits. The unit $\frac{1}{cm}$, referred to as wave number, is discussed in Ch. 6. The conversion value listed in Table 1.3 for the therm is the US, not European, accepted value [7]. A barrel, used in the measure of crude oil, is 42 US gallons [8].

1 INTRODUCTION

1 J = 6.241508 · 10^{18} electron Volt, eV	1 eV = 1.602176 · 10^{-19} J
1 J = 10^7 erg	1 erg = 10^{-7} J
1 J = 0.7375621 foot pound-force	1 foot pound-force = 1.355818 J
1 J = 0.23885 calories	1 calorie = 4.1868 J
1 J = 9.47817 · 10^{-4} British thermal units, Btu	1 Btu = 1055.056 J
1 J = 2.3885 · 10^{-4} kilocalories (food calories)	1 kilocalorie = 4186.8 J
1 J = 9.140 · 10^{-7} cubic feet of natural gas	1 cubic foot of nat. gas = 1.094 · 10^6 J
1 J = 2.778 · 10^{-7} kilowatt hour, kW·h	1 kW·h = 3.6 · 10^6 J
1 J = 6.896 · 10^{-9} gallons diesel fuel	1 gallon diesel fuel = 1.450 · 10^8 J
1 J = 9.480434 · 10^{-9} therm (US)	1 therm (US) = 1.054804 · 10^8 J
1 J = 7.867 · 10^{-9} gallons motor gasoline	1 gallon motor gasoline = 1.271 · 10^8 J
1 J = 2.390 · 10^{-10} ton of TNT	1 ton of TNT = 4.184 · 10^9 J
1 J = 1.658 · 10^{-10} barrels crude oil	1 barrel crude oil = 6.032 · 10^9 J
1 J = 4.491 · 10^{-11} metric ton of coal	1 metric ton coal = 2.227 · 10^{10} J
1 J = 1.986447 · 10^{-23} $\frac{1}{cm}$	1 $\frac{1}{cm}$ = 5.03411 · 10^{22} J

Table 1.3: Energy unit conversion factors.

1 W = 1 $\frac{J}{s}$	1 $\frac{J}{s}$ = 1 W
1 W = 1 · 10^7 $\frac{erg}{s}$	1 $\frac{erg}{s}$ = 10^{-7} W
1 W = 1.34 · 10^{-3} horsepower	1 horsepower = 7.46 · 10^2 W
1 W = 2.655224 · 10^3 foot pound-force per h	1 foot pound-force per h = 3.766161 · 10^{-4} W

Table 1.4: Power unit conversion factors.

1.5 Properties of Materials

1.5.1 Macroscopic Properties

To understand energy conversion devices, we need to understand materials both microscopically on the atomic scale and macroscopically on large scales. A *macroscopic property* is a property that applies to large pieces of the material as opposed to microscopic sized pieces.

One way to classify materials is based on their state of matter. Materials can be classified as solids, liquids, gases, or plasmas. A *plasma* is an ionized gas. Other more unusual states of matter exist such as Bose Einstein condensates, but they will not be discussed in this book.

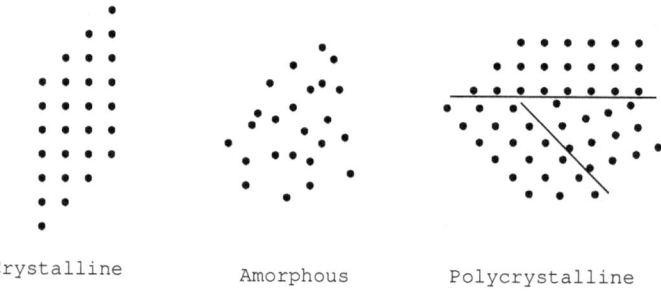

Figure 1.1: Illustration of crystalline, amorphous, and polycrystalline atomic structure.

We can further classify solids as crystalline, polycrystalline, or amorphous based on the regularity of their atomic structure [9]. Figure 1.1 illustrates these terms. In a *crystal*, the arrangement of atoms is periodic. The atoms may be arranged in a cubic array, hexagonal array, or some other way, but they are arranged periodically in three dimensions. In an *amorphous material*, the arrangement of the atoms is not periodic. The term amorphous means glassy. A *polycrystalline material* is composed of small crystalline regions. These definitions can apply to materials made of single elements or materials made of multiple elements. Many energy conversion devices are made from very pure crystalline, amorphous, or polycrystalline materials. For example, amorphous cadmium telluride is used to make solar cells, and crystalline silicon is used to make Hall effect devices. Many materials, including both silicon and silicon dioxide, can be found in all three of these forms at room temperature. In crystalline and amorphous silicon, for example, the silicon atoms may have the same number of nearest neighbors, and the density of atoms in both materials may be the same, but there is no medium-range order in the amorphous material. Electrical properties of crystalline, amorphous, and polycrystalline forms of a material may differ. Electrons can flow more easily through a pure crystalline material while electrons are more likely to be scattered or absorbed as they flow through an amorphous material, crystalline materials with impurities, or a crystalline material with crystal defects.

We can further classify crystals as either isotropic or anisotropic [10, p. 210]. A crystal is *isotropic* if its macroscopic structure and material properties are the same in each direction. A crystal is *anisotropic* if the macroscopic structure and material properties are different in different directions.

We can also classify materials based on how they behave when a voltage is applied across the material [11]. In a *conductor*, electrons flow easily in the presence of an applied voltage or electric field. In an *insulator*, also called a *dielectric*, electrons do not flow in the presence of an applied voltage or electric field. In the presence of a small external voltage or electric field, a *semiconductor* acts as an insulator, and in the presence of a strong voltage or electric field, a semiconductor acts as a conductor. Both solids and liquids can be conductors, and both solids and liquids can be insulators. For example, copper is a solid conductor while salt water is a liquid conductor.

1.5.2 Microscopic Properties

The *electron configuration* lists the energy levels occupied by electrons around an atom. The electron configuration can describe neutral or ionized atoms, and it can describe atoms in the lowest energy state or excited atoms. For example, the electron configuration of a neutral aluminum atom in the lowest energy state is $1s^2 2s^2 2p^6 3s^2 3p^1$. The electron configuration of an aluminum Al^+ ion in the lowest energy state is $1s^2 2s^2 2p^6 3s^2$, and the electron configuration of a neutral aluminum atom with an excited electron can be written as $1s^2 2s^2 2p^6 3s^2 4s^1$.

Electrons are labeled by four *quantum numbers*: the principle quantum number, the azimuthal quantum number, the magnetic quantum number, and the spin quantum number [6] [12]. No two electrons around an atom can have the same set of quantum numbers. The principle quantum number takes integer values, 1, 2, 3 and so on. All electrons with the same principle quantum number are said to be in the same *shell*. The large numbers in the electron configuration refer to principle quantum numbers. The neutral aluminum atom in the lowest energy state has two electrons in the 1 shell, eight electrons in the 2 shell, and three electrons in the 3 shell. For most atoms, especially atoms with few electrons, electrons with lower principle quantum numbers both are spatially closer to the nucleus and require the most energy to remove. However, there are exceptions to this idea for some electrons around larger atoms [13] [14].

Azimuthal quantum numbers are integers, and these values define *subshells*. For shells with principle quantum number n, the azimuthal quantum number can take values from 0 to $n-1$. In the electron configuration, values of this quantum number are denoted by lowercase letters: s=0, p=1, d=2, f=3, and so on. Magnetic quantum numbers are also integers, and these values define *orbitals*. For a subshell with azimuthal quantum number l, the magnetic quantum number takes values from $-l$ to l. In the electron configuration, superscript numbers indicate the magnetic quantum number. Spin quantum numbers of electrons can take the values $\frac{1}{2}$ and $\frac{-1}{2}$. They are not explicitly denoted in the electron configuration.

Consider again the neutral aluminum atom in the lowest energy state. This atom has electrons with principle quantum numbers $n=1$, 2, and 3. For electrons with principle quantum number 1, the only possible values for both the azimuthal quantum number and the magnetic quantum number are zero. The spin quantum number can take the values of $\frac{1}{2}$ and $\frac{-1}{2}$. Only two electrons can occupy the 1 shell, and these electrons are denoted by the $1s^2$ term of the electron configuration. For the electrons with principle quantum number 2, the azimuthal quantum number can be 0 or 1. Two electrons can occupy the 2s orbital, and six electrons can occupy the 2p orbital. For the 3 shell, the azimuthal quantum number can take three possible values: s=0, p=1, and d=2. However since aluminum only has 13 electrons, electrons do not have all of these possible values, so the 3 shell is only partially filled. The atoms in the rightmost column of the periodic table have completely filled shells. They are rarely involved in chemical reactions because adding electrons, removing electrons, or forming chemical bonds would require too much energy.

Valence electrons are the electrons that are most easily ripped off an atom. Valence electrons are the electrons involved in chemical reactions, and electrical current is the flow of valence electrons. Other, inner shell, electrons may be involved in chemical reactions or electrical current only in cases of unusually large applied energies, and these situations will not be discussed in this text. Valence electrons occupy the subshell or subshells with the highest quantum numbers, and valence electrons are not part of completely filled shells. For the example of the neutral aluminum atom in the lowest energy state, the three electrons in the 3 shell are valence electrons.

Where are the electrons around the atom spatially? This question is of interest to chemists, physicists, and electrical engineers. If we know the orbital of an electron, we have some information on where the electron

1 INTRODUCTION

is likely to be found spatially around an atom. However, identifying the location of an electron with any degree of precision is difficult for multiple reasons. First, atoms are tiny, roughly 10^{-10} m in diameter. Second, at any temperature above absolute zero, atoms and electrons are continually in motion. Third, electrons have both particle-like and wave-like properties. Fourth, according to Heisenberg's Uncertainty Principle, the position and momentum of an electron cannot simultaneously be known with complete precision. At best, you can say that an electron is most likely in some region and moves with some range of speed. Fifth, in many materials including conductors and semiconductors, valence electrons are shared by many atoms instead of bound to an individual atom [10, p. 544].

1.6 Electromagnetic Waves

1.6.1 Maxwell's Equations

In this text, V and I denote DC voltage and current respectively while v and i denote AC or time varying voltage and current. In circuit analysis, we are unconcerned with what happens outside these wires. We are only interested in node voltages and currents through wires. Furthermore, the voltages and currents in the circuit are described as functions of time t but not position (x, y, z). Devices like resistors, capacitors, and inductors too are assumed to be point-like and not extended with respect to position (x, y, z). This set of assumptions is just a model. In reality, if two nodes in a circuit have a voltage difference between them, then necessarily a force is exerted on nearby charges not in the path of the circuit. This force per unit charge is the electric field intensity $\vec{E}$. Similarly, if there is current flowing through a wire, there is necessarily a force exerted on electrons in nearby loops of wire, and this force per unit current element is the magnetic flux density $\vec{B}$. Energy can be stored in an electric or magnetic field. In later chapters, we will discuss devices, including antennas, electro-optic devices, photovoltaic devices, lamps, and lasers, that convert energy of an electromagnetic field to or from electricity.

Four interrelated vector quantities are used to describe electromagnetic fields. These vector fields are functions of position (x, y, z) and time t. The four vector fields are

$$\vec{E}(x, y, z, t) = \text{Electric field intensity in } \frac{\text{V}}{\text{m}}$$

$$\vec{D}(x, y, z, t) = \text{Displacement flux density in } \frac{\text{C}}{\text{m}^2}$$

$\vec{H}(x,y,z,t)$=Magnetic field intensity in $\frac{A}{m}$

$\vec{B}(x,y,z,t)$=Magnetic flux density in $\frac{Wb}{m^2}$

In these expressions, V represents the units volts, C represents the units coulombs, A represents the units amperes, and Wb represents the units webers. Additional abbreviations for units are listed in Appendix B.

Coulomb's law

$$\vec{F} = \frac{Q_1 Q_2 \hat{a}_r}{4\pi \epsilon r^2} \qquad (1.3)$$

tells us that charged objects exert forces on other charged objects. In this expression, Q_1 and Q_2 are the magnitude of the charges in coulombs. The quantity ϵ is the permittivity of the surrounding material in units farads per meter, and it is discussed further in Sections 1.6.3 and 2.2.3. The quantity r is the distance between the charges in meters, and $\hat{a}_r$ is a unit vector pointing along the direction between the charges. Force in newtons is represented by $\vec{F}$. Opposite charges attract, and like charges repel. Electric field intensity is force per unit charge, so the electric field intensity due to a point charge is given by

$$\vec{E} = \frac{Q \hat{a}_r}{4\pi \epsilon r^2} \qquad (1.4)$$

These vector fields can describe forces on charges or currents in a circuit as well as outside the path of a circuit. Maxwell's equations relate time varying electric and magnetic fields. Maxwell's equations in differential form are:

$$\vec{\nabla} \times \vec{E} = -\frac{\partial \vec{B}}{\partial t} \qquad \text{Faraday's Law} \qquad (1.5)$$

$$\vec{\nabla} \times \vec{H} = \vec{J} + \frac{\partial \vec{D}}{\partial t} \qquad \text{Ampere's Law} \qquad (1.6)$$

$$\vec{\nabla} \cdot \vec{D} = \rho_{ch} \qquad \text{Gauss's Law for the Electric Field} \qquad (1.7)$$

$$\vec{\nabla} \cdot \vec{B} = 0 \qquad \text{Gauss's Law for the Magnetic Field} \qquad (1.8)$$

The additional quantities in Maxwell's equations are the volume current density $\vec{J}$ in $\frac{A}{m^2}$ and the charge density ρ_{ch} in $\frac{C}{m^3}$. In this text, we will not be solving Maxwell's equations, but we will encounter references to them.

1 INTRODUCTION

The quantity $\vec{\nabla}$ is called the *del* operator. In Cartesian coordinates, it is given by

$$\vec{\nabla} = \hat{a}_x \frac{\partial}{\partial x} + \hat{a}_y \frac{\partial}{\partial y} + \hat{a}_z \frac{\partial}{\partial z}. \qquad (1.9)$$

When this operator acts on a scalar function, $\vec{\nabla} f$, it is called the *gradient*. The gradient of a scalar function returns a vector representing the spatial derivative of the function, and it points in the direction of largest change in that function. In Maxwell's equations, $\vec{\nabla}$ acts on vector, instead of scalar, functions. The operation $\vec{\nabla} \times \vec{E}$ is called the *curl*, and the operation $\vec{\nabla} \cdot \vec{E}$ is called the *divergence*. Both of these operations represent types of spatial derivatives of vector functions. The del operator obeys the identity

$$\nabla^2 = \vec{\nabla} \cdot \vec{\nabla}. \qquad (1.10)$$

The operation $\nabla^2 f$ is called the *Laplacian* of a scalar function, and it represents the spatial second derivative of that function.

1.6.2 Electromagnetic Waves in Free Space

Electromagnetic waves travel through empty space at the speed of light in free space, $c = 2.998 \cdot 10^8 \, \frac{m}{s}$, and through other materials at speeds less than c. For a sinusoidal electromagnetic wave, the speed of propagation is the product of the frequency and wavelength

$$|\vec{v}| = f\lambda \qquad (1.11)$$

where $|\vec{v}|$ is the magnitude of the velocity in $\frac{m}{s}$, f is the frequency in Hz, and λ is the wavelength in meters. In free space, Eq. 1.12 becomes

$$c = f\lambda. \qquad (1.12)$$

The speed of light in free space is related to two constants which describe free space.

$$c = \frac{1}{\sqrt{\epsilon_0 \mu_0}} \qquad (1.13)$$

The *permittivity of free space* is given by $\epsilon_0 = 8.854 \cdot 10^{-12} \, \frac{F}{m}$ where F represents farads, and the *permeability of free space* is given by $\mu_0 = 1.257 \cdot 10^{-6} \, \frac{H}{m}$ where H represents henries. (Constants specified in this section and in Appendix A are rounded to four significant digits.)

In free space, the electric field intensity $\vec{E}$ and the displacement flux density $\vec{D}$ are related by ϵ_0.

$$\vec{D} = \epsilon_0 \vec{E} \qquad (1.14)$$

Relatedly in free space, the magnetic field intensity $\vec{H}$ and the magnetic flux density $\vec{B}$ are related by μ_0.

$$\vec{B} = \mu_0 \vec{H} \qquad (1.15)$$

1.6.3 Electromagnetic Waves in Materials

Electromagnetic fields interact very differently with conductors and with insulators. Electromagnetic fields do not propagate into perfect conductors. Instead, charges and currents accumulate on the surface. While no materials are perfect conductors, commonly encountered metals like copper and aluminum are very good conductors. When these materials are placed in an external electromagnetic field, surface charges and currents build up, and the electromagnetic field in the material quickly approaches zero. Electromagnetic fields propagate through perfect insulators for long distances without decaying, and no charges or currents can accumulate on the surface because there are no electrons free from their atoms. In practical dielectrics, electromagnetic waves propagate long distances with very little attenuation. For example, optical electromagnetic waves remain strong enough to detect after propagating hundreds of kilometers through optical fibers made of pure silicon dioxide [10, p. 886].

Resistance R in ohms, capacitance C in farads, and inductance L in henries describe the electrical properties of *devices*. Resistivity ρ in Ωm, permittivity ϵ in $\frac{F}{m}$, and permeability μ in $\frac{H}{m}$ describe the electrical properties of *materials*. The quantities ρ, ϵ, and μ describe properties of materials alone while the quantities R, C, and L incorporate effects the material, shape, and size of a device.

Resistivity ρ is a measure of the inability of charges or electromagnetic waves to propagate through a material. Conductors have a very small resistivity while insulators have a large resistivity. Sometimes *electrical conductivity*, $\sigma = \frac{1}{\rho}$ in units $\frac{1}{\Omega m}$, is used in place of the resistivity. For a device made of a uniform material with length l and cross sectional area A, resistance and resistivity are related by

$$R = \frac{\rho l}{A}. \qquad (1.16)$$

Resistance is a measure of the inability of charges or electromagnetic waves to flow through a device while resistivity is a measure of the inability to flow through a material.

1 INTRODUCTION

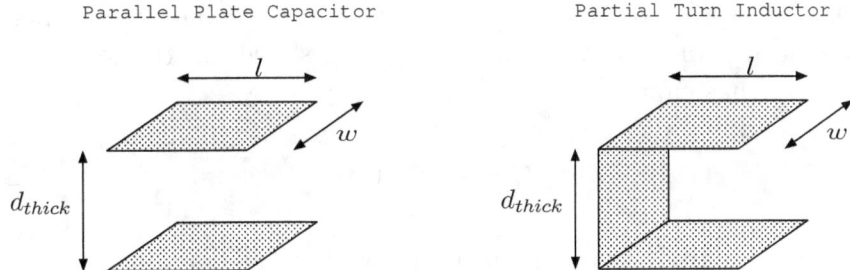

Figure 1.2: Geometry of a parallel plate capacitor and partial turn inductor.

Permeability μ is a measure of the ability of a material to store energy in the magnetic field due to currents distributed throughout the material. Materials can also be described by their relative permeability μ_r, a unitless measure.

$$\mu_r = \frac{\mu}{\mu_0} \tag{1.17}$$

While permeability describes a *material*, inductance describes a *device*. The magnetic flux density in a material is a scaled version of the magnetic field intensity.

$$\vec{B} = \mu \vec{H} \tag{1.18}$$

Often insulators have permeabilities close to μ_0 while conductors used to make permanent magnets have significantly larger permeabilities. The right part of Fig. 1.2 shows a partial turn coil in a vacuum with length l, thickness d_{thick}, and width w. The inductance and permeability of this device are related by [11, p. 311]

$$L = \frac{\mu d_{thick} l}{w}. \tag{1.19}$$

Permittivity ϵ is a measure of the ability of a material to store energy as an electric field due to charge separation distributed throughout the material. Materials can also be described by their relative permittivity ϵ_r, a unitless measure.

$$\epsilon_r = \frac{\epsilon}{\epsilon_0} \tag{1.20}$$

The displacement flux density in a material is a scaled version of the electric field intensity.

$$\vec{D} = \epsilon \vec{E} \tag{1.21}$$

Some insulators have a permittivity hundreds of times larger than the permittivity of free space. Permittivity is a measure of ability to store energy in a *material* while capacitance is a measure of the ability to store energy in in a *device*.

A uniform parallel plate capacitor with cross sectional area of plates $A = l \cdot w$ and distance between plates d_{thick}, is shown on the left part of Fig. 1.2, and it has capacitance

$$C = \frac{\epsilon A}{d_{thick}} \qquad (1.22)$$

where ϵ is the permittivity of the insulator between the plates.

Permittivity, permeability, and resistivity, depend on frequency. In some contexts, the frequency dependence can be ignored, and throughout most of this text, these quantities will be assumed to be constants. In other contexts, the frequency dependence can be quite significant. For example, the permittivity of semiconductor materials is a strong function of frequency for frequencies close to the semiconductor energy gap. The permittivity $\epsilon(\omega)$ and resistivity $\rho(\omega)$ are not independent. If one of them is known as a function of frequency and μ is assumed constant, the other can be derived. This relationship is known as the Kramers Kronig relationship [10] or occasionally as the dielectric dispersion formula [15].

When discussing electrical properties of a device, resistance, inductance, and capacitance are combined into one complex measure, the impedance. Similarly, some authors find it convenient to combine resistivity, permittivity, and permeability into a pair of complex measures of the electrical properties of materials [6]. The complex permittivity is defined $\epsilon^* = \epsilon + j\rho$, and the complex permeability is defined $\mu^* = \mu + j\rho_{mag}$. The quantity ρ represents the resistivity which is a measure of the energy converted to heat as a charge flows through a material due to an applied electrical field. The quantity ρ_{mag} represents an analogous measure of energy converted to heat from currents in an applied magnetic field. Complex permittivity and permeability will not be used in this text.

1.7 Problems

1.1. A Ford Focus produces 160 horsepower [16]. Calculate the power produced in watts, and calculate the approximate energy produced by the vehicle in one hour.

1.2. A gallon of gas contains $1.21 \cdot 10^5$ Btu and weighs 6 pounds [8]. Calculate the energy stored in the gallon of gas in joules, and calculate the specific energy in joules per kilogram.

1.3. An Oreo cookie has 53 food calories and weighs 11 grams [17]. A ton (2000 pounds) of TNT contains approximately $4.184 \cdot 10^9$ J of energy [7]. Calculate the specific energy of the cookie in joules per kilogram, and calculate the specific energy of the TNT in joules per kilogram. (Yes, the value for the cookie is higher.)

1.4. Find the electron configuration of an isolated indium atom in the lowest energy state. How many electrons are found around the atom? Repeat for a Cl^- ion.

1.5. Use a periodic table for this problem.

   (a) Which element has the electron configuration $1s^2 2s^2 2p^6 3s^2 3p^6 4s^2 3d^2$?

   (b) List two elements which have exactly two valence electrons.

Part I
Survey of Energy Conversion Devices

2 Capacitors and Piezoelectric Devices

2.1 Introduction

This chapter begins with a discussion of material polarization, and then it discusses capacitors and piezoelectric devices. The next chapter discusses pyroelectric devices and electro-optic devices. All of these devices are all constructed from a thin dielectric layer, and operation of all of these devices involves establishing a material polarization, charge build up, throughout this dielectric material. In piezoelectric materials, mechanical strain causes a material polarization. As with many energy conversion devices, piezoelectric devices can work both ways, converting mechanical energy to electricity or converting electricity to mechanical vibrations. In pyroelectric devices, a temperature gradient causes the material polarization, and in electro-optic devices, an external optical electric field causes the material polarization.

Why start the discussion of energy conversion devices with a discussion of capacitors? Capacitors are familiar to electrical engineers, and they are energy storage devices. How do capacitors work? What are the components of a capacitor? What materials are capacitors made out of? What are the differences between different types of capacitors such as mica capacitors and electrolytic capacitors? In an introductory circuits course, a capacitor is a device where the relationship between the current i and voltage v is given by

$$i = C\frac{dv}{dt} \qquad (2.1)$$

and the capacitance C is just a constant. The only difference between one capacitor and another is the capacitance value. In order to answer these questions further, we need to go beyond this model. Through this study, we will gain insights into piezoelectric devices, pyroelectric devices, and electro-optic devices too.

2.2 Capacitors

2.2.1 Material Polarization

When an external voltage is applied across an insulator, charges separate throughout the material, and this charge separation is called a *material polarization*. Material polarization can be defined more precisely in terms of the electric field intensity $\vec{E}$ and the displacement flux density $\vec{D}$, two vector fields which show up in Maxwell's equations, Eqs. 1.5 - 1.8. These vector fields are related by

$$\vec{D} = \epsilon \vec{E}. \tag{2.2}$$

Why do we define two electric field parameters when they are just scaled versions of each other? It is useful to separate the description of the electric field inside a material from the description of the field in free space. Similarly, two vector fields describe magnetic field, the magnetic field intensity $\vec{H}$ and magnetic flux density $\vec{B}$, and these fields show up in Maxwell's equations for the same reason. Material polarization, $\vec{P}$ in units $\frac{C}{m^2}$, is defined as the difference between the electric field in the material $\vec{D}$ and the electric field that would be present in free space $\vec{E}$. More specifically,

$$\vec{P} = \vec{D} - \epsilon_0 \vec{E} \tag{2.3}$$

or

$$\vec{P} = (\epsilon - \epsilon_0) \vec{E}. \tag{2.4}$$

These expressions involve the permittivity of free space ϵ_0 and the permittivity of a material ϵ which were defined in Sec. 1.6.3.

Scientists overload both the words capacitance and polarization with multiple meanings. See Appendix C for more details on the different uses of these terms.

2.2.2 Energy Storage in Capacitors

When a capacitor is charged, energy is converted from electrical energy to energy stored in a material polarization which is energy of the charge separation. When it is discharged, energy is converted from energy stored in the material polarization back to electrical energy of flowing electrons. Capacitors are made from an insulating material between conducting plates. As we supply a voltage across the insulator, charges accumulate on the plates. The voltage built up is proportional to the charge accumulated on the plates.

$$Q = Cv \tag{2.5}$$

2 CAPACITORS AND PIEZOELECTRIC DEVICES

In Eq. 2.5, Q is the charge in coulombs, v is the voltage, and the constant of proportionality is the capacitance C in farads. If we take the derivative with respect to time, we get the more familiar expression relating the current and voltage across the capacitor.

$$\frac{dQ}{dt} = i = C\frac{dv}{dt} \qquad (2.6)$$

The capacitance of a capacitor is related to the permittivity of the dielectric material between the conductors. Permittivity is a measure of the amount of energy that can be stored by a dielectric material. As described by Eq. 1.22, for a parallel plate capacitor this relationship is

$$C = \frac{\epsilon A}{d_{thick}} \qquad (2.7)$$

where A is the area of the plates and d_{thick} is the distance between the plates. The energy E stored in a capacitor as a function of voltage applied across it is given by

$$E = \frac{1}{2}Cv^2 = \frac{1}{2}Qv. \qquad (2.8)$$

The capacitance of a vacuum-filled parallel plate capacitor is described by Eq. 2.7 with permittivity $\epsilon = \epsilon_0$, the permittivity of free space. As we charge the capacitor, charges accumulate on the plates, and no change occurs to the vacuum between the plates. If we replace the vacuum with a dielectric with $\epsilon > \epsilon_0$, the capacitance becomes larger. The dielectric filled capacitor can store more energy, all else equal, because the dielectric material changes as the capacitor charges. More specifically, the material polarizes. In an insulator, electrons are bound to their atoms, and current cannot flow. Instead, the electrons in a dielectric move slightly with respect to their nuclei while still staying bound to the atoms. Electrons are always in motion for materials at temperatures above absolute zero, but when a material polarizes, the net location of electrons with respect to the nuclei changes. As the capacitor charges, the electrons are slightly displaced from their atoms, balancing the charges on the plates, and more energy is stored in the dielectric for a given voltage. We say that this process induces *electric dipoles*. The larger the permittivity, ϵ, the more the material can store energy by polarizing in this way. For this reason, capacitors are often filled with dielectric materials like tantalum dioxide Ta_2O_5 which has $\epsilon = 25\epsilon_0$ [18]. A material with $\epsilon = 25\epsilon_0$, for example, will be able to store 25 times the energy of an air filled capacitor of the same size with the same applied voltage.

2.2.3 Permittivity and Related Measures

For historical reasons, the permittivity may be expressed by different measures. The electric susceptibility χ_e, relative permittivity ϵ_r, index of refraction n, and permittivity ϵ all describe the ability of a material to store energy in the electric field. *Electric susceptibility* is a unitless measure related to the permittivity by

$$\chi_e = \frac{\epsilon}{\epsilon_0} - 1 \qquad (2.9)$$

and *relative permittivity* is another unitless measure defined by

$$\epsilon_r = \frac{\epsilon}{\epsilon_0}. \qquad (2.10)$$

With some algebra, we can write the material polarization in terms of the relative permittivity or the electric susceptibility.

$$\vec{P} = (\epsilon_r - 1)\epsilon_0 \vec{E} = \epsilon_0 \chi_e \vec{E} \qquad (2.11)$$

Scientists studying optics often use *index of refraction*, another unitless measure which represents the ratio of the speed of light in free space to the speed of light in the material.

$$n = \frac{c}{|\vec{v}|} = \frac{\text{speed of light in free space}}{\text{speed of light in material}} \qquad (2.12)$$

Since electromagnetic waves cannot travel faster than the speed of light in free space, index of refraction of a material is greater than one, n > 1. Assuming a material is a good insulator and $\mu = \mu_0$, which are typically safe assumption for optics, the relationship between index of refraction and permittivity simplifies to

$$n = \sqrt{\epsilon_r}. \qquad (2.13)$$

Table 2.1 lists relative permittivities of some insulators used to make capacitors or piezoelectric devices. The values are all approximates. See the references cited for more detailed information.

In the definitions of Section 1.6.3 and in Table 2.1, permittivity is treated as a scalar constant, but in some contexts a more complicated description is needed. In a crystalline material, a voltage applied along one crystallographic axis may induce charge separation throughout the material more easily than a voltage of the same size applied along a different axis.

Material	Relative permittivity ϵ_r	Reference
Vacuum	1.0	[3, p. 20]
Teflon	2.1	[3, p. 20]
Polyethylene	2.3	[3, p. 20]
Paper	3.0	[3, p. 20]
SiO_2	3.5	[18]
Mica	6.0	[3, p. 20]
Al_2O_3	9	[18]
AlP	10.2	[9]
$ZrSiO_4$	12.5	[19]
Si	11.8	[9]
Ge	16	[9]
Ta_2O_5	24	[20]
ZrO_2	25	[18]
HfO_2	40	[18]
TiO_2	50	[18]
PbS	161	[9]
PbSe	280	[9]
$BaSrTiO_3$	300	[18]
PbTe	360	[9]

Table 2.1: Approximate values of relative permittivity of various materials.

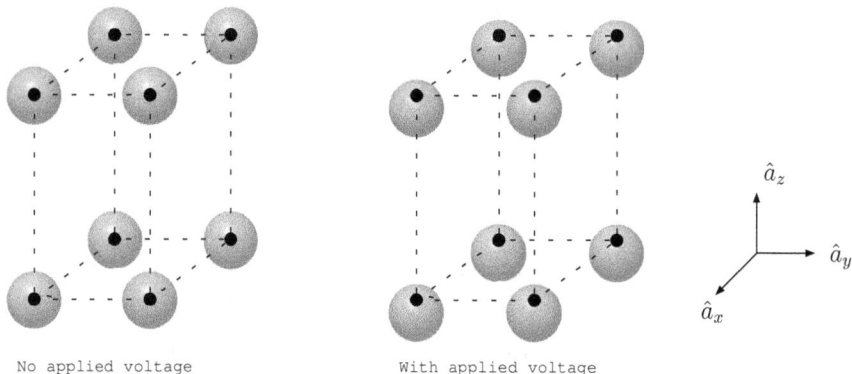

Figure 2.1: Illustration of material polarization.

In such cases, the material is called *anisotropic*. Permittivity of anisotropic materials is more accurately described by a matrix.

$$\begin{pmatrix} \epsilon_{xx} & \epsilon_{xy} & \epsilon_{xz} \\ \epsilon_{yx} & \epsilon_{yy} & \epsilon_{yz} \\ \epsilon_{zx} & \epsilon_{zy} & \epsilon_{zz} \end{pmatrix}$$

The left part of Fig. 2.1 shows some atoms of a crystal. The small black circles represent the location of the nuclei of atoms in the crystals, and the gray circles represent the electron cloud surrounding the nuclei of each atom. If an electric field is applied in the $\hat{a}_z$ direction, the material polarizes, so the electrons are slightly displaced with respect to the nuclei as shown in the figure on the right. Since the spacing of atoms is different in the $\hat{a}_x$ and $\hat{a}_y$ direction than the $\hat{a}_z$ direction, the external field required to get the same charge displacement will be different in the $\hat{a}_x$ and $\hat{a}_y$ directions than the $\hat{a}_z$ direction for this material. For this reason, the material illustrated in the figure is anisotropic, and the permittivity is best described by a matrix as opposed to a scalar quantity.

2.2.4 Capacitor Properties

Capacitors are energy conversion devices used in applications from stabilizing power supplies, to filtering communication signals, to separating out a DC offset from an AC signal. Though capacitors and batteries both store electrical energy, energy in batteries is stored in the chemical bonds of atoms of the electrodes while energy is stored in capacitors in the material polarization from bound charges shifting in a dielectric layer.

The first two measures to consider when selecting a capacitor to use in a circuit are the capacitance and the maximum voltage. A capacitor can

2 CAPACITORS AND PIEZOELECTRIC DEVICES

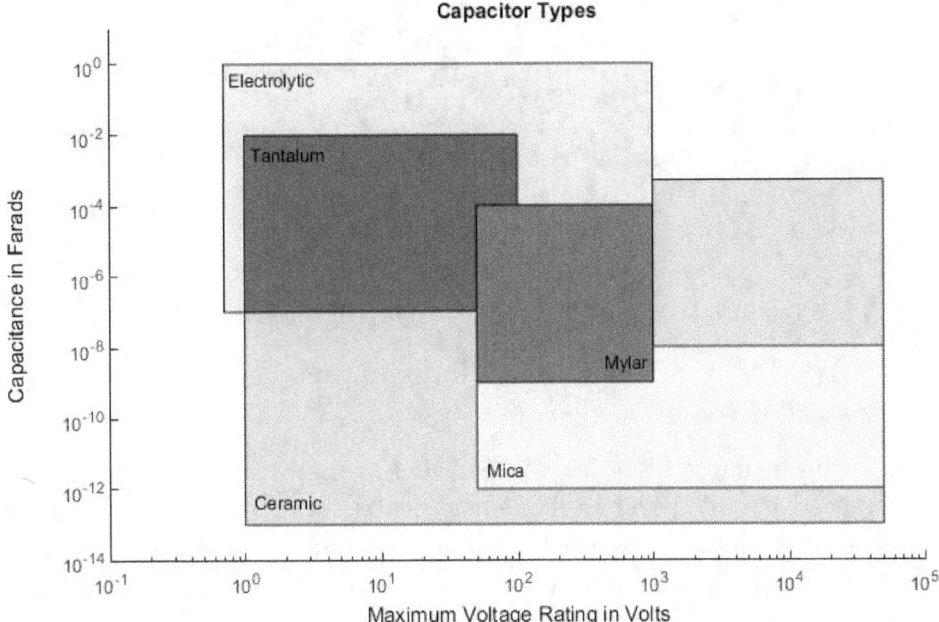

Figure 2.2: Range of capacitance and maximum voltage values for various capacitor types, following [21] and [22].

be damaged if it is placed in a circuit where the voltage across it exceeds the maximum rated value. Approximate ranges for these parameters for capacitors with different dielectric materials are shown in Fig. 2.2. Capacitance ranges are on the vertical axis, and maximum voltage ranges are on the horizontal axis. For example, electrolytic capacitors often can be found with capacitance values ranging from 10^{-7} to 1 F and maximum voltage ratings in the range of 1 to 1000 V. Similarly, ceramic capacitors can often be found with capacitance values ranging from 10^{-13} to $5 \cdot 10^{-4}$ F and maximum voltage ratings in the range of 1 to 50,000 V.

While capacitance and maximum voltage rating are important parameters to consider, they are not the only considerations. Another factor to consider is temperature stability. Ideally, the capacitance will be independent of temperature. However, all materials have a nonzero temperature coefficient. Ceramic and electrolytic capacitors tend to be more sensitive to temperature variation than polymer or vacuum capacitors [22]. Accuracy, or precision, is also important. Just as resistors are labeled with tolerances, capacitors may have tolerances of, for example, ±5% or ±10%. Another factor to consider is equivalent series resistance [23, ch. 1]. All materials have some resistivity, so all capacitors have some finite resistance. To account for the internal resistance, we can model any physical capacitor

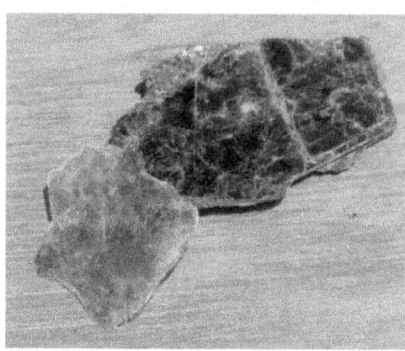

Figure 2.3: Natural mica.

as an ideal capacitor in series with an ideal resistor, and the value of the resistor used is called the equivalent series resistance. Also, leakage of a capacitor should be considered [22]. If a capacitor is able to retain its stored charge for a long period of time, the capacitor has small leakage. If the capacitor discharges quickly even when disconnected from a circuit, it has large leakage. An ideal capacitor has no leakage [22]. Capacitors are also differentiated by their lifetime. An ideal capacitor operates for decades without degradation. However, some types of capacitors, such as electrolytic capacitors, are not designed to have long lifetimes [22]. Other factors to consider include cost, availability, size, and frequency response [22].

Ceramics, glasses, polymers, and other materials are used as the dielectric [22]. Often capacitors are classified by the dielectric material they contain [22]. Ceramic capacitors are small, cheap, and readily available [22]. They can often tolerate large applied voltages [22]. They typically have small capacitance values, poor accuracy, poor temperature stability and moderate leakage [22]. They have low equivalent series resistance and can withstand a lot of current, but they can cause transient voltage spikes, [23, ch. 1]. Some ceramic capacitors are piezoelectric. If these capacitors are vibrated, or even tapped with a pencil, noise will be introduced in the circuit due to piezoelectricity [23, ch. 12].

Mica is an interesting material which is used as a dielectric in capacitors. Figure 2.3 shows naturally occurring mica collected at Ruggles Mine near Grafton, New Hampshire. Mica comes in different natural forms including biotite and muscovite $KAl_2(AlSi_3O_{10})(OH)_2$ [24]. Mica is a flaky mineral with a layered structure [24], so mica capacitors can be made with very thin dielectric layers. Mica capacitors often have good accuracy and small leakage [22].

Capacitor dielectrics have been made from many types of polymers in-

Figure 2.4: Through-hole size capacitors.

cluding polystyrene, polycarbonate, polyester, polypropylene, Teflon, and mylar [22]. These capacitors often have good accuracy, temperature stability, and leakage characteristics [22].

Not all capacitors have solid dielectrics. A vacuum is a dielectric. Capacitors with a vacuum dielectric are used in applications which involve high voltage or which require very low leakage [22]. Capacitors with liquid dielectrics made of oil are used in similar situations [22]. Electrolytic capacitors often have dielectrics which are a combination of solid materials with liquid electrolytes. An *electrolyte* is a liquid through which some charges can flow more easily than others. Electrolytic capacitors are *polarized,* meaning that they have positive and negative terminals, so, similar to a diode, the orientation of the capacitor in a circuit is important. Inside an electrolytic capacitor is a junction of multiple materials. The initial application of voltage in the factory chemically creates an oxide layer which is the dielectric. Reversing the voltage will dissolve the dielectric and destroy the capacitor. One advantage of electrolytic capacitors is that a small device can have a large capacitance. However, they often have poor accuracy, temperature stability, and leakage [22]. Also, electrolytic capacitors have a finite lifetime because the liquid can degrade over time.

2.3 Piezoelectric Devices

Can we induce a material polarization in an insulator in a way that does not involve applying a voltage? If so, then this method can charge a capacitor, and we can discharge the capacitor as usual to produce electricity. Any device that accomplishes this task is an energy conversion device.

Piezoelectric, pyroelectric, and electro-optic devices all involve this type of energy conversion, and they are all currently available as sensors and as other products. In piezoelectric devices, discussed in this section, a mechanical stress causes a material polarization.

If a large enough strain is exerted on a material, the crystal structure will change. For example, at high enough temperature and pressure, coal will crystallize into diamond, and when the pressure is removed, the material stays in diamond form. Steel can be hardened by repeatedly hitting it in a process called *shot peening*. A significant amount of energy is needed to permanently change the crystal structure of a material. In this section, we are not discussing this effect. Instead, we are discussing an effect that typically requires little energy. When a mechanical strain is exerted on a piezoelectric device, a material polarization is established. The valence electrons are displaced, but the nuclei of the material and other electrons do not move. When we release the stress, the material polarization goes away.

2.3.1 Piezoelectric Strain Constant

We can describe the material polarization of a piezoelectric insulating material by incorporating a term which depends on the applied mechanical stress, [25].

$$\vec{P} = \vec{D} - \epsilon_0 \vec{E} + d\vec{\varsigma} \qquad (2.14)$$

In this equation, $\vec{P}$ is material polarization in $\frac{C}{m^2}$, $\vec{D}$ is displacement flux density in $\frac{C}{m^2}$, ϵ_0 is the permittivity of free space in $\frac{F}{m}$, $\vec{E}$ is the applied electric field intensity in $\frac{V}{m}$, d is the *piezoelectric strain constant* in $\frac{m}{V}$, and $\vec{\varsigma}$ is the stress in pascals. Stress can also be given in other units.

$$1 \text{ Pa} = 1 \frac{J}{m^3} = 1 \frac{N}{m^2} \qquad (2.15)$$

For many materials, the piezoelectric strain constant d is zero, and for many other materials, d is quite small. Barium titanate is used to make piezoelectric sensors because it has a relatively large piezoelectric strain coefficient, $d \approx 3 \cdot 10^{-10} \frac{m}{V}$ [25, p. 408]. Additional example coefficients are given in the next chapter in Table 3.1.

Mechanical strain is a unitless measure of deflection or deformation while *stress* has units pascals. Without an external electric field, these quantities are related by Young's elastic modulus which has units $\frac{N}{m^2}$.

$$\text{strain} = \left(\frac{1}{\text{Young's elastic modulus}} \right) \cdot \text{stress} \qquad (2.16)$$

If an electric field is also applied, stress and strain are related by

$$\text{strain} = \left(\frac{1}{\text{Young's elastic modulus}}\right) \cdot \text{stress} + \vec{E} \cdot d \qquad (2.17)$$

where d is the piezoelectric strain constant.

The energy stored in a piezoelectric device under stress $\vec{\varsigma}$ is given by

$$E = |\vec{\varsigma}| \cdot A \cdot l \cdot \eta_{eff} \qquad (2.18)$$

where A is the cross sectional area of a device in m^2, l is the deformation in m, and η_{eff} is the efficiency. Devices which are bigger, are deformed more, or are made from materials with larger piezoelectric constants store more energy.

According to Eq. 2.14, the material polarization of an insulating crystal is linearly proportional to the applied stress. While this accurately describes many materials, it is a poor description of other materials. For other piezoelectric crystals, the material polarization is proportional to the square of the applied stress

$$\left|\vec{P}\right| = \left|\vec{D}\right| - \epsilon_0 \left|\vec{E}\right| + d|\vec{\varsigma}| + d_{quad}|\vec{\varsigma}|^2 \qquad (2.19)$$

where d_{quad} is another piezoelectric strain constant. To model the material polarization in other materials, terms involving higher powers of the stress are needed.

2.3.2 Piezoelectricity in Crystalline Materials

To understand which materials are piezoelectric, we need to introduce some terminology for describing crystals. Crystalline materials may be composed of elements, such as Si, or compounds, such as NaCl. By definition, atoms in crystals are arranged periodically. Two components are specified to describe the arrangement of atoms in a crystal: a lattice and a basis [25, p. 4]. A *lattice* is a periodic array of points in space. An n-dimensional lattice is specified by n lattice vectors for integer n. We can get from one lattice point to every other lattice point by traveling an integer number of lattice vectors. Three vectors, $\vec{a_1}$, $\vec{a_2}$, and $\vec{a_3}$, are used to describe physical lattices in three-space. The choice of lattice vectors is not unique. Lattice vectors which are as short as possible are called *primitive lattice vectors*. A *cell* of a lattice is the area (2D) or volume (3D) formed by lattice vectors. A *primitive cell* is the area or volume formed by primitive lattice vectors, and it is the smallest possible repeating unit which describes a lattice.

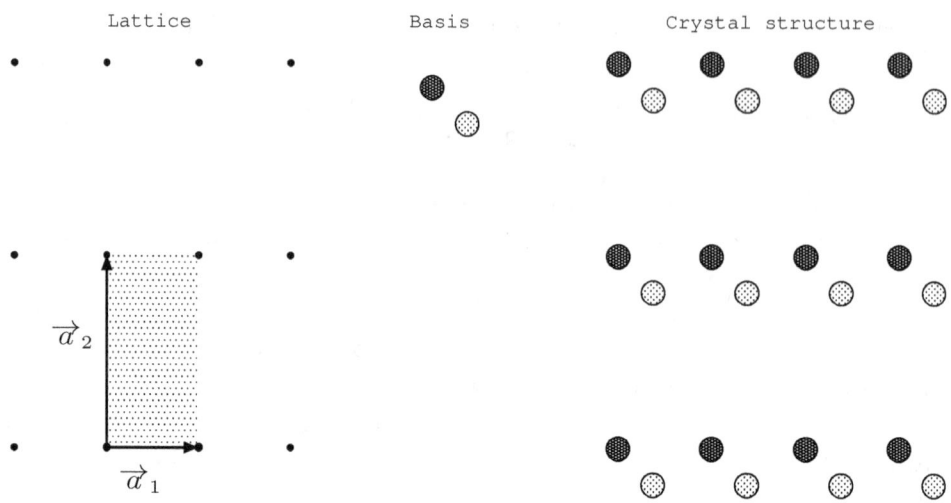

Figure 2.5: Two dimensional illustration of the terms lattice, basis, crystal structure, and primitive lattice vector.

To specify the structure of a material, we attach one or more atoms to every point in the lattice. This arrangement of atoms is called a *crystal basis*. The lattice and crystal basis together define the *crystal structure* [25]. Figure 2.5 shows a two dimensional example of a lattice, crystal basis, and crystal structure. Since this example is two dimensional, only two lattice vectors are needed to specify the lattice. Two primitive lattice vectors are shown, and a primitive cell is shaded.

There are 14 possible three dimensional lattice types, and these are called *Bravais lattices* [25]. Each of these possible lattices has a descriptive name. Figure 2.6 shows four of the possible Bravais lattices: simple cubic, body centered cubic, face centered cubic, and asymmetric triclinic. In the simple cubic lattice, all angles between line segments connecting nearest neighbor points are right angles, and all lengths between nearest neighbor points are equal. In the asymmetric triclinic lattice, none of these angles are right angles, and none of these lengths between nearest neighbor points are equal. Figure 2.6 shows lattice cells, but the cells for the body centered cubic and face centered cubic lattices are not primitive cells because smaller repeating units can be found.

Consider some example lattices and crystal structures. The crystal structure of sodium chloride, for example, involves a face centered cubic lattice and a basis composed of one sodium and one chlorine atom. Another example is silicon which crystallizes in what is known as the diamond structure [25]. This crystal structure involves a face centered cubic lattice and a basis composed of two silicon atoms, at location $(0,0,0)$ and $\left(\frac{l}{4}, \frac{l}{4}, \frac{l}{4}\right)$

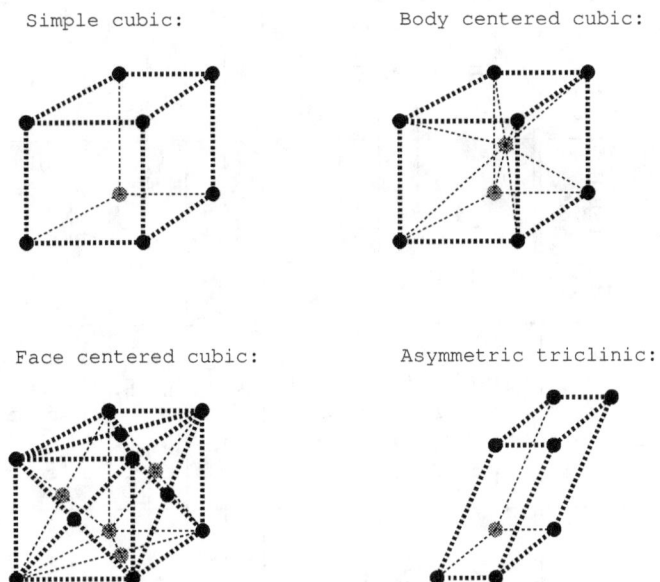

Figure 2.6: Illustration of some Bravais lattices.

where l is the length of the primitive cell. Carbon, Si, Ge, and Sn all crystallize in this diamond structure with cell lengths of $l = 0.356, 0.543, 0.565,$ and 0.646 nm respectively [25].

While there are only 14 possible three dimensional lattices, there are significantly more possible crystal structures because the crystal structure also incorporates the basis. It is not possible to list all possible crystal structures. Instead, they are classified based on the symmetries they contain. Possible symmetry operations are 2-fold, 3-fold, 4-fold, and 6-fold rotations, horizontal and vertical mirror planes, and inversion. Crystal structures are grouped based on the symmetry elements they contain into classes called *crystal point groups*. There are 32 possible crystal point groups, and they are listed in the Table 2.2.

Some authors classify crystal structures into *crystal space groups* instead of *crystal point groups* [6] [26]. While there are 32 crystal point groups, there are 230 crystal space groups. Crystal space groups are based on symmetry transformations which can incorporate not only rotations and mirror planes but also combination of translations along with rotations and mirror planes. Crystal space groups will not be discussed further in this text.

Hermann-Mauguin Notation	Schoenflies Notation	Crystal System	Angles of Primitive Lattice Cell	Lengths of Primitive Lattice Cell	Piezoelec., Pockels Electro-optic, No Inversion Symmetry	Pyro-electric
1	C_1	triclinic	$\alpha, \beta, \gamma \neq 90°$	$a \neq b \neq c$	y	y
$\bar{1}$	S_2	triclinic	$\alpha, \beta, \gamma \neq 90°$	$a \neq b \neq c$	n	n
2	C_2	monoclinic	$\alpha, \gamma = 90°, \beta \neq 90°$	$a \neq b \neq c$	y	y
m	C_{1h}	monoclinic	$\alpha, \gamma = 90°, \beta \neq 90°$	$a \neq b \neq c$	y	y
$\frac{2}{m}$	C_{2h}	monoclinic	$\alpha, \gamma = 90°, \beta \neq 90°$	$a \neq b \neq c$	n	n
222	D_2, V	orthorhombic	$\alpha, \beta, \gamma = 90°$	$a \neq b \neq c$	y	n
$2mm$	C_{2v}	orthorhombic	$\alpha, \beta, \gamma = 90°$	$a \neq b \neq c$	y	y
$\frac{2}{m}\frac{2}{m}\frac{2}{m}$	D_{2h}, V_h	orthorhombic	$\alpha, \beta, \gamma = 90°$	$a \neq b \neq c$	n	n
4	C_4	tetragonal	$\alpha, \beta, \gamma = 90°$	$a = b \neq c$	y	y
$\bar{4}$	S_4	tetragonal	$\alpha, \beta, \gamma = 90°$	$a = b \neq c$	y	n
$\frac{4}{m}$	C_{4h}	tetragonal	$\alpha, \beta, \gamma = 90°$	$a = b \neq c$	n	n
422	D_4	tetragonal	$\alpha, \beta, \gamma = 90°$	$a = b \neq c$	y	n
$4mm$	C_{4v}	tetragonal	$\alpha, \beta, \gamma = 90°$	$a = b \neq c$	y	y
$\bar{4}2m$	D_{2d}, V_d	tetragonal	$\alpha, \beta, \gamma = 90°$	$a = b \neq c$	y	n
$\frac{4}{m}\frac{2}{m}\frac{2}{m}$	D_{4h}	tetragonal	$\alpha, \beta, \gamma = 90°$	$a = b \neq c$	n	n
3	C_3	trigonal	$\alpha, \beta, \gamma \neq 90°$	$a = b = c$	y	y

Table 2.2: Summary of crystal point groups.

Hermann-Mauguin Notation	Schoenflies Notation	Crystal System	Angles of Primitive Lattice Cell	Lengths of Primitive Lattice Cell	Piezoelec., Pockels Electro-optic, No Inversion Symmetry	Pyro-electric
$\bar{3}$	S_6	trigonal	$\alpha, \beta, \gamma \neq 90°$	$a = b = c$	n	n
32	D_3	trigonal	$\alpha, \beta, \gamma \neq 90°$	$a = b = c$	y	n
$3m$	C_{3v}	trigonal	$\alpha, \beta, \gamma \neq 90°$	$a = b = c$	y	y
$3\frac{2}{m}$	D_{3d}	trigonal	$\alpha, \beta, \gamma \neq 90°$	$a = b = c$	n	n
6	C_6	hexagonal	$\alpha = \beta = 90°, \gamma = 120°$	$a = b \neq c$	y	y
$\bar{6}$	C_{3h}	hexagonal	$\alpha = \beta = 90°, \gamma = 120°$	$a = b \neq c$	y	n
$\frac{6}{m}$	C_{6h}	hexagonal	$\alpha = \beta = 90°, \gamma = 120°$	$a = b \neq c$	n	n
622	D_6	hexagonal	$\alpha = \beta = 90°, \gamma = 120°$	$a = b \neq c$	y	n
$6mm$	C_{6v}	hexagonal	$\alpha = \beta = 90°, \gamma = 120°$	$a = b \neq c$	y	y
$\bar{6}m2$	D_{3h}	hexagonal	$\alpha = \beta = 90°, \gamma = 120°$	$a = b \neq c$	y	n
$\frac{6}{m}\frac{2}{m}\frac{2}{m}$	D_{6h}	hexagonal	$\alpha = \beta = 90°, \gamma = 120°$	$a = b \neq c$	n	n
23	T	cubic (isometric)	$\alpha, \beta, \gamma = 90°$	$a = b = c$	y	n
$\frac{2}{m}\bar{3}$	T_h	cubic (isometric)	$\alpha, \beta, \gamma = 90°$	$a = b = c$	n	n
432	O	cubic (isometric)	$\alpha, \beta, \gamma = 90°$	$a = b = c$	y	n
$\bar{4}3m$	T_d	cubic (isometric)	$\alpha, \beta, \gamma = 90°$	$a = b = c$	y	n
$\frac{4}{m}\bar{3}\frac{2}{m}$	O_h	cubic (isometric)	$\alpha, \beta, \gamma = 90°$	$a = b = c$	n	n

Table 2.2 continued: Summary of crystal point groups.

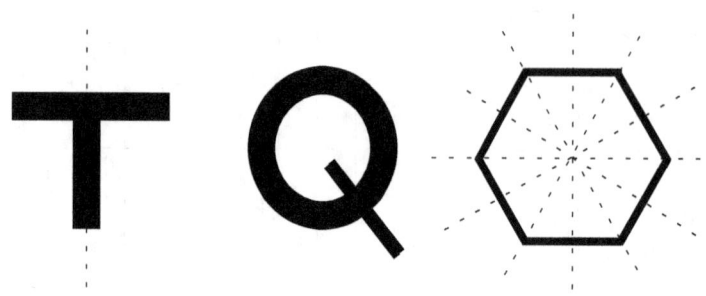

Figure 2.7: Shapes used to illustrate symmetry elements.

As an example of identifying symmetry elements, consider the 2D shapes in Fig. 2.7. The T-shaped figure has one symmetry element, a mirror plane symmetry. The shape looks the same if it is reflected over the mirror plane shown in the figure by a dotted line. The Q-shape has no symmetry elements. The hexagon has multiple symmetry elements. It contains 2-fold rotation because it looks the same when rotated by 180°. It also has 3-fold and 6-fold rotation symmetries because it looks the same when rotated by 60° and 30° respectively. It also has multiple mirror planes shown by dotted lines in the figure. In this example, symmetry elements of 2D shapes are identified, but material scientists are interested in identifying symmetries of 3D crystal structures to gain insights in the properties of materials. Materials are classified into categories called crystal point groups based on the symmetries of their crystal structures.

We generalize about crystalline materials based on whether or not their crystal structure possesses inversion symmetry. What is the inversion operation? In 2D, inversion is the same as a rotation by 180°. In 3D, a shape or crystal structure contains *inversion symmetry* if it is identical when rotated by 180° and inverted through the origin [24, p. 269]. More specifically, draw a vector $\vec{V}$ from the center of the shape to any point on the surface. If the shape has inversion symmetry, then for any such vector $\vec{V}$, the point a distance $-\vec{V}$ from the origin is also on the surface of the shape. The example on the left of Fig. 2.8 has inversion symmetry because for any such vector $\vec{V}$ from the center of the shape to a point on the surface, there is a point on the surface a vector $-\vec{V}$ away from the origin too. The example on the right does not contain inversion symmetry as illustrated by the vector $\vec{V}$ shown by the arrow.

If a crystal structure has inversion symmetry, we say the crystal has a *center of symmetry* otherwise we say it is *noncentrosymmetric*. Crystal structures are classified into classes called crystal point groups, and twenty-

With Inversion Symmetry   Without Inversion Symmetry

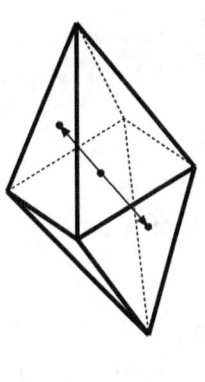 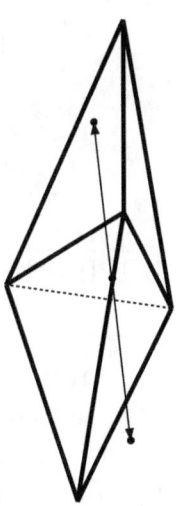

Figure 2.8: The shape on the left contains inversion symmetry while the shape on the right does not.

one of the 32 point groups have no center of symmetry thus do not contain inversion symmetry [24, p. 35]. Twenty of these crystal point groups have a *polar axis*, some axis in the crystal with different forms on opposite ends of the axis. These twenty one crystal point groups are specified as noncentrosymmetric in the sixth column of Table 2.2. If we mechanically stress these materials along the polar axis, different amounts of charges will build up on the different sides of the axis. Dielectric crystalline materials whose crystal structure belongs to any one of these 21 of these noncentrosymmetric crystal point groups are piezoelectric [24].

Table 2.2 lists all crystal point groups and summarizes whether crystalline materials whose crystal structure belongs to each group can be piezoelectric, pyroelectric, and Pockels electro-optic. Pyroelectricity and electro-optics are discussed in the next chapter. Information in the table comes from references [24] [26] [27] [28]. The left two column list the 32 possible crystal point groups. There are two different, but equivalent, ways of labeling the crystal point groups. The first column names the crystal point groups using Hermann-Mauguin notation. This notation dates to the 1930s and is used by chemists, mineralogists, and some physicists. The second column names the crystal point groups using Schoenflies notation. Schoenflies notation dates from 1891 [29], and it is used by mathematicians, spectroscopists, and other physicists.

The third column of Table 2.2 lists the *crystal system*. As shown in Fig.

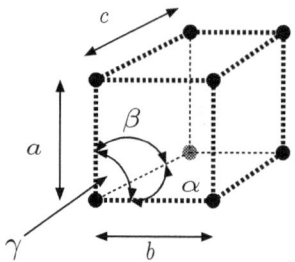

Figure 2.9: Labels on a primitive cell of a lattice.

2.9, the angles of a primitive cell of a lattice are labeled α, β, and γ, and the lengths of the sides are labeled a, b, and c. Crystal point groups can be classified based on the angles and lengths of the primitive cell of the lattice which belongs to that group. The literature contains multiple subtly different ways of defining crystal systems [30]. The information in the third column follows reference [28]. The fourth column gives relationships between the angles of the primitive cell. The fifth column gives relationships between the lengths of sides of the primitive cell. Combinations of angles and lengths are not unique to a specific row. For example, classes C_2 and C_{1h} both have $\alpha = 90°$, $\beta \neq 90°$, $\gamma = 90°$, and $a \neq b \neq c$. However, crystal structures belonging to these crystal point groups contain different symmetry elements. For more details on specifically which symmetry elements are contained in which crystal point group, see [24] [26] [27] [28]. The sixth column lists whether or not the crystal point group has inversion symmetry. Crystal structures with no inversion symmetry or center of symmetry, called *noncentrosymmetric*, are both piezoelectric and Pockels electro-optic. The last column lists whether or not crystalline materials whose crystal structure belongs to the various crystal point groups can be pyroelectric.

It is possible to start with a crystal structure of a material, derive the symmetry elements it contains, derive whether or not the material is piezoelectric, and derive whether or not the material is pyroelectric. Furthermore, it is possible to derive along which axes piezoelectricity or pyroelectricity can occur in the material. However, this derivation is beyond the scope of this book. For further details, see [6] [27] [28] [31].

To predict whether or not a dielectric crystalline material is piezoelectric, identify its lattice and crystal basis to identify its crystal structure. Identify the symmetries of the crystal structure to classify its crystal structure into a particular crystal point group. If that crystal structure contains inversion symmetry, the material can be piezoelectric. We often do not have to go through all of these steps because the crystal point group for many crystalline materials is tabulated [32]. Even if the crystal structure for a material contains inversion symmetry, the piezoelectric effect and the piezoelectric strain coefficient d may be too small to measure.

The effect may only occur when you stress the material along some particular axis, and it may not occur for a mechanical stress of an arbitrary orientation with respect to the direction of the crystal axes. There is only one crystal point group, called asymmetric triclinic, where a random stress will produce a material polarization [24]. For all other point groups, only stresses along certain axes will produce a material polarization [24]. Furthermore, in most crystals a given amount of stress along one axis of the crystal will produce a different amount of material polarization than the same amount of stress applied along a different crystallographic axis. Qualitatively, compressing a crystal along one axis may cause more charge displacement than compressing a crystal along a different axis. For this reason, it is more accurate to treat the piezoelectric strain coefficient as a matrix. This 3x6 matrix has elements

$$d_{ik} = \left(\frac{\partial \text{strain along k}}{\partial \text{electric field along i}} \right) \bigg|_{\text{for a given stress}} . \quad (2.20)$$

where electric field has x, y, and z components, and the stress can be applied along the xx, xy, xz, yy, yz, or zz directions.

2.3.3 Piezoelectricity in Amorphous and Polycrystalline Materials and Ferroelectricity

The previous section discussed piezoelectricity in crystals. We can discuss symmetries of the crystal structure of crystalline materials, but we cannot even define a crystal structure for amorphous materials. However, it is possible to make piezoelectric devices out of polycrystalline and amorphous materials. In a dielectrics, if we apply an external electric field, a material polarization is induced. Electric dipoles form because the electrons and nuclei of the atoms displace slightly from each other. Coulomb's law tells us that charge buildups, such as these electric dipoles, induce an electric field. So, if we apply an external electric field to a dielectric, this primary effect induces a material polarization, and this material polarization will, as a secondary effect, induce additional material polarization in the material. Once one atom polarizes forming an electric dipole, nearby atoms will polarize. Small regions of the same material polarization are called *electrical domains*.

In certain dielectric materials, an external mechanical stress induces a local material polarization. The charge buildup of this material polarization induces a material polarization in nearby atoms forming electrical domains [23]. This piezoelectric effect can occur whether the original material is crystalline, amorphous, or polycrystalline [23]. In noncrystalline materials, this effect is necessarily nonlinear, so these materials are not well described by Eqs. 2.14 or 2.19.

The nonlinear process of a material polarization of one atom inducing a material polarization of nearby atoms causing the formation of electrical domains is called *ferroelectricity*. Ferroelectric materials may be crystalline, amorphous, or polycrystalline. We will see in the next chapter that materials can be ferroelectric pyroelectric and ferroelectric electro-optic in addition to ferroelectric piezoelectric. The ferroelectric effect is limited by temperature. For many ferroelectric materials, these effects occur only below some temperature, called the *Curie temperature*. When the materials are heated above the Curie temperature, the ferroelectric effect goes away [33]. The material polarization of a ferroelectric material may depend on whether or not a material polarization has previously been induced. If the state of a material depends on its past history, we say that the material has *hysteresis*. Ferroelectric materials may have a material polarization even in the absence of an external mechanical stress or electric field if a source of energy has previously been applied.

While the prefix ferro- means iron, most ferroelectric materials do not contain iron, and most iron containing materials are not ferroelectric. The word *ferroelectric* is used as an analogy to the word *ferromagnetic*. Some iron containing materials are ferromagnetic. If an external magnetic field is applied across a *ferromagnetic* material, an internal magnetic field is set up in the material. Ferromagnetic materials can have a permanent magnetic dipole even in the absence of an applied magnetic field. We can model an electric dipole as a pair of charges. We can model a magnetic dipole as a small current loop. Ferromagnetic materials exhibit hysteresis, and they have magnetic domains where the magnetic dipoles are aligned.

Originally, a piezoelectric ferroelectric material has randomly aligned electrical domains and no net material polarization, so it starts out as neither piezoelectric or ferroelectric. The process of causing a material to exhibit piezoelectricity and ferroelectricity is called *poling*. To pole a material, place it in a strong external electric field [23], for example, across the poles of a battery, hence the term. Poling does not change the atomic structure, so if the material was originally amorphous, it will remain amorphous. During this process, electrical domains form, and these domains remain even when the external field is removed. A material that is

2 CAPACITORS AND PIEZOELECTRIC DEVICES

piezoelectric due to this type of poling is sometimes called an *electret* [15, p. 297]. After the material is poled, it may have a net material polarization throughout. Furthermore after poling, it is piezoelectric and ferroelectric, so an external mechanical stress induces a material polarization locally and throughout the material.

2.3.4 Materials Used to Make Piezoelectric Devices

What makes a good material for a piezoelectric sensor or piezoelectric energy conversion device? First piezoelectric devices are made from electrical insulators. When an external voltage is applied across a conductor, valence electrons are removed from their atoms, so no material polarization accumulates. Second, piezoelectric devices are made from materials with large piezoelectric strain constants. The piezoelectric strain constant is so small that it cannot be detected in many crystals with crystal structures from one of the 21 crystal point groups known to be piezoelectric, and it is zero in crystals from the other crystal point groups. Third, piezoelectric devices should be made from materials that are not brittle so that they can withstand repeated stressing without permanent damage. Thermal properties may also be important [33]. There is no material that is best in all applications.

Quartz, crystalline SiO_2, was the first material in which piezoelectricity was studied. Pierre and Jacques Curie discovered the effect in quartz in the 1880s [3]. Today, many piezoelectric devices, including crystal oscillators, are made from quartz. Lead zirconium titanate is another material used due to its relatively high piezoelectric strain constant [3] [34]. In applications which require flexibility and the ability to withstand repeated mechanical stress without damage, polymers such as polyvinyldenfluoride are used [25]. Piezoelectricity has also been studied in materials including barium titanate $BaTiO_3$, lithium niobate, tourmaline

$$(Na,Ca)(Li,Mg,Al)_3(Al,Fe,Mn)_6(BO_3)_3(Si_6O_8)(OH)_4,$$

and Rochelle salt

$$KNaC_4H_4O_6 \cdot 4H_2O$$

[3] [23] [24] [34].

Manufacturers of piezoelectric devices do not often label their products to say whether they are made from crystalline, amorphous, or polycrystalline materials, but there are advantages and disadvantages to the different types of materials. An advantage of making piezoelectric devices from polycrystalline or amorphous materials is that the devices can be made more easily into different shapes such as cylinders and spheres [33]. However, the materials used often have lower melting temperatures, higher temperature expansion coefficients, and are more brittle [33]. Crystalline materials, such as quartz, have the advantages of being harder and having a higher melting temperature [33].

2.3.5 Applications of Piezoelectricity

A number of electrical components involve piezoelectricity. When a voltage is applied across a piece of piezoelectric material, it mechanically bends and deforms. When the voltage is released, it springs back at a natural resonant frequency. This material can be integrated with a feedback circuit to produce oscillations at a precise frequency. Electrical oscillators of this type are often made from crystalline quartz. A more recent application is the piezoelectric transformer. These devices are used in the cold cathode fluorescent lamps which are used as backlight for LCD panels [23, p. 289]. The lamps require around a thousand volts to turn on and hundreds of volts during use. Transformers made of magnets and coils can achieve these high voltages, but piezoelectric transformers are much smaller, small enough to be mounted on a printed circuit board. A traditional transformer involves a pair of coils, and it converts AC electricity to magnetic energy to AC electricity at a different voltage. Similarly, a piezoelectric transformer also involves multiple energy conversion processes. In such a device, AC electricity is converted to mechanical vibrations and then to AC electricity at a different voltage. Energy is conserved in these devices, so they can produce high voltages with low currents. Figure 2.10 shows a piezoelectric transformer that can convert an input of 8 to 14 V to an output up to 2 kV [35]. Figure 2.11 shows an example of some small piezoelectric circuit components. Starting in the upper left and going clockwise, a microphone, ultrasonic transmitter and receiver, vibration sensor, and oscillator are shown.

Efficiency of energy conversion devices is hard to discuss because every author makes different assumptions. However by any measure, efficiency of a commercial piezoelectric device is low, often 6% or less [36]. Due to this low efficiency, many piezoelectric devices are used as sensors. Regardless of this low efficiency, other devices are used for energy harvesting. For example, one train station embedded piezoelectric devices in the platforms to generate electricity. Piezoelectric devices also have been used to convert the energy from the motion of a fluid or from wind directly to electricity [36].

There is interest in using piezoelectric devices for biomedical applica-

2 CAPACITORS AND PIEZOELECTRIC DEVICES

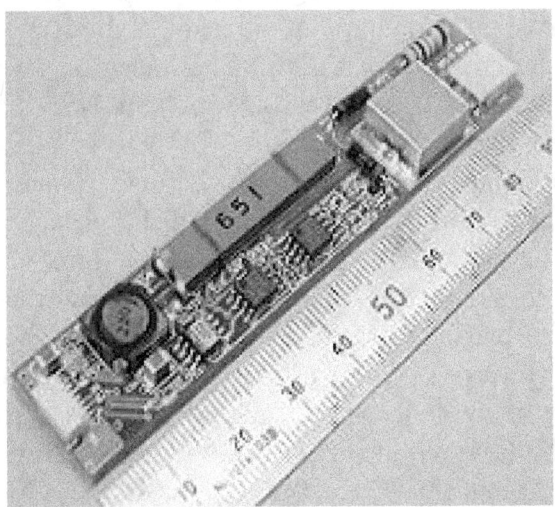

Figure 2.10: A piezoelectric transformer that takes an input of 8-14 V and produces and output of up to 2 kV. This picture is used with permission from [35].

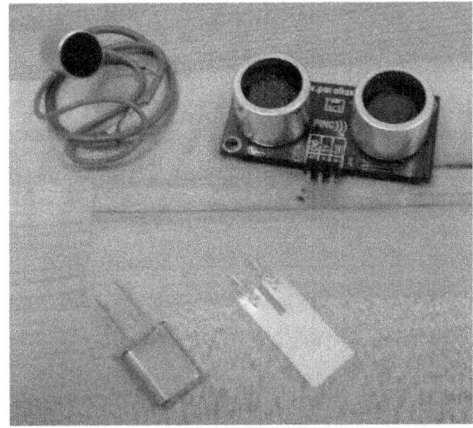

Figure 2.11: Example small piezoelectric devices. Clockwise from top left: electret microphone, ultrasonic distance sensor, vibration sensor, oscillator crystal.

tions. Quartz is piezoelectric, and it is durable, readily available, and nontoxic. Engineers have developed piezoelectric devices designed for use outside of the body and to be implanted inside the body. Some piezoelectric devices are used as sensors. For example, piezoelectric sensors can monitor knees or other joints [3]. Also, ultrasonic imaging is a common diagnostic technique. Piezoelectric devices are used both to generate the ultrasonic vibrations and to detect them [33]. Other biomedical piezoelectric devices are used as a source of electrical power. Artificial hearts, pacemakers, and other devices require electricity, and they are often limited by battery technology available to supply the energy [36]. Piezoelectric generators have no moving parts to wear out, and they can avoid the problem of needing to change the batteries. Physical activity can be classified as continuous, such as breathing, or discontinuous, such as walking. Both types of physical activity can be used as a source of mechanical energy for piezoelectric devices [36]. The amount of power required for different biomedical devices varies quite a bit. For example, an artificial heart may require around 8 W while a pacemaker may require only a few microwatts [36]. Piezoelectric devices may be able to capture energy from typical physical activity and convert it into electrical energy to power the device. A piezoelectric device in an artificial knee has produced 0.85 mW [36], and a device in a shoe has generated 8.4 mW from walking [36].

Piezoelectric devices are used in other types of imaging systems besides biomedical imaging systems. One of the earlier applications was in sonar systems. Around the time of WWI, the military actively developed sonar systems to detect boats and submarines. Today, sonar systems are used to detect fish and to measure the depth of bodies of water [33]. Sonar imaging is also used to analyze electrical circuits and to detect imperfections and cracks in steel and in welds [33].

Piezoelectric devices are used in a variety of other applications too. Piezoelectric sensors are used in some buttons and keyboards [36]. Piezoelectric devices are used to make accelerometers [37, p. 353], and they are used to measure pipe flow [33]. Speakers, microphones and buzzers can all be made from piezoelectric devices, and they can operate at both audio and ultrasonic frequencies. Piezoelectric devices that generate ultrasonic signals can be used to emulsify dyes, paints, and food products like peanut butter [33]. Also, they are used in some barbecue grill ignitions where mechanical stress induces an electric spark [23, ch. 15].

2.4 Problems

2.1. A parallel plate capacitor has a capacitance of $C = 10$ pF. The plates have area 0.025 cm^2. A dielectric layer of thickness $d_{thick} = 0.01$ mm separates the plates. For the dielectric layer, calculate the permittivity ϵ, the relative permittivity ϵ_r, and the electric susceptibility χ_e.

2.2. We often assume that the capacitance of a capacitor and the permittivity of a material are constants. However, sometimes these quantities are better described as functions of frequency. Consider a capacitor made from parallel plates of area 0.025 cm^2 separated by 0.01 mm. Assume that for $\omega \lesssim 10^6 \frac{\text{rad}}{\text{s}}$, the capacitance is well modeled by

$$C(\omega) = 8 \cdot 10^{-11} + 3 \cdot 10^{-15} \omega$$

in farads. For the dielectric material between the plates of the capacitor, calculate the permittivity $\epsilon(\omega)$, the relative permittivity $\epsilon_r(\omega)$, and the electric susceptibility $\chi_e(\omega)$.

2.3. A cylindrical sandwich cookie has a radius of 0.75 in. The cookie is made from two wafers, each of thickness 0.15 in, which are perfect dielectrics of relative permittivity $\epsilon_r = 2.8$. Between the wafers is a layer of cream filling of thickness 0.1 in which is a perfect dielectric of relative permittivity $\epsilon_r = 2.2$. Find the overall capacitance of the cookie.
Hint: Capacitances in series combine as $\frac{1}{\frac{1}{C_1} + \frac{1}{C_2}}$.

2.4. A parallel plate capacitor has a capacitance of 3 μF.

(a) Suppose another capacitor is made using the same dielectric material and with the same cross sectional area. However, the thickness of the dielectric between the plates of the capacitor is double that of the original capacitor. What is its capacitance?

(b) Suppose a third capacitor is made with the same cross sectional area and thickness as the first capacitor, but from a material with twice the permittivity. What is its capacitance?

2.5. A piezoelectric material has a permittivity of $\epsilon = 3.54 \cdot 10^{-11} \, \frac{F}{m}$ and has a piezoelectric strain constant of $d = 2 \cdot 10^{-10} \, \frac{m}{V}$. If the material is placed in an electric field of strength $|\vec{E}| = 70 \, \frac{V}{m}$ and is subjected to a stress of $|\vec{\varsigma}| = 3.5 \, \frac{N}{m^2}$. Calculate the material polarization.

2.6. A piezoelectric material has permittivity $\epsilon_r = 2.5$. If the material is placed in an electric field of strength $|\vec{E}| = 2 \cdot 10^3 \, \frac{V}{m}$ and is subjected to a stress of $|\vec{\varsigma}| = 200 \, \frac{N}{m^2}$, the material polarization of the material is $3.2 \cdot 10^{-8} \, \frac{C}{m^2}$. Calculate d, the piezoelectric strain constant.

2.7. Consider two piezoelectric devices of the same size and shape. The dielectric material of the first device has a permittivity of $\epsilon = 2.21 \cdot 10^{-11} \, \frac{F}{m}$ and a piezoelectric strain constant of $d = 8 \cdot 10^{-11} \, \frac{m}{V}$. The dielectric material of the second device has an electric susceptibility of $\chi_e = 3.2$ and a piezoelectric strain constant of $d = 2 \cdot 10^{-10} \, \frac{m}{V}$.

   (a) Find ϵ_r, the relative permittivity, for each device.
   (b) Find $\frac{C_1}{C_2}$, the ratio of the capacitance of the first device to the capacitance of the second device.
   (c) The devices are placed in an external electric field of strength $|\vec{E}| = 32 \, \frac{V}{m}$. No stress is placed on the devices. Calculate the material polarization, $\vec{P}$ for each device.
   (d) The devices are placed in an external electric field of strength $|\vec{E}| = 32 \, \frac{V}{m}$, and a stress of $|\vec{\varsigma}| = 100 \, \frac{N}{m^2}$ is applied to the devices. Calculate the material polarization, $\vec{P}$ for each device.
   (e) Which device would you expect is able to store more energy? Explain your answer.

2.8. A particular piezoelectric device has a cross sectional area of 10^{-5} m². When a stress of $800 \, \frac{N}{m^2}$ is applied, the device compresses by 10 μm. Under these conditions, the device can generate $2.4 \cdot 10^{-9}$ J. Calculate the efficiency of the device.

2.9. A particular piezoelectric device has a cross sectional area of 10^{-5} m² and an efficiency of 5%. When a stress of $1640 \, \frac{N}{m^2}$ is applied to the device, it oscillates with an average velocity of $0.01 \, \frac{m}{s}$. Calculate the power that can be generated from the device.

2 CAPACITORS AND PIEZOELECTRIC DEVICES

2.10. A piezoelectric device is placed in an electric field of strength $|\vec{E}| = 500 \, \frac{V}{m}$. The device is tested twice. In the first test, a stress of $|\vec{\varsigma}| = 1000 \, \frac{N}{m^2}$ was put on the device, and the material polarization was measured to be $|\vec{P}| = 2.75 \cdot 10^{-8} \, \frac{C}{m^2}$. In the second test also with $|\vec{E}| = 500 \, \frac{V}{m}$, a stress of $|\vec{\varsigma}| = 100 \, \frac{N}{m^2}$ was put on the device, and the material polarization was measured to be $|\vec{P}| = 6.50 \cdot 10^{-9} \, \frac{C}{m^2}$. Find the piezoelectric strain constant d, and find the relative permittivity of the material ϵ_r.

2.11. According to the data sheet, a piezoelectric device is 3% efficient. A coworker says that energy is not conserved in the device because 97% of the energy is lost when it is used. Explain what is wrong with your coworker's explanation.

2.12. Match the material property with its definition. (Not all definitions will be used.)

1. A mechanical stress will cause a (material) polarization in this type of material.	A. Amorphous
2. This type of material is glassy and noncrystalline.	B. Dielectric
3. Charges do not easily flow through this type of material	C. Ferroelectric
4. In the presence of a weak external voltage, charges do not flow in this type of material. In the presence of a strong external voltage, charges flow easily.	D. Piezoelectric
5. A material polarization in one atom induces material polarization in nearby atoms in this type of material.	

2.13. Consider a piezoelectric material in an external electric field $\vec{E}$ in units $\frac{V}{m}$. The figure shows the magnitude of the material polarization, $|\vec{P}|$ in units $\frac{C}{m^2}$, as a function of the strength of the external electric field when no mechanical stress is applied. The material has a piezoelectric strain constant of $d = 5 \cdot 10^{-10} \frac{m}{V}$.

(a) Find the relative permittivity ϵ_r, and find the electric susceptibility χ_e.

(b) Find and plot an expression for the magnitude of the material polarization as a function of the external electric field strength when a stress of 1000 $\frac{N}{m^2}$ is applied. Label the axes of your plot well.

(c) This material is used to make a piezoelectric device with a cross sectional area of 1 cm². When this device is compressed a distance of 1 mm, an energy of $2 \cdot 10^{-10}$ J is stored. Find the efficiency of the device.

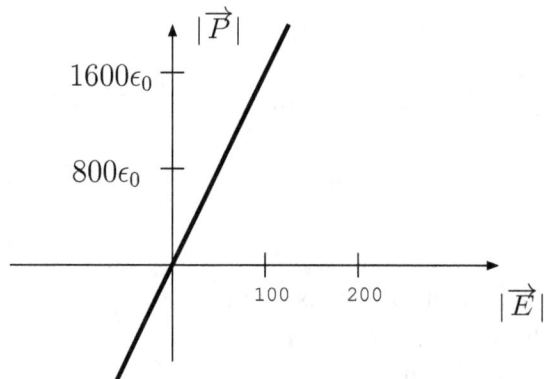

2.14. Consider the 2D crystal structure, shown in the figure, composed of a lattice and a crystal basis. The crystal basis is composed of two atoms of type A and one atom of type B.

   (a) Sketch the crystal basis.
   (b) Sketch the 2D lattice.
   (c) Draw two primitive vectors $\vec{a}_1$ and $\vec{a}_2$ on your sketch.

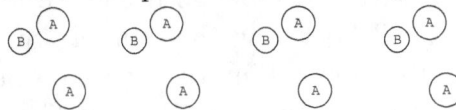

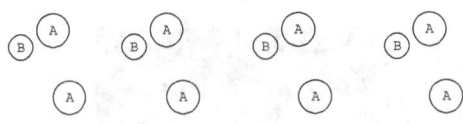

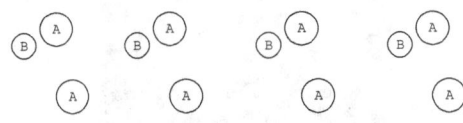

2.15. Consider the illustrations of the crystal structure of two 2D materials where X and O represent the location of different types of atoms. Do the materials have the same crystal structure? basis? lattice? crystal point group? Answer yes or no, and explain.

Material 1:

```
X O    X O    X O    X O




X O    X O    X O    X O




X O    X O    X O    X O
```

Material 2:

```
XX     XX     XX     XX
XX     XX     XX     XX



XX     XX     XX     XX
XX     XX     XX     XX



XX     XX     XX     XX
XX     XX     XX     XX
```

2.16. The figure below illustrates two possible crystal lattices: a face centered cubic lattice and a body centered cubic lattice. The solid arrows represent lattice vectors, but not primitive lattice vectors, and the cells shown are not primitive cells. The dotted vectors in the figure show primitive lattice vectors. In the case of the face centered cubic lattice, the primitive lattice vectors go from a corner point to a point on in the middle of one of the faces of the cube. In the body centered cubic lattice, the primitive lattice vectors go from a corner point to a point in the center of a cell bordering that corner. Suppose that the solid vectors have length 0.4 nm. Find the length of the primitive lattice vector in the face centered cubic lattice, and find the length of the primitive lattice vector in the body centered cubic lattice.

Face Centered Cubic   Body Centered Cubic

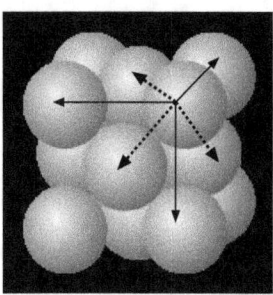
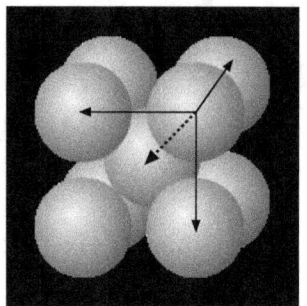

3 Pyroelectrics and Electro-Optics

3.1 Introduction

Electrical engineers interested in materials often focus their study on semiconductors or occasionally conductors. However, energy conversion devices are made out of all types of materials. In the last chapter we discussed capacitors and piezoelectric devices. Both are constructed from a layer of insulating material between conductors. The properties of this dielectric layer determine the properties of the resulting devices. This chapter discusses two additional types of devices that involve material polarization of insulators, pyroelectric devices and electro-optic devices. As with other types of energy conversion devices, these can operate two ways. *Pyroelectric devices* can convert a temperature difference to a material polarization and therefore electricity, or they can convert a material polarization to a temperature difference. *Electro-optic devices* can convert optical electromagnetic radiation to a material polarization or vice versa. As with the devices studied in the last chapter, these devices are constructed around a dielectric layer, and the choice of material in the dielectric layer determines the behavior of the device. Studying these devices is worthwhile even though they are encountered significantly less often than capacitors and piezoelectric devices because this study illustrates the variety of energy conversion devices that engineers have produced.

If a solid is heated enough, it melts. Some materials have multiple crystal structures that are stable at room temperature. These materials may be converted from one crystal structure to another by heating and cooling. Similar effects can occur if energy is supplied by shining a strong enough laser on the material instead of heating it. This chapter is *not* concerned with effects involving melting or thermally changing the crystal structure from one phase to another. Instead, we consider the case when a small amount of energy is supplied, by heat or by electromagnetic radiation. The energies involved are enough to change the material polarization and the internal momentum of electrons but not the location of the nuclei of the material, for example.

3.2 Pyroelectricity

3.2.1 Pyroelectricity in Crystalline Materials

Pyroelectric devices are energy conversion devices which convert a temperature difference to or from electricity through changes in material polarization. The pyroelectric effect was first studied by Hayashi in 1912 and by

Material	Chemical composition	Piezoelectric strain const. d in $\frac{m}{V}$ from [38] [39]	Pyroelectric coeff. $|\vec{b}|$ in $\frac{C}{m^2 K}$ from [38] [39]	Pockels electro-optic coeff. γ in $\frac{m}{V}$ from [27]
Sphalerite	ZnS	$1.60 \cdot 10^{-12}$	$4.34 \cdot 10^{-7}$	$1.6 \cdot 10^{-12}$
Quartz	SiO$_2$	$2.3 \cdot 10^{-12}$	$1.67 \cdot 10^{-6}$	$0.23 \cdot 10^{-12}$
Barium Titanate	BaTiO$_3$	$2.6 \cdot 10^{-10}$	$12 \cdot 10^{-6}$	$19 \cdot 10^{-12}$

Table 3.1: Example piezoelectric strain constants, pyroelectric coefficients, and Pockels electro-optic coefficients. Values for sphalerite assume the $\overline{4}3m$ crystal structure. Pockels coefficients assume a wavelength of $\lambda = 633$ nm. Average values specified in the references are given. See the cited references for additional assumptions. The Pockels electro-optic coefficient γ is defined in Sec. 3.3.1.

Rontgen in 1914 [3] [40]. This effect occurs in insulators, so it is different from the thermoelectric effect. The thermoelectric effect, to be discussed in Chapter 8, is a process that converts between energy of a temperature difference and electricity and occurs because heat and charges flow at different rates through junctions.

If an insulating crystal is placed in an external electric field, the material will polarize. The electrons will displace slightly forming electric dipoles, and energy can be stored in this material polarization. In some pyroelectric materials, heating or cooling will also cause the material to polarize. We can model the material polarization by adding a term to Eq. 2.14 to account for the temperature dependence [3, p. 327].

$$\vec{P} = \vec{D} - \epsilon_0 \vec{E} + \vec{b} \Delta T. \qquad (3.1)$$

As in Eq. 2.14, $\vec{P}$ represents material polarization in $\frac{C}{m^2}$, $\vec{D}$ represents displacement flux density in $\frac{C}{m^2}$, $\vec{E}$ represents electric field intensity in $\frac{V}{m}$, and ϵ_0 is the permittivity of free space in $\frac{F}{m}$. The *pyroelectric coefficient* $\vec{b}$ has units $\frac{C}{m^2 \cdot K}$, and ΔT represents the change in temperature. The coefficient $\vec{b}$ is a vector because the material polarization may be different along different crystal directions. Table 3.1 lists example values for the pyroelectric coefficient as well as for other coefficients. (Note that this definition of $\vec{b}$ is similar but not identical to the definition in [3].) In some materials, the material polarization depends linearly on the temperature as described by Eq. 3.1. In other materials, more terms are needed to describe the dependence of the material polarization on temperature.

$$\vec{P} = \vec{D} - \epsilon_0 \vec{E} + \vec{b}\Delta T + \vec{b_{quad}}(\Delta T)^2 \dots \quad (3.2)$$

Many materials exhibit pyroelectricity only below a temperature known as the *pyroelectric Curie temperature*.

In the last chapter, we saw that we could determine whether or not a crystalline material was piezoelectric from its crystal structure. To do so, we identified the symmetries of the crystal structure. Crystal structures are grouped into 32 classes called *crystal point groups* based on the symmetries they contain. Crystal structures in the 21 of the crystal point groups that do not have a center of symmetry can be piezoelectric. We can use a similar technique to determine if a crystalline material is or is not pyroelectric. All pyroelectric crystals are piezoelectric, but not all piezoelectric crystals are pyroelectric. To determine if a crystalline material can be pyroelectric, identify its crystal structure and determine the corresponding crystal point group. Crystals in the 10 crystal point groups listed in Table 2.2 are pyroelectric [3, p. 366] [26, p. 557].

3.2.2 Pyroelectricity in Amorphous and Polycrystalline Materials and Ferroelectricity

In the Sec. 2.3.3 we saw that some materials, called ferroelectric piezoelectric materials, had a material polarization that depended nonlinearly on the mechanical stress applied. These materials could be crystalline, amorphous, or polycrystalline. When a charge separation occurred in one atom, the charges from that electric dipole induce dipoles to form in nearby atoms, and electrical domains with aligned material polarization form in the material. This effect depends on the mechanical stress applied to the material previously, and the dependence on past history is called hysteresis.

Materials can also be *ferroelectric pyroelectric*, and these materials can be crystalline, amorphous, or polycrystalline. In these materials, the material polarization depends nonlinearly on the temperature, as opposed to the mechanical stress. As with the piezoelectric version of this effect, polarization of one atom induces a material polarization in nearby atoms. Such materials can have a material polarization even when no temperature gradient is applied, and they can exhibit hysteresis.

3.2.3 Materials and Applications of Pyroelectric Devices

Pyroelectricity has been studied in a number of materials including barium titanate $BaTiO_3$, lead titanate $PbTiO_3$, and potassium hydrogen phosphate

KH_2PO_4 [25] [26]. It has also been studied in chalcogenide glasses which are sulfides, selenides, and tellurides such as GeTe [25] [26]. When selecting a pyroelectric material for an application, the pyroelectric coefficient should be considered. Thermal properties are important too. The material should be able to withstand repeated heating and cooling, and it should have a relatively high melting temperature to be useful.

The pyroelectric effect does not have many applications. Some optical detectors designed to detect infrared light are made from pyroelectric materials [41] [42]. However, most optical detectors are photovoltaic devices made from semiconductor junctions, and this technology will be discussed in Chapter 6. While sensors using the pyroelectric effect could be used to measure temperature, other types of temperature sensors, such as thermocouples, are typically used. Thermocouples, which operate based on the thermoelectric effect which is discussed in Chapter 8, are more convenient to build and operate. Additionally, in many pyroelectric materials, the effect is nonlinear while linear sensors are easier to work with and calibrate.

3.3 Electro-Optics

3.3.1 Electro-Optic Coefficients

Typically, the magnitude of material polarization in a dielectric is proportional to the strength of an applied electric field.

$$\vec{P} = \vec{D} - \epsilon_0 \vec{E} = \epsilon_0 \chi_e \vec{E} \tag{3.3}$$

In this equation χ_e is the *electric susceptibility*, and it is unitless. It is defined in Sec. 2.2.3 and related to permittivity by Eq. 2.9. However in other materials, the material polarization depends nonlinearly on the applied electric field. Materials for which the material polarization depends linearly on the external electric field are called linear materials while others are called nonlinear or electro-optic materials. The *electro-optic effect* occurs when an applied external electric field induces a material polarization in a material where the amount of polarization depends nonlinearly on the external field. The name involves the word optic because the external field is often due to a visible laser beam. However, the external field can be from any type of source at any frequency, and a material polarization will occur even with a constant applied electric field. A large enough external electric field will cause a material to melt or to crystallize in a different phase, but this effect is not the electro-optic effect. Instead, the electro-optic effect only involves a change in the material polarization, not the crystal structure, and the change involved is not permanent.

3 PYROELECTRICS AND ELECTRO-OPTICS

We can write the magnitude of the material polarization as a function of powers of the applied external field.

$$|\vec{P}| = \epsilon_0 \chi_e |\vec{E}| + \epsilon_0 \chi^{(2)} |\vec{E}|^2 + \epsilon_0 \chi^{(3)} |\vec{E}|^3 + ... \quad (3.4)$$

The quantity $\chi^{(2)}$ is called the *chi-two coefficient*, and it has units $\frac{m}{V}$. The quantity $\chi^{(3)}$ is called the *chi-three coefficient*, and it has units $\frac{m^2}{V^2}$ [27] [42, ch. 1].

If an infinite number of terms are included on the right side of Eq. 3.4, any arbitrary material can be described. In most materials, only the first term of Eq. 3.4 is needed while $\chi^{(2)}$, $\chi^{(3)}$, and all higher order coefficients are negligible, and these materials are not electro-optic. Materials with $\chi^{(2)}$, $\chi^{(3)}$ or other coefficients nonzero are called electro-optic. It is rare to need more coefficients than χ_e, $\chi^{(2)}$, and $\chi^{(3)}$ to describe a material.

The effect due to the $\epsilon_0 \chi^{(2)} |\vec{E}|^2$ term is called the *Pockels effect* or linear electro-optic effect. It was first observed by Friedrich Pockels in 1893 [3, p. 382] [10]. In this case the material polarization depends on the square of the external field. The effect due to the $\epsilon_0 \chi^{(3)} |\vec{E}|^3$ term is called the *Kerr effect* or the quadratic electro-optic effect. In this case, the material polarization depends on the cube of the external electric field. John Kerr first described this effect in 1875 [3, p. 382] [10].

While some authors use the coefficients χ_e, $\chi^{(2)}$ and $\chi^{(3)}$, this effect is most often studied by optics scientists who instead prefer index of refraction n, a unitless measure introduced in Sec. 2.2.3. In electro-optic materials, the index of refraction is a nonlinear function of the strength of the external electric field. Instead of expanding the material polarization in a power series as a function of the external field strength as in Eq. 3.4, the index of refraction is expanded. Pockels and Kerr coefficients are defined as terms of this expansion.

As described by Eq. 2.3, material polarization is the difference in $\frac{C}{m^2}$ between an external electric field in a material and the field in the absence of the material.

$$|\vec{P}| = |\vec{D}| - \epsilon_o |\vec{E}| \quad (3.5)$$

With some algebra, we can identify the displacement flux density component and the overall index of refraction. Add two terms which sum to zero to Eq. 3.4.

$$|\vec{P}| = \epsilon_0 \chi_e |\vec{E}| + \epsilon_0 |\vec{E}| + \epsilon_0 \chi^{(2)} |\vec{E}|^2 + \epsilon_0 \chi^{(3)} |\vec{E}|^3 - \epsilon_0 |\vec{E}| \quad (3.6)$$

The first two terms can be combined, and $\epsilon_0|\vec{E}|$ can be distributed out.

$$|\vec{P}| = \left[(\chi_e + 1) + \chi^{(2)}|\vec{E}| + \chi^{(3)}|\vec{E}|^2 + ...\right]\epsilon_0|\vec{E}| - \epsilon_0|\vec{E}| \qquad (3.7)$$

The first term is the displacement flux density.

$$\vec{D} = \epsilon_{r\ eo}\vec{E} = \left[(\chi_e + 1) + \chi^{(2)}|\vec{E}| + \chi^{(3)}|\vec{E}|^2 + ...\right]\epsilon_0|\vec{E}| \qquad (3.8)$$

The quantity in brackets in Eq. 3.8 is the relative permittivity, $\epsilon_{r\ eo}$. Since we are considering electro-optic materials, it depends nonlinearly on the applied external field. Assuming the material is a perfect dielectric with $\mu = \mu_0$, the index of refraction is the square root of this quantity. It represents the ratio of the speed of light in free space to the speed of light in this material, and it also depends nonlinearly on the applied external field.

$$n_{eo} = \sqrt{\epsilon_{r\ eo}} \qquad (3.9)$$

The index of refraction must be larger than one because electromagnetic waves in materials cannot go faster than the speed of light, so the quantity $\frac{1}{\epsilon_{r\ eo}}$ must be less than one.

Some authors expand the term $\frac{1}{\epsilon_{r\ eo}}$ in a Taylor expansion instead of the material polarization, and electro-optic coefficients are defined with respect to this expansion [42].

$$\frac{1}{\epsilon_{r\ eo}} = \frac{1}{\epsilon_{r\ x}} + \gamma|\vec{E}| + s|\vec{E}|^2 + ... \qquad (3.10)$$

The coefficient γ is called the *Pockels coefficient*, and it has units $\frac{m}{V}$. The coefficient s is called the *Kerr coefficient*, and it has units $\frac{m^2}{V^2}$. In the absence of nonlinear electro-optic contributions, we can denote the relative permittivity as $\epsilon_{r\ x}$ and the index of refraction as n_x where

$$\epsilon_{r\ x} = n_x^2 = \chi_e + 1. \qquad (3.11)$$

The expansion of Eq. 3.10 is guaranteed to converge because $\frac{1}{\epsilon_{r\ eo}} < 1$. Example values of the Pockels electro-optic coefficient are listed in Table 3.1.

With some algebra, the overall index of refraction n_{eo} can be written in terms of the Pockels and Kerr coefficients. Equations 3.9 and 3.10 can be combined.

3 PYROELECTRICS AND ELECTRO-OPTICS

$$n_{eo} = \left(\frac{1}{\epsilon_{r\;x}} + \gamma\left|\vec{E}\right| + s\left|\vec{E}\right|^2 + ...\right)^{-1/2} \quad (3.12)$$

$$n_{eo} = \left[\frac{1}{\epsilon_{r\;x}}\left(1 + \gamma\epsilon_{r\;x}\left|\vec{E}\right| + s\epsilon_{r\;x}\left|\vec{E}\right|^2 + ...\right)\right]^{-1/2} \quad (3.13)$$

$$n_{eo} = n_x \left[1 + \gamma n_x^2 \left|\vec{E}\right| + s n_x^2 \left|\vec{E}\right|^2 + ...\right]^{-1/2} \quad (3.14)$$

The quantity of Eq. 3.14 in brackets can be approximated using the binomial expansion and keeping only the first terms.

$$\left(1 + \gamma n_x^2 \left|\vec{E}\right| + s n_x^2 \left|\vec{E}\right|^2 + ...\right)^{-1/2} \approx \left(1 - \frac{1}{2}\gamma n_x^2 \left|\vec{E}\right| - \frac{1}{2} s n_x^2 \left|\vec{E}\right|^2\right) \quad (3.15)$$

Finally, the overall index of refraction can be written as a polynomial expansion of the strength of the external electric field [10, p. 698].

$$n_{eo} \approx n_x \left(1 - \frac{1}{2}\gamma n_x^2 \left|\vec{E}\right| - \frac{1}{2} s n_x^2 \left|\vec{E}\right|^2\right) \quad (3.16)$$

The Pockels electro-optic effect is called the linear electro-optic effect while the Kerr effect is called the quadratic effect due to the form of the equation above.

3.3.2 Electro-Optic Effect in Crystalline Materials

As with the piezoelectric effect, we can determine which crystalline insulating materials will exhibit the Pockels effect by looking at the symmetries of the material. To determine if a crystal can show the Pockels effect, determine the crystal structure, identify the symmetries, and determine its crystal point group. The Pockels effect occurs in noncentrosymmetric materials, materials with a crystal structure with no inversion symmetry. Of the 32 crystal point groups, 21 of these groups may exhibit the Pockels electro-optic effect. For materials in these crystal point groups, $\chi^{(2)}$ and the Pockels coefficient γ are nonzero. These 21 groups are also the piezoelectric crystal point groups [10, ch. 18], and they are listed in Table 2.2. In some crystalline materials which belong to these crystal point groups, the Pockels effect is nonzero but too small to be measurable.

From Table 2.2 we can see that all materials that are piezoelectric are also Pockels electro-optic and vice versa. Also, all materials that are pyroelectric are piezoelectric but not the other way around. Thus, if a device is used as an electro-optic device, and the device is accidentally mechanically stressed or vibrated, the material polarization will be induced by piezoelectricity. In many devices, these effects simultaneously occur, and it can be difficult to identify the primary cause of a material polarization when multiple effects simultaneously occur.

Tables of Pockels electro-optic coefficients for crystals can be found in reference [27] and [42].

The Kerr electro-optic effect can occur in crystals whether or not they belong to a crystal point group which has a center of symmetry, so some materials exhibit the Kerr effect but not the Pockels effect. In many materials, the Kerr effect is quite small.

3.3.3 Electro-Optic Effect in Amorphous and Polycrystalline Materials

Table 2.2 only applies to crystalline materials because only crystalline materials have a specific crystal structure and can be classified into to a crystal point group. However, crystalline, polycrystalline, and amorphous materials can all be electro-optic. In amorphous and polycrystalline materials, the electro-optic effect is necessarily nonlinear. When an external electric field, for example from a laser, is applied, a material polarization develops. The charge separation in that region induces a material polarization in nearby atoms. Just as materials can be ferroelectric piezoelectric and ferroelectric pyroelectric, amorphous and polycrystalline materials can be ferroelectric electro-optic.

3.3.4 Applications of Electro-Optics

Some controllable optical devices are made from electro-optic materials. Examples of such devices include controllable lenses, prisms, phase modulators, switches, and couplers [10]. Operation of these devices typically involves two laser beams. One of these beams controls the material polarization of the device. The intensity, phase, or electromagnetic polarization of the second optical beam is altered as it travels through the device [10, p. 698-700]. Combinations of these electro-optic devices are used to make controllable optical logic gates and interconnects for optical computing applications [10, ch. 21] [31, ch. 20].

Most memory devices are not made from electro-optic materials, but some creative memory device designs involve electro-optic materials. For example, electro-optic materials are used for some rewritable memory [10, p. 712] [27, p. 534] and for hologram storage [10, ch. 21] [27, ch. 20].

Also, electro-optic materials are used in liquid crystal displays [10, ch. 18]. Liquid crystals are electro-optic materials because an external voltage alters their material polarization [10, ch. 18].

Electro-optic materials are also used to convert an optical beam at one frequency to an optical beam at a different frequency. Second harmonic generation involves converting an optical beam with photons of energy E to a beam with photons at energy $\frac{1}{2}E$ [10, ch. 19] [27, ch. 18] [31, ch. 16]. Electro-optic materials are used in the second harmonic generation process as well as in the related processes of third harmonic generation, three wave mixing, four wave mixing, optical parametric oscillation, and stimulated Raman scattering [10, ch. 19].

3.4 Notation Quagmire

This text attempts to use notation consistent with the literature. However, consistency is a challenge because every author seems to have a different name for the same physical phenomena. Furthermore, the same term used by different authors may have completely different meanings. For example, as described by Eq. 2.19, in some materials, a mechanical stress induces a material polarization proportional to the square of that stress. This text calls this phenomenon piezoelectricity, or to be more specific, quadratic piezoelectricity. However, references [3] and [6] call this phenomenon electrostriction. To make matters worse, reference [33] calls this effect ferroelectricity. Some authors make different assumptions when using terms too. For example, when reference [26] uses the term ferroelectricity, it assumes crystalline materials, but it makes no assumptions about whether the effect is linear or not.

Table 3.2 summarizes the notation used in this text to describe energy conversion processes involving material polarization. The first column lists the name used here to describe the effect. The second column lists what effect causes a material polarization. The third column describes whether the effect occurs in crystals only. The fourth column describes whether the material polarization varies linearly or not with the parameter described in the second column. The next column lists references which call this effect ferroelectricity. The last column gives names used by other references to describe this particular phenomenon. The last two columns are quite incomplete because a thorough literature survey was not done. However, these columns show quite a variety to the terminology even for the small fraction of the literature reviewed.

You might think that you can avoid confusion of terminology by looking for Greek or Latin roots. While many of the terms introduced in the preced-

Notation in this text	$\vec{P}$ induced by ...	Crystalline? Amorphous? Polycrystalline?	Linear?	Who calls this ferroelectricity?	What others call this quantity
(Linear) Piezoelectricity	Mechanical stress, $\overset{\curvearrowright}{\zeta}$	Crystalline	Linear		
(Quadratic) Piezoelectricity	Mechanical stress $\overset{\curvearrowright}{\zeta}$	Crystalline	Quadratic	[33]	Electrostriction [3, p. 327] [6], photoelasticity [31]
Ferroelectric Piezoelectricity	Mechanical stress $\overset{\curvearrowright}{\zeta}$	All	Nonlinear	[25, p. 408]	
(Linear) Pyroelectricity	Temperature differential, ΔT	Crystalline	Linear	[26, p. 556], [42, p. 50], [43]	Thermal nonlinear optical effects [42]
(Quadratic) Pyroelectricity	Temperature differential, ΔT	Crystalline	Quadratic	[26, p. 556], [43]	Thermal nonlinear optical effects [42]
Ferroelectric Pyroelectricity	Temperature differential, ΔT	All	Nonlinear	[3, p. 366], [26, p. 556], [43]	Thermal nonlinear optical effects [42]
Linear (Pockels) Electro-optic Effect	Optical Electromag. radiation $\vec{E}$	Crystalline	Linear		Electronic polarizablility [25, p. 390]
Quadratic (Kerr) Electro-optic Effect	Optical Electromag. radiation $\vec{E}$	Crystalline	Quadratic		
Ferroelectric Electro-optic Effect	Optical Electromag. radiation $\vec{E}$	All	Nonlinear		Photoinduced anisotropy, photodarkening [44] [45], intimate valence alternation pair state [44]

Table 3.2: Terminology related to processes involving material polarization.

ing chapters do have etymological roots, looking at the roots of the words does not help and sometimes makes matters worse. As discussed above, the prefix *ferro-* means iron. However, the ferroelectric effect has nothing to do with iron, and ferroelectric materials rarely contain iron. This name is an analogy to ferromagnetics. Some forms of iron are ferromagnetic. In ferromagnetics, an external magnetic field changes the permeability of a material. In ferroelectrics, an external electric field influences the permittivity. To make matters worse, iron has the periodic table symbol Fe while iridium has the symbol Ir. In this text, the term *pyroelectric effect* follows Roentengen's terminology which dates 1914 [3]. The root *pyro-*, showing up in pyroelectricity, also shows up in pyrite and pyrrhotite which are iron containing compounds.

Sometimes the terms *phase change* and *photodarkening* are applied to the electro-optic effect in amorphous materials, but not crystalline materials. More specifically, sulfides, selenides, and tellurides, referred to as *chalcogenides*, are sometimes called *phase change* materials. Examples include GeAsS, GeInSe, and so on. The word chalcogenide is itself a misnomer. The prefix *chalc-* comes from the Greek root meaning copper [24]. They are named in analogy to CuS, chalcosulfide. The name phase change material was popularized by a company that made CDs and battery components. While crystalline materials can also be electro-optic, the name phase change is not typically applied to crystals.

Sometimes the terminology used in the literature can be quite different from the terminology of this text. For example, reference [44] describes material polarization in chalcogenide glasses by saying that when exposed to external optical electric fields, a material stores energy by "a transient exciton which can be visualized as a transient intimate valence alternation defect pair." ... "This means essentially that macroscopic anisotropies result from geminate recombination of electron-hole pairs, which do not diffuse out of the microscopic entity in which they were created by absorbed photons." An *exciton* is a bound electron-hole pair. In other words, the material polarizes. When an external optical electric field is applied, electric dipoles form throughout the material. When reading the literature related to piezoelectricity, pyroelectricity, and electro-optics, be aware that there is not much consistency in the terminology used.

3.5 Problems

3.1. For each of the three crystalline materials below

- Find the crystal point group to which it belongs.
  (Hint: use http://www.mindat.org)
- Using Table 2.2, determine whether or not the material is piezo-electric.
- Using Table 2.2, determine whether or not the material is pyro-electric.
- Using Table 2.2, determine whether or not the material is Pockels electro-optic.

(a) ZnS (sphalerite)

(b) HgS (cinnabar)

(c) Diamond

3.2. Cane sugar, also called saccharose, has chemical composition $C_{12}H_{22}O_{11}$ and belongs to the crystal point group given by 2 in Hermann-Maguin notation [38]. Reference [38] lists values specified in cgse units for its piezoelectric constant as $10.2 \cdot 10^{-8} \frac{\text{esu}}{\text{dyne}}$ and its pyroelectric coefficient as $0.53 \frac{\text{esu}}{\text{cm}^{2}\cdot{}^\circ\text{C}}$. Convert these values to the SI units of $\frac{\text{m}}{\text{V}}$ and $\frac{\text{C}}{\text{m}^2 \cdot \text{K}}$ respectively.
Hint: The electrostatic unit or statcoulomb is a measure of charge [7] where
$$1 \text{ esu} = 1 \text{ statC} = 3.335641 \cdot 10^{-10} \text{ C}$$
and the dyne is a measure of force where $1 \text{ dyne} = 10^{-5} \text{ N}$.

3.3. A material has relative permittivity $\epsilon_{r\,x}$ when no external electric field is applied. The coefficient $\chi^{(2)}$ is measured in the presence of an external electric field of strength $|\vec{E}|$. Assume that $\chi^{(3)}$ and all higher order coefficients are zero. Find the Pockels coefficient γ as a function of the known quantities $\epsilon_{r\,x}$, $\chi^{(2)}$, and $|\vec{E}|$.

3.4. The first figure below shows the displacement flux density $\left|\vec{D}\right|$ as a function of the strength of an applied electric field intensity $\left|\vec{E}\right|$ in a non-electro-optic material. The second figure below shows the displacement flux density $\left|\vec{D}\right|$ as a function of the strength of an applied electric field intensity $\left|\vec{E}\right|$ in a ferroelectric electro-optic material. The solid line corresponds to an unpoled material. The dotted line corresponds to the material after it has been poled in the $\hat{a}_z$ direction, and the dashed line corresponds to the material after it has been poled in the $-\hat{a}_z$ direction.

(a) For the non-electro-optic material, find the relative permittivity, ϵ_r. Also find the magnitude of the material polarization, $\vec{P}$.

(b) Assume the ferroelectric electro-optic material is poled by a strong external electric field, and then the field is removed. Find the magnitude of the material polarization $\left|\vec{P}\right|$ after the external field is removed.

(c) Assume the ferroelectric material is poled in the $-\hat{a}_z$ direction by a strong external field, and then the field is removed. A different external electric field given by $\vec{E} = 100\hat{a}_z \, \frac{V}{m}$ is applied. Find the approximate relative permittivity of the material.

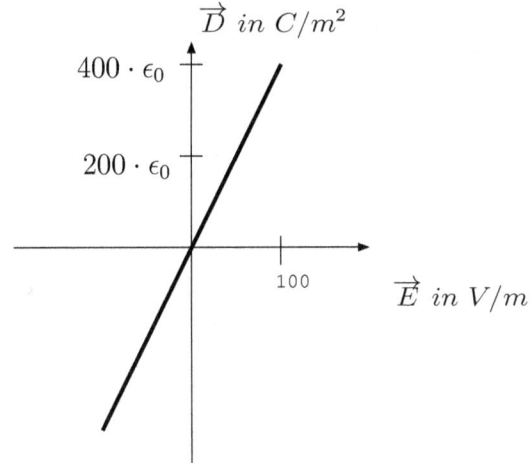

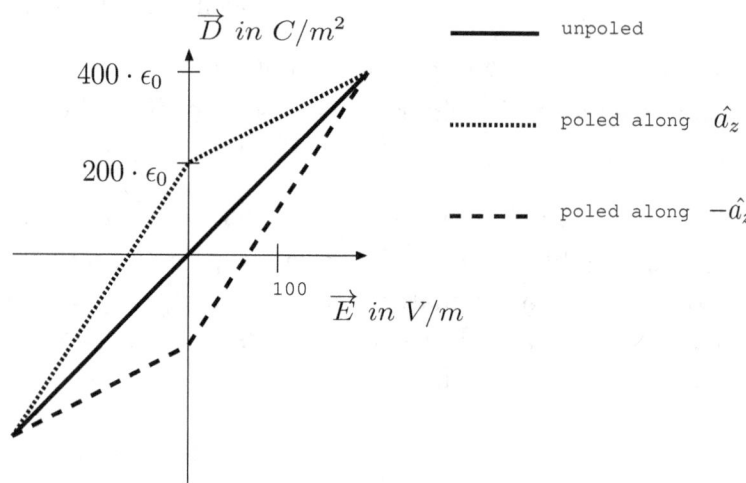

3.5. A crystalline material is both piezoelectric and pyroelectric. When an external electric field of $|\vec{E}| = 100 \, \frac{V}{m}$ is applied, the material polarization is determined to be $|\vec{P}| = 1500\epsilon_0 \, \frac{C}{m^2}$. When both a stress of $|\vec{\varsigma}| = 30 \, \frac{N}{m^2}$ and an external electric field of $|\vec{E}| = 100 \, \frac{V}{m}$ are applied, the material polarization is determined to be $|\vec{P}| = 6.0123 \cdot 10^{-6} \, \frac{C}{m^2}$. When a temperature gradient of $\Delta T = 50 \, ^\circ C$, a stress of $|\vec{\varsigma}| = 30 \, \frac{N}{m^2}$, and an external electric field of $|\vec{E}| = 100 \, \frac{V}{m}$ are applied, the material polarization is determined to be $|\vec{P}| = 6.3 \cdot 10^{-6} \, \frac{C}{m^2}$. Find:

- The relative permittivity of the material
- The piezoelectric strain constant
- The magnitude of the pyroelectric coefficient

4 Antennas

4.1 Introduction

In the previous two chapters we discussed energy conversion devices which are made from insulators and which are related to capacitors. In Chapters 4 and 5 we discuss energy conversion devices involving conductors and related to inductors. Maxwell's equations say that time varying electric fields induce magnetic fields and time varying magnetic fields induce electric fields. If a permanent magnet moves near a coil of wire, the time varying magnetic field will induce a current in the coil of wire. This idea is the basis behind motors and electrical generators, which are some of the most common energy conversion devices. However, they are outside the scope of this text because they involve magnets and coils. Instead, we will study two other types of energy conversion devices based on this same principle. In this chapter we discuss antennas, and in the next chapter we will discuss Hall effect devices.

Antennas are energy conversion devices that convert between electrical energy and electromagnetic energy. Antennas can act as both transmitters and receivers. Transmitters convert electrical energy of the flow of electrons to energy of electromagnetic waves. Receivers convert energy from electromagnetic waves to the electrical energy of electrons in a circuit. The same physical antenna can operate in both ways depending on how it is used.

Antennas are all around us. Cell phones and laptops have antennas, and antennas are mounted on the roofs of most cars. Antennas relay information about the electrical grid to the local power utility, and antennas on satellites transmit weather maps to weather stations on earth. Antennas are even built into RFID tags on shirts in stores, and these tags are used to track inventory and prevent theft.

Electrical engineers study both electrical energy and electromagnetic energy, and the words used to describe these phenomena are similar. Is this really an energy conversion process? The answer is yes. Electrical energy involves the flow of electrons through a wire. We often think of electrons as particles. We often use the term electromagnetic wave to describe the flow of electromagnetic energy transmitted by an antenna. However, electrons have both wave-like and particle-like properties. Similarly, electromagnetic waves have both wave-like and particle-like properties. The wavelengths involved are orders of magnitude apart, so it is convenient to only discuss either the wave-like or the particle-like properties. There are fundamental differences between electricity and electromagnetic waves. *Fermions* are

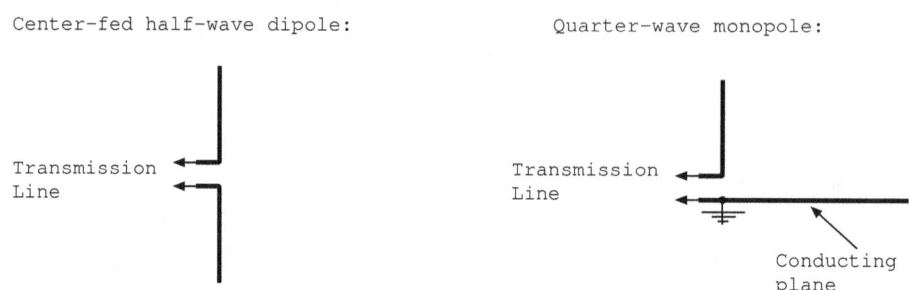

Figure 4.1: Center-fed half-wave dipole and quarter-wave monopole antennas.

elementary particles with half integer spin quantum numbers and with quantum mechanical wave functions which are antisymmetric when two particles are interchanged [46, p. 391]. *Bosons* are elementary particles with integer spin quantum numbers and with wave functions which are symmetric when two particles are interchanged [46, p. 391]. Electrons are fermions while electromagnetic waves are bosons. So, antennas are energy conversion devices. A complete discussion of the differences between fermions and bosons requires the study of quantum mechanics and quantum field theory which are beyond the scope of this book.

An antenna may be as simple as a single metal rod, it may be a copper trace on a printed circuit board, it may be a cone shaped horn, or it may be a complicated arrangement of multiple wires. Some antennas even resemble planar or volume fractals [47] [48]. Hundreds of types of antennas have been developed. Seventy five types are discussed in [49], and 91 types are discussed in [50].

The simplest antenna is just a piece of wire. It may be straight and taut, or it may be carelessly strung from a tree. For an antenna designed to operate at wavelength λ, the length of the antenna is often approximately $\frac{\lambda}{2}$. A straight antenna of length $\frac{\lambda}{2}$ with signal supplied to the center is called a *center-fed half-wave dipole* or a $\frac{\lambda}{2}$ dipole. Some antennas are placed above a conducting plate, or above a conductive surface, which acts as a reflector. A straight antenna of around length $\frac{\lambda}{4}$ supplied by a signal at one end with a reflector beneath is called a *quarter-wave monopole* or a $\frac{\lambda}{4}$ monopole. Figure 4.1 illustrates both dipole and monopole antennas. While a random wire will act as an antenna, an antenna with frequency response, impedance, radiation pattern, and electromagnetic polarization designed for the specific application will perform much more efficiently, and these factors are discussed below in Sec. 4.4.

4 ANTENNAS

4.2 Electromagnetic Radiation

4.2.1 Superposition

The physics of antenna operation is described by Maxwell's equations. Ampere's law, one of Maxwell's equations, was introduced in Section 1.6.1.

$$\vec{\nabla} \times \vec{H} = \vec{J} + \frac{\partial \vec{D}}{\partial t} \qquad (4.1)$$

In Eq. 4.1, $\vec{H}$ is the magnetic field intensity in $\frac{A}{m}$, $\vec{D}$ is the displacement flux density in $\frac{C}{m^2}$, and $\vec{J}$ is the current density in $\frac{A}{m^2}$. In the case of a transmitting antenna, the current density in the antenna comes from a known source, and the electromagnetic field, described by $\vec{D}$ and $\vec{H}$, can be derived.

Using Maxwell's equations, we can algebraically derive the electromagnetic field only for very simple antennas. The simplest antenna is an infinitesimal dipole antenna, also known as a *Hertzian dipole*. References [11] derives the electric field intensity, $\vec{E}$ in units $\frac{V}{m}$, for an infinitesimal dipole antenna with length dl and sinusoidal current $I_0 \cos(\omega t)$. The result is given in spherical coordinates is

$$\vec{E} = \frac{2 I_0 \cdot dl \cdot \cos\theta}{4\pi\epsilon\omega} \left[\frac{\sin(\omega t - \frac{2\pi}{\lambda} r)}{r^3} + \frac{\frac{2\pi}{\lambda} \cos(\omega t - \frac{2\pi}{\lambda} r)}{r^2} \right] \hat{a}_r \qquad (4.2)$$
$$+ \frac{I_0 \cdot dl \cdot \sin\theta}{4\pi\epsilon\omega} \left[\frac{\sin(\omega t - \frac{2\pi}{\lambda} r)}{r^3} + \frac{\frac{2\pi}{\lambda} \cos(\omega t - \frac{2\pi}{\lambda} r)}{r^2} - \frac{\left(\frac{2\pi}{\lambda}\right)^2 \sin(\omega t - \frac{2\pi}{\lambda} r)}{r} \right] \hat{a}_\theta .$$

In this expression, ω is frequency in $\frac{rad}{s}$, λ is the wavelength in meters, ϵ is the permittivity of the material surrounding the antenna in $\frac{F}{m}$, and (r, θ, ϕ) are the coordinates of a point specified in spherical coordinates.

For complicated antennas, superposition is used to make the computation feasible. To derive the electromagnetic radiation from a complicated antenna, small straight antenna segments are considered [15, ch. 10]. The electromagnetic radiation from each piece is found, and the *principle of superposition* is the idea that the radiation from the entire antenna is the sum of these pieces. The same idea applies to linear circuits. If a circuit has a complicated input, the input can be broken up into simpler components. Any voltage in the circuit can be found by finding the contribution due to each of these components then summing.

4.2.2 Reciprocity

Reciprocity is the idea that the behavior of an antenna as a function of angle is the same regardless of whether the antenna is being used to send

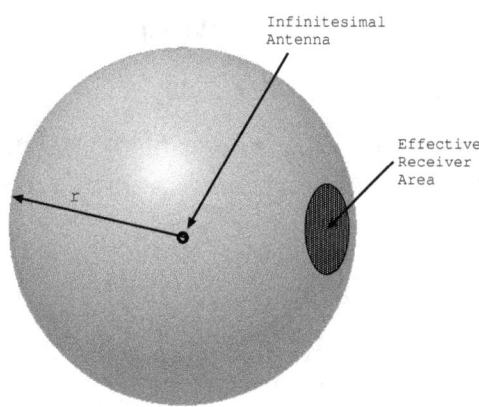

Figure 4.2: Illustration of power radiating from an isotropic antenna.

or receive a signal [15, ch. 10]. A plot of the strength of the field radiated from a transmitter as a function of the angles θ and ϕ is called a *radiation pattern plot*. Similarly, a plot of the strength of the signal received by a receiving antenna as a function of angles θ and ϕ assuming a uniform field strength is also called a *radiation pattern plot*. Consider two identical antennas, one being used as a transmitter and the other as a receiver. The radiation pattern plots will be the same for these two antennas.

Regardless of the idea of reciprocity, it is often a bad idea to swap the transmitting and the receiving antennas of a system because a transmitter may be designed to handle much more power than a receiver [15, p. 479]. A receiving antenna of effective area A at a distance r from an antenna which transmits uniformly in all directions receives at most only the fraction $\frac{A}{4\pi r^2}$ of the transmitted power [49, p. 4].

$$P_{rec} = P_{trans} \frac{A}{4\pi r^2} \qquad (4.3)$$

For example, consider an antenna that transmits 20 kW of power uniformly in all directions. Assume a receiving antenna has an effective area of 10 cm^2 and covers a portion of a spherical shell as shown in Fig. 4.2. What is the power received assuming that the antenna is at a distance of $r = 1$ m, and what is the power received assuming a distance of $r = 1$ km?

The surface area intercepted by the receiver is 10 cm^2 = 10^{-3} m^2. In the first case, this surface area is the fraction $\frac{10^{-3}}{4\pi \cdot 1^2}$ of the surface sphere of radius 1 m. At most, the antenna can receive this fraction of the power.

$$P = 20 \cdot 10^3 \cdot \frac{10^{-3}}{4\pi \cdot 1^2} = 1.6 \text{ W} \qquad (4.4)$$

4 ANTENNAS

In the second case, this surface area is the fraction $\frac{10^{-3}}{4\pi \cdot (10^3)^2}$ of the surface of the sphere of radius 1 km. At most, this antenna can receive

$$P = 20 \cdot 10^3 \cdot \frac{10^{-3}}{4\pi \cdot (10^3)^2} = 1.6 \; \mu W. \tag{4.5}$$

From this example, we can already see some of the advantages and challenges in using electromagnetic waves for communication, and we can see some of the consequences of antenna design. The transmitted power in this example is orders of magnitude larger than the received power. In such a situation, the transmitting circuitry and receiving circuitry will look very different due to the amount of power and current expected during operation. The antennas used will likely also look very different. An antenna transmitting kilowatts of power may need to be mounted on a tower while a receiving antenna that receives milliwatts of power may be built into a portable hand held device.

A typical radio station may want to transmit throughout a city, a radius much larger than 1 km. Furthermore, no energy conversion device is 100% efficient. The electrical power at the receiver 1 m away is therefore going to be less than 1.6 W, and the power at the receiver 1 km away is going to be less than 1.6 μW. Also, all radio receivers are limited by noise. Suppose, for example, that this transmitter is placed at the center of a city of radius 1 km and the receiver can only successfully receive signals with power above 1 μW due to 1 μW of background noise. A receiver placed 1 km away at the edge of the city may be able to receive the signal successfully, while a receiver further away in the suburbs may not. However, many receivers placed 1 km away with this surface area of 10 cm^2 could simultaneously detect the radio signal.

If no building in the city is taller than 10 stories, no receivers are likely to be found at a height over 30 m, for example, above the surface of the earth. However, the transmitter in this example radiates power uniformly in all directions including up. We can design antennas which radiate power in some directions more than others. If we could focus all power from this antenna at altitudes below 30 m, the power at a particular receiver may be larger than we calculated above, so a receiver farther away may be able to detect the signal. The radiation pattern of an antenna is the spatial distribution of the power from the antenna. Radiation pattern plots are discussed further in Sec. 4.4.3.

This example also provides some insights on the safety of working with antennas. The 10 cm^2 surface area in this example is, to an order of magnitude, the surface area of a human hand. A typical microwave oven uses less power than the transmitter in this example. Kilowatts of power are enough to cook with, so for this reason, it would be dangerous to touch or even, depending on the frequency, be close to the transmitting antenna. The antenna in this example needs to be mounted on an antenna tower not only for mechanical reasons but also for safety reasons. The amount of power through this surface depends on distance from the transmitter as $\frac{1}{r^2}$, so the danger level is strongly dependent on distance from the antenna.

4.2.3 Near Field and Far Field

The region within about a wavelength of an antenna is called the *near field region*. The region beyond multiple wavelengths from an antenna is called the *far field* or *Fraunhofer region*. For aperture antennas, instead of wire antennas, distances larger than $\frac{2(\text{aperture size})^2}{\lambda}$ are considered in the far field [15, p. 498]. The radiation pattern in the near field region and in the far field region are quite different. Near field electromagnetic radiation is used for some specialized applications including tomographic imaging of very small objects [51]. However, receiving antennas used for communication signals almost always operate in the far field region from transmitting antennas. As an example of the difference between near field and far field behavior of an antenna, consider the infinitesimal dipole antenna. The electric field intensity is given in Eq. 4.2. The near field electric field from this infinitesimal antenna is found by taking the limit as $r \to 0$.

$$\vec{E} = \frac{I_0 \cdot dl \cdot \cos\theta}{4\pi\epsilon\omega} \cdot \frac{\sin\left(\omega t - \frac{2\pi}{\lambda}r\right)}{r^3} (2\hat{a}_r + \hat{a}_\theta) \qquad (4.6)$$

The far field electric field is found by taking the limit as $r \to \infty$.

$$\vec{E} = \frac{-I_0\omega \cdot dl \cdot \sin\theta}{4\pi\epsilon} \cdot \frac{\sin\left(\omega t - \frac{2\pi}{\lambda}r\right)}{r}\hat{a}_\theta \qquad (4.7)$$

4.2.4 Environmental Effects on Antennas

The electromagnetic radiation from an antenna is affected by the environment surrounding the antenna, specifically nearby large conductors. Sometimes conductors are purposely placed nearby to make an antenna directional. Other times, the conductors, like metal roofs or bridges, just happen to be nearby. If an antenna is placed near a salty lake, the lake surface will reflect the electromagnetic radiation. In other cases, the electrical properties of soil underneath an antenna will affect the electromagnetic radiation [50, ch. 8] [15, p. 635].

Environment	Conductivity σ in $\frac{1}{\Omega m}$	Relative permittivity ϵ_r
Industrial city	0.001	5
Sand	0.002	10
Rich soil	0.01	14
Fresh water	0.001	80
Salt water	5	80

Table 4.1: Conductivity and relative permittivity of different environments, [50, ch. 8].

Numerical simulations are used to understand how an antenna behaves near metal roofs, nearby lakes, or other objects. The effects of the environment are modeled by assigning nearby materials an electrical conductivity σ, permittivity ϵ, and permeability μ. Often the surroundings have $\mu \approx \mu_0$, but the other parameters can vary widely. Table 4.1 lists values of electrical conductivity and relative permittivity used to model different environments as suggested by reference [50, ch. 8]. The values listed are approximates due to the variety of environments within each category. Additionally, the conductivity can vary from day to day. For example, electromagnetic waves may interact with farmland very differently on a snowy winter day, after a spring rainfall, and during a dry spell in summer. Also, even for a single uniform material, conductivity and permittivity are functions of frequency.

4.3 Antenna Components and Definitions

Antennas used for radio frequency communication are made from conducting wire elements. These elements may be classified as driven or parasitic [50]. All antennas have at least one *driven element*. In a transmitting antenna, power is supplied to the driven element. Current flowing through the antenna induces an electromagnetic field around the antenna. In a receiving antenna, the driven element is connected to the receiving circuitry. Some antennas also have *parasitic elements*. These elements affect the antenna's radiation pattern, but they are not connected to the power supply or receiving circuitry [50]. The electric field inside a perfect conductor is zero, so putting a good conductor near an antenna influences the antenna's radiation pattern. Parasitic elements may be included in the antenna to focus the electromagnetic field in a particular direction, alter the bandwidth of the antenna, or for other reasons. Antennas are often mounted on a metal rod for mechanical support, and this rod is called a *boom*.

Antennas may be used individually or as part of an *array*. Arrays may also be driven or parasitic. In a *driven array*, all elements are connected to the power supply or receiving circuitry [50]. In a *parasitic array*, one or more of the elements are parasitic and not connected [50]. Arrays are also classified based on the direction of radiation compared to the axis of the array. In a *broadside array*, radiation is mostly perpendicular to the axis of the array while in an *end fire array*, radiation is mostly along the direction of the axis of the array [50].

A *transmission line* is a pair of conductors which is used to transmit a signal and which is very long compared to the wavelength of the signal being sent. Communications engineers and power systems engineers both use the term transmission line, but they make different assumptions. To a communications engineer, it is a long pair of conductors over which a signal is sent. To a power systems engineer, it is a cable that is part of the power grid. The communications definition will be used in this text. The conductors of a transmission line may be a pair of parallel wires, they may be a waveguide formed by a pair of parallel plates, they may be a coax cable, or they may have another geometry. Coax cable is formed by a wire and cylindrical tube separated by an insulator, both with the same axis, so they are coaxial. For example, a coax cable connecting a transmitter operating at a frequency of $f = 88$ MHz on the first floor of a building and an antenna on the top of the tenth floor of the building is a transmission line because the length of the cable is long compared to the wavelength of $\lambda = 3.4$ m. As another example, a pair of wires connecting a transmitting circuit operating at $f = 4$ GHz on one end of a printed circuit board and an antenna on the other end 25 cm away is also a transmission line because the length of the wires is long compared to the wavelength of $\lambda = 7.5$ cm.

Some antennas have a *balun*. Balun is a contraction for balanced/unbalanced. It is used between balanced loads and unbalanced transmission lines [15, p. 406] [50]. A typical transmission line, made up of a coax cable, is constructed from an inner conductor and an outer conductor. These conductors have different radii, so they have different impedances. The transmission line is called unbalanced due to this impedance difference. Suppose that this transmission line is connected to a dipole antenna formed from two symmetric conductors. The impedance of the two arms of the dipole are equal, so we say that it is a balanced load. A balun can used in this type of situation when a balanced antenna is connected to an unbalanced transmission line. By properly choosing the impedance of a balun, reflections at the interface between the antenna and transmission line can be reduced so that more energy gets to or from the antenna and less remains stored in the transmission line.

Frequency	Abbreviation	Name
30-3000 Hz	ELF	Extremely Low Frequency
3-30 kHz	VLF	Very Low Frequency
30-300 kHz	LF	Low Frequency
300 kHz -3 MHz	MF	Medium Frequency
3-30 MHz	HF	High Frequency
30-300 MHz	VHF	Very High Frequency
300 MHz-3 GHz	UHF	Ultra High Frequency
3-30 GHz	SHF	Super High Frequency
30-300 GHz	EHF	Extremely High Frequency

Table 4.2: Names of electromagnetic frequency ranges [15] [54].

4.4 Antenna Characteristics

Four main factors which differentiate antennas are frequency response, impedance, directivity, and electromagnetic polarization. When selecting an antenna for a particular application, these factors should be considered. In this section, these and other factors which influence antenna selection are discussed.

4.4.1 Frequency and Bandwidth

Electromagnetic waves of a wide range of frequencies are used for communication. Different names are given to electromagnetic signals at different frequency ranges. Table 4.2 lists the name used to refer to various frequency bands for which antennas are used.

Electromagnetic waves are rarely used for communication at the lowest frequency band listed in Table 4.2. However, one example was Project ELF (short for Extremely Low Frequency). It was a US military radio system used to communicate with submarines, and it operated at 76 Hz [52]. The array involved 84 miles of antennas spread out near a transmitting facilities in northern Wisconsin and the upper peninsula of Michigan [52], and it operated from 1988 to 2004 [53]. It had an input power of 2.3 MW, but only 2.3 W of electromagnetic radiation was transmitted due to the fact that the length of the antenna elements used was a small fraction of the wavelength. The few watts transmitted were able to reach submarines under the ocean throughout the world [52]. Three letter messages took 15-20 minutes to transmit or receive [52].

Antennas are commonly used to transmit and receive electromagnetic radiation in the frequency range from $3 \text{ kHz} \lesssim f \lesssim 3 \text{ THz}$. However, an antenna designed to operate at 3 kHz looks quite different from an antenna designed to operate at 3 THz. Wire-like antennas are used for signals roughly in the frequency range 3 kHz $\lesssim f \lesssim$ 3 GHz. Solid cone, plate-like, or aperture antennas are used to transmit and receive signals in the frequency range 3 GHz $\lesssim f \lesssim$ 3 THz [15, ch. 15]. To understand the need for different techniques, consider the wavelengths involved. A signal with frequency $f = 30$ kHz, for example, has a wavelength $\lambda = 1.00 \cdot 10^4$ m. The length of an antenna is often of the same order of magnitude as the wavelength. While we can construct wire antennas of this length, they not portable. As another example, a wifi signal which operates at 2.5 GHz has a wavelength of $\lambda = 12.5$ cm. Wire antennas which are this length are easy to build and transport. However, wire antennas designed for signals at higher frequencies can be difficult to construct accurately due to their small size. For this reason, wire antennas are typically used at lower frequencies while cone or plate-like antennas are used higher frequencies.

A human eye can detect electromagnetic radiation with frequencies and wavelengths in the range

$$4.6 \cdot 10^{14} \text{ Hz} \lesssim f \lesssim 7.5 \cdot 10^{14} \text{ Hz} \quad \text{or} \quad 400 \text{ nm} \lesssim \lambda \lesssim 650 \text{ nm}$$

Antennas are not used to receive and transmit optical signals due to the small wavelengths involved even though optical signals obey the same fundamental physics as radio frequency electromagnetic radiation. Green light, for example, has a wavelength near $\lambda = 500$ nm and a frequency near $6 \cdot 10^{14}$ Hz. An antenna designed to transmit and detect this light would need to be approximately of length $\frac{\lambda}{2} \approx 250$ nm. An atom is around 0.1 nm in length, so an antenna designed for green light would be only approximately 2500 atoms long. Antennas of this size would be impractical for many reasons. Another reason that different techniques are needed to transmit and receive optical signals is that electrical circuits cannot operate at the speed of optical frequencies. Techniques for transmitting and detecting optical signals are discussed in Chapters 6 and 7.

When selecting an antenna, the range of frequencies that will be transmitted or received as well as their bandwidth should be considered. Some antennas are designed to operate over a narrow range of frequencies while other antennas are designed to operate over a broader band of frequencies. An antenna with a narrow bandwidth would be useful in the case when an antenna is used to receive signals only in a specific frequency band while an antenna with a broad bandwidth would be useful when an antenna is to receive signals over a wider frequency range. For example, an antenna designed to receive over the air television signals in the US should be designed for the broad range from 30 MHz - 3 G Hz because television signals fall in the VHF and UHF ranges.

Like all sensors, antennas detect both signal and noise. Noise in a radio receiver may be internal to the receiving circuitry or due to external sources such as other nearby transmitters [49, p. 4]. An antenna with a broad bandwidth will receive more noise due external sources than an antenna with a narrow bandwidth. Noise characteristics of an antenna influence the ability to receive weak signals, so they should be considered in selecting an antenna for an application [50].

4.4.2 Impedance

Both antennas and transmission lines have a *characteristic impedance*. The term transmission line is defined in Sec. 4.3 as a long pair of conductors. If the length of the conductors is long compared to the wavelength of signal transmitted, the voltage and current may vary along the length of the line, and energy may be stored in the line. For this reason, transmission lines are described by a characteristic impedance in ohms. The characteristic impedance gives the ratio of voltage to current along the line, and it provides information on the ability of the transmission line to store energy in the electric and magnetic field. Typical values for the impedance of transmission lines used for communications are 50 or 75 Ω. Similarly, each antenna has its own characteristic impedance, measured in ohms, which represents the ratio of voltage to current in the antenna.

Why is the impedance important? Transmitting antennas are often physically removed from the signal source and connected by a transmission line. Similarly, receiving antennas are often in a different location than receiving circuitry and connected by a transmission line. To efficiently transmit a signal between transmitting or receiving circuitry and an antenna, the impedance between the antenna and transmission line should be *matched*. In this case, where the characteristic impedance of the line and antenna are equal, energy flows along the transmission line between the circuitry and the antenna. Transmission lines are made from good, but not perfect, conductors. A small amount of energy may be converted to heat due to the resistance in the lines, but this amount of energy is often trivial. However, if there is an impedance mismatch between the antenna and the transmission line, reflections will be set up at the transmission line antenna interface. Less energy will be transmitted to or from the antenna because energy will be stored in the line, and the amount of energy involved may be significant. In a properly designed system were the impedances of the antenna and the transmission line are matched, no reflection occurs, so as much energy as possible is transmitted to or from the antenna.

Impedance of an antenna is a function of frequency. Antennas transmit and receive communications signals which are almost never sinusoids of a single frequency. Often, however, the signals contain only components with frequencies within a narrow band. For example, a radio station may have a carrier frequency of 100 MHz, and it may transmit signals with frequency components 99.99 MHz $< f <$ 100.01 MHz. In this case, the impedance of the antenna may be approximated by the impedance at 100 MHz.

4.4.3 Directivity

Antennas can be designed to radiate energy equally in all directions. Alternatively, antennas can be designed to radiate energy primarily along a single direction. *Directivity D* is a unitless measure of the uniformity of the radiation pattern plot. It is defined as the ratio of the maximum power density over the average power density.

$$D = \frac{\text{Maximum power density radiated by antenna}}{\text{Average power density radiated by antenna}} \quad (4.8)$$

An antenna which radiates equally in all directions is called *isotropic*. An antenna that radiates equally in two, but not the third, direction is called *omnidirectional* [15]. For example, an omnidirectional antenna may radiate equally in all horizontal directions but not the vertical direction. Isotropic antennas have $D = 1$ while all other antennas have $D > 1$. Some applications require an isotropic antenna. For example, a radio station in the center of a town might use an isotropic or omnidirectional antenna to transmit to all of the town. In other cases, a directional antenna is preferred. A stationary weather station that transmits data to a fixed base station would be wasting energy using an isotropic antenna because it could use less transmitted power with the same received power using a directional antenna.

Received power may be larger than given by Eq. 4.3 if directional antennas are used instead of isotropic antennas. For a transmitting antenna with gain G_{trans} and a receiving antenna with gain G_{rec} compared to an isotropic antenna, Eq. 4.3 becomes

$$P_{rec} = P_{trans} G_t G_r \left(\frac{\lambda}{4\pi r}\right)^2 \quad (4.9)$$

where the effective area is assumed to be related to the receiver gain by

$$G_r = \frac{4\pi A}{\lambda^2}. \quad (4.10)$$

Equation 4.9 is known as the Friis equation [55]. Received power will be less than given by Eq. 4.3 or 4.9 due to losses in the air or other material through which the signal travels and due to a difference in electromagnetic polarization between the transmitter and receiver [49, p. 4].

Directivity is a rough measure of an antenna. A more accurate measure is a *radiation pattern plot*. The radiation pattern plot is a graphical representation of intensity of radiation with respect to position throughout space. A radiation pattern plot may be a 3D plot or a pair of 2D plots. In the case where two 2D plots are used, one of the plots is an azimuth plot and the other is an elevation plot. The *azimuth plot* shows a horizontal slice of the 3D radiation pattern, parallel to the xy plane. The *elevation plot* shows a vertical slice, perpendicular to the xy plane. Most radiation pattern plots, including all shown in this text, are labeled by the amplitude of the electric field [15] [56]. However, occasionally they are labeled by the amplitude of the power instead. The radiation pattern of an antenna is quite different in the near field, at a distance less than about a wavelength, and in the far field, with distances much greater than a wavelength. Radiation pattern plots illustrate the far field behavior only.

Figure 4.3 shows the radiation pattern plot for a half-wave dipole antenna in free space, and it was plotted using the software EZNEC [56]. The acronym NEC stands for Numerical Electromagnetics Code. The figure in the upper left is the azimuth plot, the figure in the upper right is the elevation plot, the figure in the lower left is a 3D radiation pattern plot, and the figure in the lower right is the antenna layout.

Figure 4.4 shows the radiation pattern plots for a 15-meter quad antenna. Distinct lobes and nulls are apparent.

Front to back ratio (F/B ratio) is a measure related to directivity that can be found from the azimuth radiation pattern plot. By definition, it is the ratio of the strength of the power radiated in the front to the back. Often, the front direction is chosen to be the direction of largest magnitude in the radiation pattern plot, and the back direction is the opposite direction. F/B ratio can be specified either on a log scale in units of dB or on a linear scale which is unitless. It can also be defined either as a ratio of the strength of the electric field intensities or as a ratio of the strengths of the powers, but most often power is used.

$$\text{F/B ratio} = \left[\frac{P_{front}}{P_{back}}\right]_{dB} = 10\log_{10}\left[\frac{P_{front}}{P_{back}}\right]_{lin} = 20\log_{10}\left[\frac{|\vec{E}_{front}|}{|\vec{E}_{back}|}\right]_{lin} \quad (4.11)$$

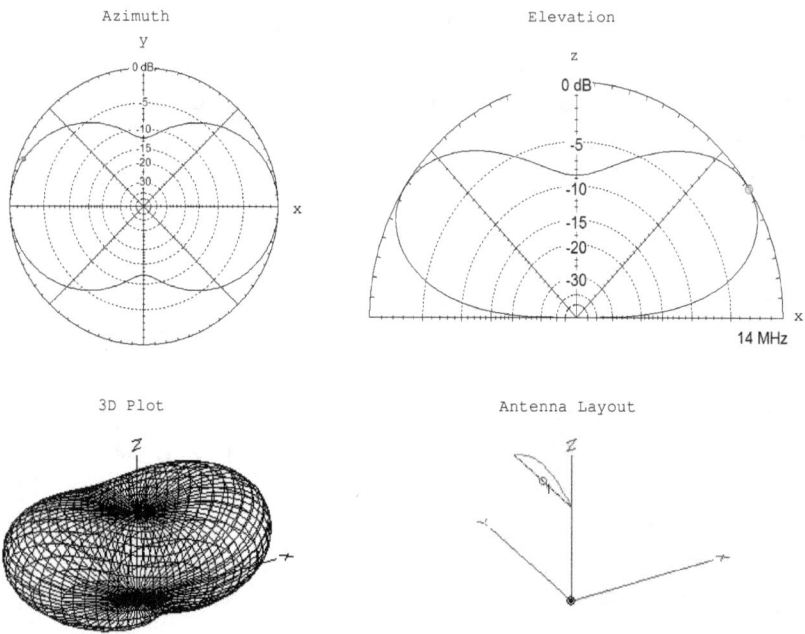

Figure 4.3: Radiation pattern plots for a half-wave dipole antenna.

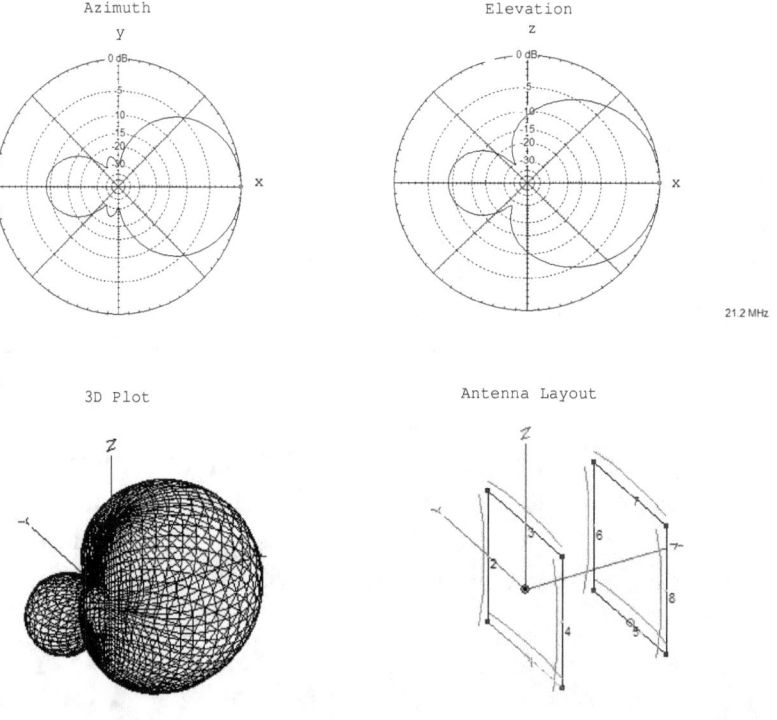

Figure 4.4: Radiation pattern plots for a 15-meter quad antenna.

4 ANTENNAS

$$\text{F/B ratio} = \left[\frac{P_{front}}{P_{back}}\right]_{dB} = 2\left[\frac{|\vec{E}_{front}|}{|\vec{E}_{back}|}\right]_{dB} \qquad (4.12)$$

The F/B ratio for the example of Fig. 4.4 can be calculated from the azimuth plot. The strength of the field in front direction is 9 dB stronger than the strength of the field in the back direction.

$$\left[\frac{|\vec{E}_{front}|}{|\vec{E}_{back}|}\right]_{dB} = 9 \text{ dB} \qquad (4.13)$$

From this information, we can calculate the strength of the field in the front direction to the strength of the field on a linear scale.

$$\left[\frac{|\vec{E}_{front}|}{|\vec{E}_{back}|}\right]_{dB} = 10\log_{10}\left[\frac{|\vec{E}_{front}|}{|\vec{E}_{back}|}\right]_{lin} \qquad (4.14)$$

$$\left[\frac{|\vec{E}_{front}|}{|\vec{E}_{back}|}\right]_{lin} = 10^{\frac{1}{10} \cdot \left[\frac{|\vec{E}_{front}|}{|\vec{E}_{back}|}\right]_{dB}} \qquad (4.15)$$

$$\left[\frac{|\vec{E}_{front}|}{|\vec{E}_{back}|}\right]_{lin} = 10^{\frac{9}{10}} = 7.94 \qquad (4.16)$$

If this antenna is being used as a transmitter, signal in the front direction is 7.9 times as strong as the signal in the back direction. The front to back ratio specifies the power ratio, and for this antenna, it is 18 dB.

$$\text{F/B ratio} = \left[\frac{P_{front}}{P_{back}}\right]_{dB} = 2\left[\frac{|\vec{E}_{front}|}{|\vec{E}_{back}|}\right]_{dB} = 18 \text{ dB}. \qquad (4.17)$$

When selecting an antenna, many decisions related to the antenna directivity are needed. A particular application may require an isotropic or a directional antenna. If a directional antenna is needed, the magnitude of the directivity must be decided. Additionally, the orientation of the antenna must be decided so that nodes and nulls are in the appropriate directions. Both the azimuth angle and the elevation angle of the nodes and nulls should be considered [50, p. 22-1].

4.4.4 Electromagnetic Polarization

The electromagnetic wave emanating from a transmitting antenna is described by an electric field $\vec{E}$ and a magnetic field $\vec{H}$. The wave necessarily has both an electric field and a magnetic field because, according to Maxwell's equations, time varying electric fields induce time varying magnetic fields, and time varying magnetic fields induce electric fields. At any point in space and at any time, the direction of the electric field, the direction of the magnetic field, and the direction of propagation of the wave are all mutually perpendicular. More specifically,

$$\left(\text{Direction of } \vec{E}\right) \times \left(\text{Direction of } \vec{H}\right) = \left(\text{Direction of propagation}\right). \tag{4.18}$$

An electromagnetic wave which varies with position in the same way that it varies with time is called a *plane wave* because planar wavefronts propagate at constant velocity in a given direction. For example, a sinusoidal plane wave which travels in the positive $\hat{a}_z$ direction is described by

$$\vec{E} = E_0 \cos\left(10^6 t - 300z\right) \hat{a}_x. \tag{4.19}$$

For this plane wave, $\vec{E}$ is directed along $\hat{a}_x$, $\vec{H}$ is directed along $\hat{a}_y$, and the wave propagates in the $\hat{a}_z$ direction. As another example, consider the plane wave described by

$$\vec{E} = E_0 \cos\left(10^6 t - 300z\right) \left(\frac{\hat{a}_x + \hat{a}_y}{\sqrt{2}}\right). \tag{4.20}$$

For this plane wave, the direction of $\vec{E}$ is 45^0 from the $\hat{a}_x$ axis, the direction of $\vec{H}$ is 45^0 from the $\hat{a}_y$ axis, and again it propagates in the $\hat{a}_z$ direction. Both of these electric fields describe sinusoidal plane waves because the electric field varies with position as it does with time, sinusoidally in both cases.

We can classify plane waves by their *electromagnetic polarization*. Plane waves can be classified as linearly polarized, left circularly polarized, right circularly polarized, left elliptically polarized, or right elliptically polarized. In a previous chapter, we encountered the distinctly different idea of material polarization. Appendix C discusses overloaded terminology including the term polarization.

Both of the electromagnetic waves described by Eq. 4.19 and by Eq. 4.20 are *linearly polarized*. In both cases, the direction of the electric field remains constant as the wave propagates with respect to both position and time. If the direction of the electric field rotates uniformly around the axis formed by the direction of propagation, the wave is called *circularly polarized*. If the direction of the electric field rotates nonuniformly, the wave is called *elliptically polarized*. For circularly polarized waves, the projection of the wave on a plane perpendicular to the axis formed by the direction of propagation is circular. For elliptical waves, the projection is elliptical. To determine if the polarization is left or right, point your right thumb in the direction of propagation, and compare the rotation of the electric field to the rotation of your fingers. If the rotation is along the direction of the fingers of your right hand, the wave is right polarized. Otherwise, it is left polarized. For example, the wave described by

$$\vec{E} = E_0 \cos\left(10^6 t - 300z\right) \frac{\hat{a}_x}{\sqrt{2}} + E_0 \sin\left(10^6 t - 300z\right) \frac{\hat{a}_y}{\sqrt{2}} \qquad (4.21)$$

is right circularly polarized. As another example, the wave

$$\vec{E} = E_0 \cos\left(10^6 t - 300z\right) \frac{\hat{a}_x}{2} + E_0 \sin\left(10^6 t - 300z\right) \frac{\hat{a}_y \sqrt{3}}{2} \qquad (4.22)$$

is right elliptically polarized. The wave

$$\vec{E} = E_0 \cos\left(10^6 t - 300z\right) \frac{\hat{a}_x}{\sqrt{2}} - E_0 \sin\left(10^6 t - 300z\right) \frac{\hat{a}_y}{\sqrt{2}} \qquad (4.23)$$

is left circularly polarized. These definitions are illustrated in the Fig. 4.5.

What does electromagnetic polarization have to do with antennas? Antennas may be designed to transmit linearly, circularly, or elliptically polarized signals. Antennas designed to transmit or receive circularly polarized signals often contain wires that coil in the corresponding direction around an axis. If a signal is transmitted with an antenna designed to transmit linearly polarized waves, the best antenna to use as a receiver will be one that is also designed for linearly polarized waves. The signal can be detected by an antenna designed for signal of a different electromagnetic polarization, but the received signal will be noisier or weaker. Similarly, if a signal is transmitted with an antenna designed for right circular polarization, the best receiving antenna to use will be one also designed for right circular polarization.

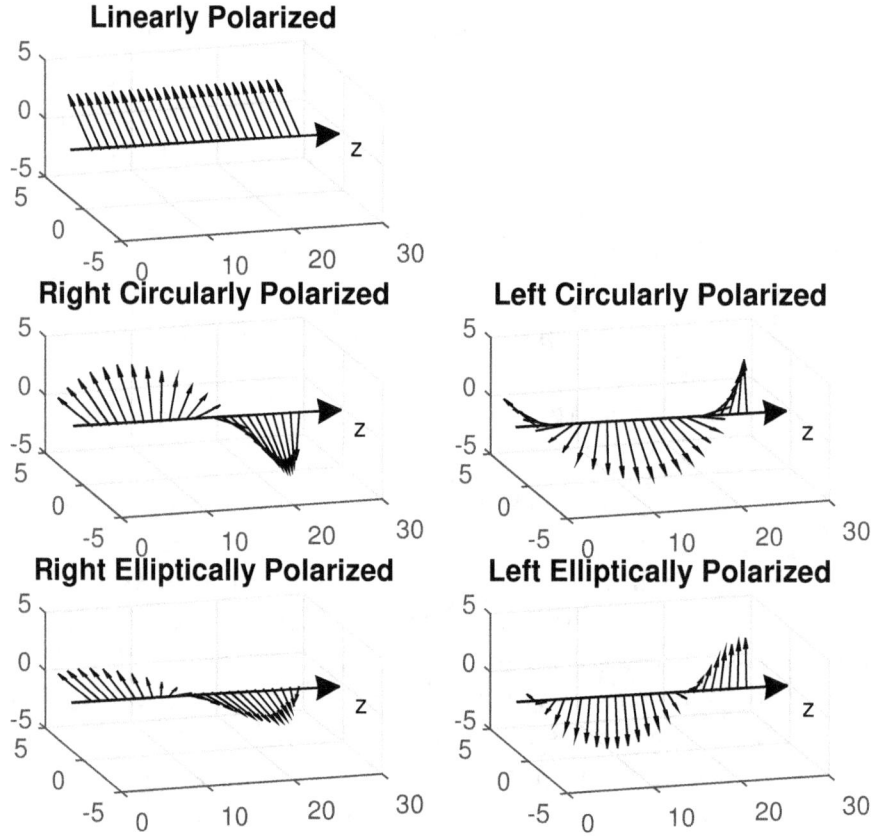

Figure 4.5: Illustration of types electromagnetic polarization for a plane wave traveling in the $\hat{a}_z$ direction.

4.4.5 Other Antenna Considerations

Antennas are made from good conductors. In Chapters 2 and 3, we saw that the materials that make up many energy conversion devices strongly influence the behavior. While the conductivity of conductors vary, overall the material that an antenna is made from does not significantly affect its behavior. In addition to bandwidth, impedance, directivity, and electromagnetic polarization, other factors, such as size, shape and configuration, distinguish one antenna from another. Mechanical factors should be considered too. An ideal antenna may be one that is easy to construct or mount in the desired location, is portable, or requires little maintenance [50]. If an antenna is to be mounted outside, the antenna should be able to withstand snow, wind, ice, and other extreme weather [50]. While Maxwell's equations are useful for predicting the radiation pattern of an antenna, they do not provide information about these other factors.

There is no perfect antenna. In one case, the best antenna may be a Yagi which is very directional and designed to operate within a narrow frequency band. In another application, the best antenna may be mechanically strong and mounted in a way to withstand extreme wind [50, p. 17-29]. In another case, the best antenna may be portable and easy to set up by one person regardless of its radiation pattern [50, p. 21-26]. In another case, the best antenna may be a wire of an arbitrary length hanging from a tree because it was the easiest and quickest to construct. As with any branch of engineering, antenna design involves trade offs. For example, the best antenna to detect an 800 MHz linearly polarized signal is an antenna that is designed to detect 800 MHz signals, is designed to detect linearly polarized signals, is oriented in the proper direction, and has an impedance matched to the impedance of the transmission line used. The signal can still be detected using an antenna designed for a different frequency, designed for a different electromagnetic polarization, improperly directed, or with mismatched impedance. However, in all of these cases, a less intense signal will be received.

Figure 4.6: A snow covered dish antenna.

4.5 Problems

4.1. An antenna is designed to operate between 4.98 GHz and 5.02 GHz, for a bandwidth of $\Delta f = 0.04$ GHz. Find $\Delta \lambda$, the wavelength range over which the antenna is designed to operate.
Hint: The answer is NOT 7.5 m.

4.2. Use the figure to find the following information. (Wires connecting to receiver or transmitter are not shown.)

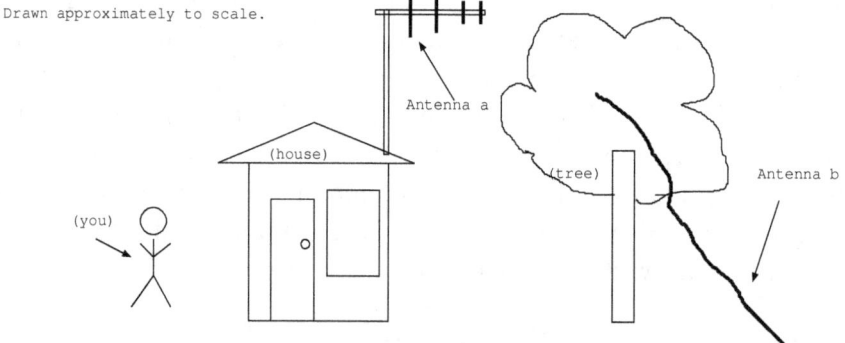

(a) Approximate the wavelength that antenna a is designed to operate at.

(b) Approximate the frequency that antenna b is designed to operate at.

(c) Which antenna most likely has parasitic elements: antenna a, antenna b, both, or neither? Explain your choice.

(d) Which antenna do you expect to be more isotropic: antenna a, antenna b, or would they be about the same? Explain your choice.

(e) Which antenna is more likely to be used as a receiver than a transmitter: antenna a, antenna b, or both antennas about equal? Explain your choice.

4.3. Some speculate that alien civilizations might be able to watch TV programs that escape the earth's atmosphere. To get an idea of the likelihood for this to occur, consider an isotropic antenna in outer space transmitting a 200 MHz TV signal.

Assume that the alien civilization uses an antenna with surface area 0.5 m^2 and has the technology to detect a signal with power as low as $5 \cdot 10^{-22}$ W. What is the minimum power that must be transmitted for detection to occur at a distance of 1.0 light year?

4.4. Project ELF, described in Sec. 4.4.1, was an extremely low frequency, 76 Hz, radio system set by the military to communicate with submarines. It had facilities near Clam Lake, Wisconsin and Republic, Michigan, 148 miles apart [52]. Because these facilities were located a fraction of a wavelength apart, antennas at these locations acted as part of a single array. The length of all antenna elements was 84 miles [52]. Assume it took 18 minutes to transmit a three letters message using 8 bit ASCII, and assume signals travel close to the speed of light in free space.

(a) Calculate the ratio of the distance between the transmitting facilities to the wavelength.

(b) Calculate the ratio of the length of all antenna elements to the wavelength.

(c) What was the speed of communication in bits per second?

(d) How many wavelengths long were each bit?

4.5. Match the following plots or antenna descriptions with their azimuth plots.
   1. An antenna with 3D plot shown below

   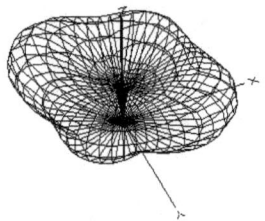

   2. An isotropic antenna
   3. An antenna with nulls at $\pm 90^0$
   4. An antenna with a gain of around 19dB

   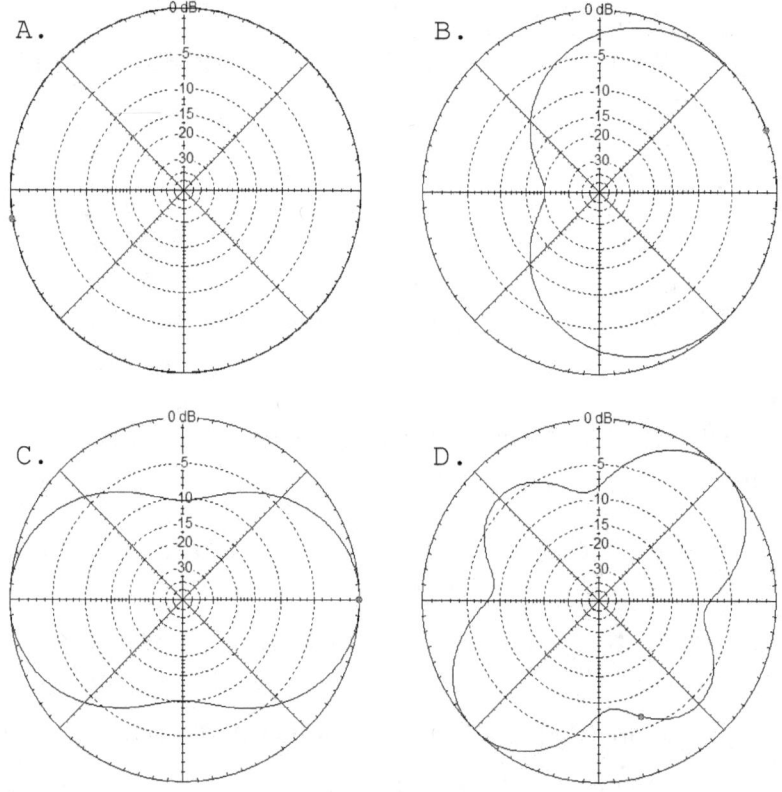

4.6. Radiation pattern plots are for a particular transmitting antenna are shown. They were plotted with EZNEC. The azimuth plot is on the left, and the elevation plot is on the right. The antenna is designed to operate at at 360 MHz. Use the plots to answer the following questions.

4 ANTENNAS

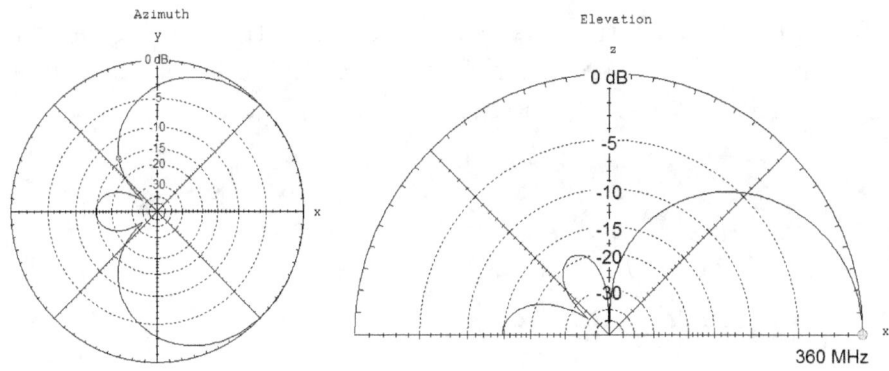

360 MHz

(a) Assume a person 100 m away and receiving signal from the antenna in the front direction (along the $\hat{a}_x$ axis) receives a signal of 15 W. Approximately how strong of a signal would the person receive by standing 100 m away from the transmitter along the $\hat{a}_y$ axis (in watts)?

(b) Find the (power) F/B ratio in dB.

(c) According to the azimuth plot, at approximately what angle are the nulls for this antenna?

(d) What wavelength is this antenna designed to operate at?

4.7. Figure 4.4 show the radiation pattern plots for a quad antenna designed to operate at $f = 21.2$ MHz. The upper left plot shows the azimuth plot, the upper right plot shows the elevation plot. The lower left plot shows the 3D radiation pattern, and the lower right plot shows the antenna elements. They were plotted with EZNEC software.

(a) Find the wavelength the antenna is designed to operate at.

(b) Find $\left[\frac{|\vec{E}_{front}|}{|\vec{E}_{back}|}\right]_{dB}$, the field front to back ratio of the antenna in dB.

(c) Find $\left[\frac{P_{front}}{P_{back}}\right]_{dB}$, the power front to back ratio in dB.

(d) Find $\left[\frac{P_{front}}{P_{back}}\right]_{lin}$, the power front to back ratio on a linear scale.

(e) Assume the electric field intensity 50 m away measured along the $\phi = 45^0$ axis (in the $z = 0$ plane) is 5 $\frac{V}{m}$. Find the electric field intensity 50 m away measured along the $\phi = 135^0$ axis (in the $z = 0$ plane).

4.8. Radiation pattern plots for a particular transmitting antenna are shown. They were plotted with EZNEC. The azimuth plot is on the left, and the elevation plot is on the right.

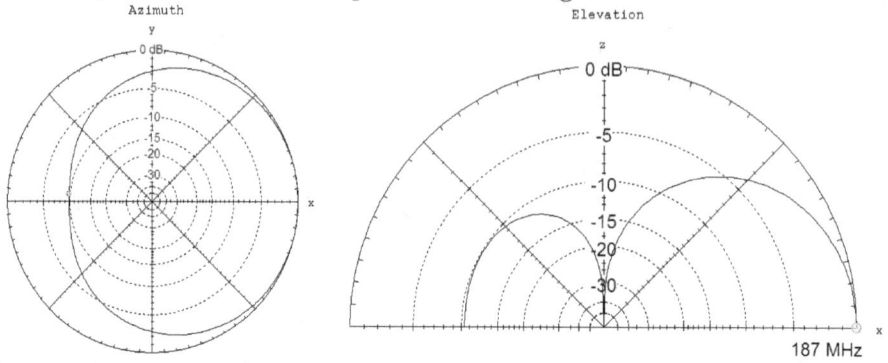

187 MHz

(a) Is this antenna isotropic? Justify your answer.

(b) The antenna is designed to operate at a frequency of 187 MHz. What is the corresponding wavelength?

(c) Find the (power) F/B ratio in dB.

(d) The signal 100 m from the transmitting antenna in the front direction ($\phi = 0$) is measured to be $|\vec{E}| = 50\ \frac{V}{m}$. What is the electric field strength of the signal in $\frac{V}{m}$ at 100 m from the antenna in the $\phi = 45^0$ direction?

(e) Radiation pattern plots do not apply for all distances from the antenna. Roughly for what distances away are the radiation plots valid?

4.9. Determine if the following electromagnetic waves are linearly polarized, right circularly polarized, left circularly polarized, right elliptically polarized, or left elliptically polarized. All of these waves travel in the $\hat{a}_z$ direction, and ω is a constants. (This is a modified version of P3.34 from [11].)

(a) $\vec{E} = 10\cos(\omega t - 8z)\,\hat{a}_x + 10\sin(\omega t - 8z)\,\hat{a}_y$

(b) $\vec{E} = 10\cos\left(\omega t - 8z + \frac{\pi}{4}\right)\hat{a}_x + 10\cos\left(\omega t - 8z + \frac{\pi}{4}\right)\hat{a}_y$

(c) $\vec{E} = 10\cos(\omega t - 8z)\,\hat{a}_x - 20\sin(\omega t - 8z)\,\hat{a}_y$

(d) $\vec{E} = 10\cos(\omega t - 8z)\,\hat{a}_x - 10\sin(\omega t - 8z)\,\hat{a}_y$

5 Hall Effect

5.1 Introduction

In this chapter we discuss another type of inductive energy conversion device, the Hall effect device. While these devices may be made from conductors, they are more often made from semiconductors, like silicon, which are easily integrated into microelectronics. The Hall effect was discovered using gold by Edwin Hall in 1879 [57]. The first practical devices were produced in the 1950s and 1960s when uniform semiconductor materials were first manufactured [57].

Hall effect sensors are used to measure some hard to observe quantities. Without external tools, humans cannot detect magnetic field. However, a small, inexpensive Hall effect sensor can act as a magnetometer. Also, the Hall effect can be used to determine if a semiconductor is n-type or p-type. One of the first applications of Hall effect devices was in computer keyboard buttons [57]. Today, Hall effect devices are used to measure the rotation speed of a motor, as flow rate sensors, in multiple types of automotive sensors, and in many other applications.

5.2 Physics of the Hall Effect

Hall effect devices are direct energy conversion devices that convert energy from a magnetic field to electricity. The physics behind these devices is described by the Lorentz force equation. This discussion follows references [3] and [9]. If we place a charge in an external electric field, it will feel a force parallel to the applied electric field. If we place a moving charge in an external magnetic field, it will feel a force perpendicular to the applied magnetic field. The Lorentz force equation

$$\vec{F} = Q\left(\vec{E} + \vec{v} \times \vec{B}\right) \qquad (5.1)$$

describes the forces on the moving charge due to the external electric and magnetic fields. In the above equation, $\vec{F}$ represents force in newtons on a charge moving with velocity $\vec{v}$ in units $\frac{m}{s}$. The quantity $\vec{E}$ represents the electric field intensity in units $\frac{V}{m}$, and $\vec{B}$ represents the magnetic flux density in units $\frac{Wb}{m^2}$. Charge in coulombs is denoted by Q. Notice that the force on the charge due to the electrical field points in the same direction as the electrical field while the force on the charge due to the magnetic field points perpendicularly to both the velocity of the charge and the direction of the magnetic field.

The Hall effect occurs in both conductors and semiconductors. In conductors, electrons are the charge carriers responsible for the effect while in semiconductors, both electrons and holes are the charge carriers responsible for the effect [9]. A *hole* is the absence of an electron. Consider a piece of semiconductor oriented as shown in Fig. 5.1a. Assume the length is specified by l, the width is specified w, and the thickness is specified by d_{thick}. For a typical Hall effect device, these dimensions may be in the millimeter range. Furthermore, assume the semiconductor is p-type with hole concentration p in units m^{-3}. The *charge concentration* represents the net, or excess, charge density above a neutral material. Materials with a net negative charge, excess valence electrons, will have a positive value for the electron concentration n and are called *n-type*. Materials with a net positive charge, an excess of holes, will have a positive value for the hole concentration p which represents the density of holes in the material and are called *p-type*. Overall *charge density* is related to n and p by

$$\rho_{ch} = -qn + qp \tag{5.2}$$

where q is the magnitude of the charge of an electron.

Assume the semiconductor is placed in an external magnetic field oriented in the $\hat{a}_z$ direction, with magnetic flux density

$$\vec{B} = B_z \hat{a}_z.$$

Also assume a current is supplied through the semiconductor in the $\hat{a}_x$ direction. The positive charge carriers in the semiconductor, holes, move with velocity $\vec{v} = v_x \hat{a}_x$ because current is the flow of charge per unit time. These measures are illustrated in Fig. 5.1b. Hall effect devices are typically used as sensors as opposed to energy harvesting devices because power must be supplied from this external current and because the amount of electricity produced is typically quite small.

The force on the charges can be found from the Lorentz force equation. The force due to the external magnetic field on a charge of magnitude q is given by

$$q\vec{v} \times \vec{B} = qv_x \hat{a}_x \times B_z \hat{a}_z = -qB_z \hat{a}_y \tag{5.3}$$

and is oriented in the $-\hat{a}_y$ direction. Positive charges accumulate on one side of the semiconductor as shown in Fig. 5.1c. This charge build up causes an electric field oriented in the $\hat{a}_y$ direction which opposes further charge build up. Charges accumulate until an equilibrium is reached when the forces on the charges in the $\hat{a}_y$ direction are zero.

$$\vec{F} = 0 = Q\left(\vec{E} + \vec{v} \times \vec{B}\right) \tag{5.4}$$

5 HALL EFFECT

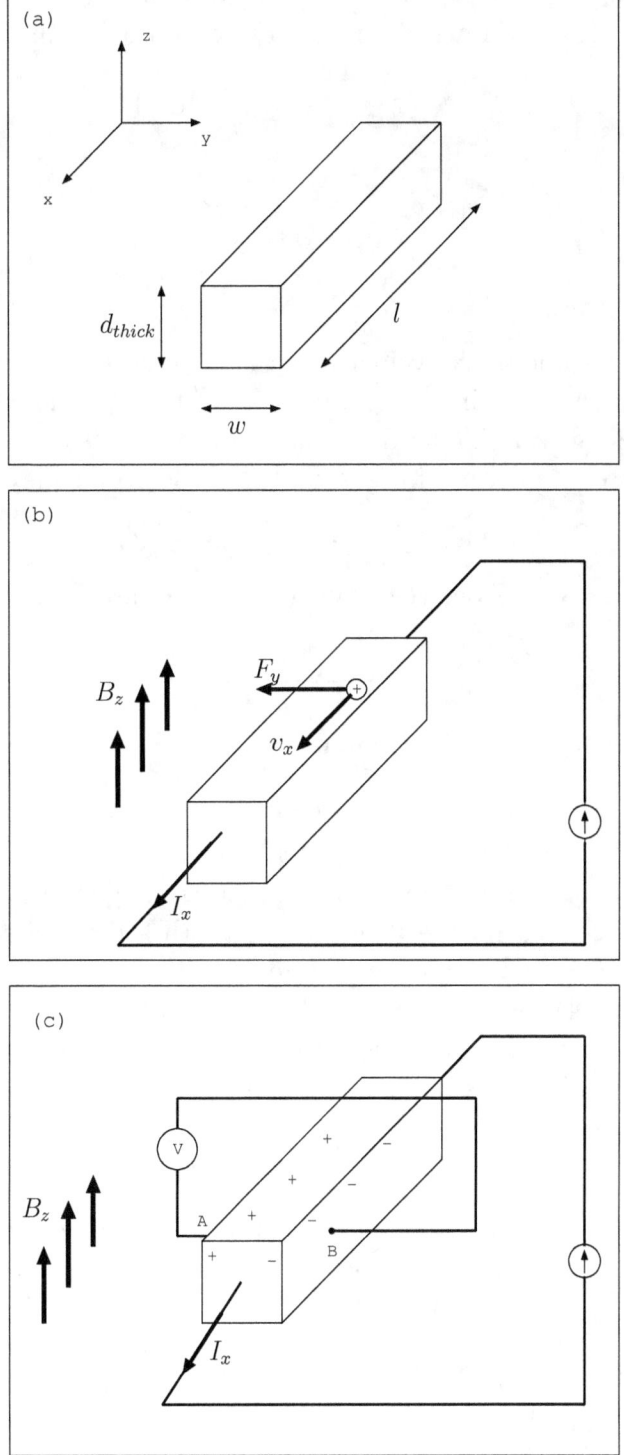

Figure 5.1: Illustration of Hall effect.

The electric field intensity can be expressed as a function of the voltage V_{AB} measured across the width of the device, in the $\hat{a}_y$ direction.

$$\vec{E} = \frac{V_{AB}}{w}\hat{a}_y \qquad (5.5)$$

$$q\vec{E} = -q\vec{v} \times \vec{B} \qquad (5.6)$$

$$\frac{V_{AB}}{w} = v_x B_z \qquad (5.7)$$

While the magnitude of the velocity of the charges v_x is often not known, the applied current, I_x, in units amperes, is known. The current density through a cross section of the device is the product of the charge concentration, the strength of the charges, and the velocity of the charges.

$$\text{current density} = \frac{I_x}{w \cdot d_{thick}} = q \cdot v_x \cdot p \qquad (5.8)$$

From the above expression, velocity can be expressed in terms of the current.

$$v_x = \frac{I_x}{w \cdot d_{thick} \cdot q \cdot p} \qquad (5.9)$$

Equations 5.7 and 5.9 can be combined.

$$V_{AB} = \frac{w \cdot I_x \cdot B_z}{w \cdot d_{thick} \cdot q \cdot p} \qquad (5.10)$$

A *magnetometer* is a device that measures magnetic field. To use a Hall effect device as a magnetometer, start with a piece of semiconductor of known dimensions and known charge concentration, and then apply a current. If the voltage perpendicular to the current is measured, the magnetic field can be calculated. The measured voltage is proportional to the strength of the external magnetic flux density.

$$B_z = \frac{d_{thick} \cdot q \cdot p \cdot V_{AB}}{I_x} \qquad (5.11)$$

Voltage is easily measured with a voltmeter, so no specialized tools are needed. To reliably measure this voltage, it is often amplified.

Alternatively, if the strength of an external magnetic field is known, the Hall effect can be used to measure the concentrations of holes or electrons in a piece of semiconductor. With some algebra, we can write the hole concentration as a function of the dimensions of the semiconductor, the known magnetic field strength, the applied current, and the measured voltage.

$$p = \frac{I_x \cdot B_z}{d_{thick} \cdot q \cdot V_{AB}} \qquad (5.12)$$

5 HALL EFFECT

An analogous expression can be found if electrons instead of holes are the dominant charge carrier. The sign of this measured voltage is also used to determine whether a piece of semiconductor is n-type or p-type [58].

The *Hall resistance* R_H is a parameter inversely proportional to the charge concentration, and it has the units of ohms [9] [59]. For the assumptions above, the Hall resistance is defined as

$$R_H = \frac{B_z}{qp} \cdot \frac{w}{l \cdot d_{thick}}. \tag{5.13}$$

By combining Eqs. 5.12 and 5.13, it can be written in terms of the measured voltage and applied current.

$$R_H = \frac{V_{AB}}{I_x} \cdot \frac{w}{l} \tag{5.14}$$

As an example, suppose that a piece of silicon with a hole concentration of $p = 10^{17}$ cm^{-3} is used as a Hall effect device. The device has dimensions $l = 1$ cm, $w = 0.2$ cm, and $d_{thick} = 0.2$ cm, and it is oriented as shown in Fig. 5.1. The material has a resistivity of $\rho = 0.9$ Ω·cm. A current of $I = 1$ mA is applied in the $\hat{a}_x$ direction. The device is in an external magnetic field of $\vec{B} = 10^{-5} \hat{a}_z \frac{\text{Wb}}{\text{cm}^2}$. If a voltmeter is connected as shown in the figure, what voltage V_{AB} is measured?

$$V_{AB} = \frac{I_x B_z}{q \cdot d_{thick} \cdot p} = \frac{1^{-3} \cdot 10^{-5}}{1.6 \cdot 10^{-19} \cdot 0.2 \cdot 10^{17}} = 3.1 \cdot 10^{-6} \text{ V} \tag{5.15}$$

Signals in the millivolt range are easily detected with a standard voltmeter, yet signals in the microvolt range often can be measured with some amplification. What output power is generated by this device? We can calculate the resistance along the $\hat{a}_y$ direction. The resistivity of the silicon was given in the problem, and resistance R and resistivity ρ are related by

$$R = \frac{\rho \cdot \text{length}}{\text{area}}. \tag{5.16}$$

The resistance across the width of the device is

$$R_{width} = \frac{\rho w}{l d_{thick}} = \frac{0.09 \cdot 0.2}{1 \cdot 0.2} = 0.09 \text{ } \Omega \tag{5.17}$$

We can use this calculated resistance and the measured voltage to find the power converted from the magnetic field to electrical power of the device.

$$P = \frac{V_{AB}^2}{R_{width}} = 1.1 \cdot 10^{-11} \text{ W} \tag{5.18}$$

This amount of power is tiny. While this device can make a useful sensor, it will not make a useful energy harvesting device. It generates tens of picowatts of power, and a 1 mA current must be supplied to generate the power.

5.3 Magnetohydrodynamics

A *magnetohydrodynamic device* converts magnetic energy to or from electrical energy through the use of a conductive liquid or plasma. Similar to the Hall effect, the fundamental physics of the *magnetohydrodynamic effect* is described by the Lorentz force equation, Eq. 5.1. The difference is that the magnetohydrodynamic effect occurs in conductive liquids or plasmas while the Hall effect occurs in solid conductors or solid semiconductors. Another related effect, which is also described by the Lorentz force equation, is the *electrohydrodynamic effect*, discussed in Sec. 10.6. The difference is that the magnetohydrodynamic effect involves magnetic fields while the electrohydrodynamic effect involves electric fields.

Matter can be found in solid, liquid, or gas state. A *plasma* is another possible state of matter. A plasma is composed of charged particles, but a plasma has no net charge. When a solid is heated, it melts into a liquid. When a liquid is heated, it evaporates into a gas. When a gas is heated, the particles will collide with each other so often that the gas becomes ionized. This ionized gas is a plasma [3]. When ions in either a conductive liquid or a plasma flow in the presence of a magnetic field perpendicular to the flow of ions, a voltage is produced.

This magnetohydrodynamic effect was first observed by Faraday in 1831 [3]. In the 1960s, there was interest in building magnetohydrodynamic devices where the conducting medium was a plasma. These devices typically operated at high temperatures, in the range of 3000-4000 K [60]. Progress was limited, however, because few materials can withstand such high temperatures. More recently, engineers have used this principle to build pumps, valves, and other devices for microfluidic systems [61] [62]. These room temperature devices can control the flow of conducting liquids through the use of an external magnetic field.

5.4 Quantum Hall Effect

Around a hundred years after the discovery of the Hall effect, the *quantum Hall effect* was discovered. Klaus von Klitzing discovered the integer quantum Hall effect in 1980 and won the physics Nobel prize for it in 1985 [63]. In 1998, Robert Laughlin, Horst Störmer, and Daniel Tsui won the physics Nobel prize for the discovery of the fractional quantum Hall effect [64]. The integer quantum Hall effect is observed in two dimensional electron gases which can occur, for example, in an inversion layer at the interface between the semiconductor and insulator in a MOSFET [59]. As in the Hall effect, a current is applied in one direction, and the Hall voltage is measured in the perpendicular direction. Following Fig. 5.1, assume that a current is applied along the $\hat{a}_x$ direction in the presence of an external magnetic field in the $\hat{a}_z$ direction. The voltage V_{AB} is measured, and Hall resistance R_H is calculated. The quantum Hall effect is observed at low temperatures and in the presence of strong applied magnetic fields. In such situations, the Hall resistance has the form

$$R_H = \frac{h}{q^2 \cdot \mathfrak{n}} \qquad (5.19)$$

where $h = 6.626 \cdot 10^{-34}$ J·s is the Planck constant and $\mathfrak{n}$ is an integer [59]. This effect is called the quantum Hall effect because R_H can take only discrete values corresponding to integer values. Values of the Hall resistance can be measured extremely accurately, to 2.3 parts in 10^{10} [59]. The fractional quantum Hall effect is observed in highly ordered two dimensional electron gases in the presence of very strong magnetic fields, and it involves quantum mechanical electron-electron interactions [65].

The formal definition of the ohm relies on definitions of the meter, kilogram, and second. The kilogram is defined with respect to the weight of a physical object made of platinum and iridium housed in the International Bureau of Weights and Measures in France [59]. Multiple national labs, including the National Institute of Standards and Technology in the United States, have come up with an experimental means of defining the ohm involving the quantum Hall effect. This standardized definition of the ohm is accurate to one part in 10^9 which is more accurate than previous definitions involving the kilogram, meter, and second [59]. Because of the high accuracy with which the integer quantum Hall effect can be measured, scientists have proposed using experiments involving it to standardize the measurement of the Planck constant and the definition of the kilogram instead of relying on a definition involving a physical object. These new standards have not been adopted yet, but they may be implemented as early as 2019 [66].

5.5 Applications of Hall Effect Devices

A Hall effect device is a simple device. It is essentially a piece of semiconductor with leads connected and calibrated for use. For this reason, Hall effect devices are inexpensive, small, and readily available. As with most integrated circuits, these devices are durable and long lasting because they have no mechanical moving parts [57].

Hall effect devices are available in two types: analog and digital. Analog Hall effect devices are typically integrated with an amplifier and circuitry to make the output more linear [57]. Some devices also contain circuitry to make the devices stable over a wider temperature range because the output of Hall effect sensors may be slightly temperature dependent [57]. The operating output voltage range of these devices is often limited by the amplifier circuit as opposed to the Hall effect sensor [57]. Digital Hall effect devices contain the Hall effect sensor integrated with additional circuitry such as a comparator to produce a digital output [57].

Analog Hall effect devices are used to sense magnetic field, temperature, current, pressure, position, and other parameters [57]. To make a Hall effect temperature sensor, for example, a magnet is mounted on a material that contracts or expands in the presence of a temperature change. As the magnet moves, it changes the magnetic field in a nearby Hall effect device and thereby generates a voltage across the Hall effect device. The same effect can be used to measure pressure or other parameters using a material that expands or contracts when the pressure changes or other parameter changes. Current flowing through a wire generates a magnetic field surrounding the wire. For this reason, the Hall effect can be used to make an ammeter that can be mounted nearby, as opposed to in the path of, the current.

Digital Hall effect devices are used as switches or as buttons in a keyboard. If a small magnet is mounted in a button, a Hall effect device can be used to sense when that magnet is pressed down near the Hall effect sensor. Hall effect devices can also be used as proximity sensors to detect the presence of nearby ferromagnetic objects [57]. Additionally, digital Hall effect devices are used in magnetic card readers [57]. One of the most common applications is in tachometers, devices that measure rotation speed. To measure the rotation speed of a motor for example, the Hall effect sensor is mounted near a ferromagnetic gear. See Fig. 5.2. As a gear tooth passes the sensor, the magnetic field at the sensor changes, and a voltage is induced across the Hall effect device. Hall effect sensors are used to measure rotation speed of motors, fans, tape machines, and disk drives [57]. Relatedly, Hall effect devices are used as flow rate sensors. These

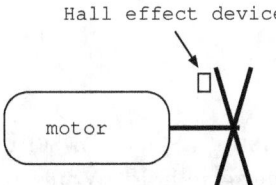

Figure 5.2: Placement of Hall effect sensor used as a tachometer.

sensors are found in devices ranging from water softeners to ocean current monitors [57]. To detect flow rate, a blade is mounted so that it rotates in the water flow. Magnets are mounted on the blade, and the Hall effect sensor is mounted nearby. When the blade passes the sensor, the magnetic field at the sensor changes and induces a voltage in the Hall effect sensor. Following the same principle, Hall effect sensors are used to measure the speed of paper flow in copiers, needles in sewing machines, drill bits in drilling machines, and bottles in bottling factories [57].

Multiple types of Hall effect devices are used in cars. Hall effect sensors are used as rotation sensors to detect transmission speed [57]. They are used as proximity sensors to detect the shift lever position, crank shaft position, and throttle position [57]. They are also used in door interlocks, in brake skidding detection, and in traction control systems [57].

5.6 Problems

5.1. Suppose that you are using a piece of semiconductor as a Hall effect device to measure a magnetic field. You supply a DC current through the device. You would like to replace the piece of semiconductor with another one that will give a larger output for the same external magnetic field. List two ways you can change the piece of semiconductor so that the output would increase. (Specify both the property and whether it would need to be increased or decreased.)

5.2. A piece of p-type semiconductor is used as a Hall effect device. The device has a thickness of $d_{thick} = 1$ mm. It is placed in an external magnetic field of $|\vec{B}| = 10^{-5} \frac{\text{Wb}}{\text{cm}^2}$. A Hall voltage of 5 μV is measured when a current of 3 mA is applied. Calculate p, the charge (hole) concentration in units $\frac{1}{\text{cm}^3}$.

5.3. A Hall effect device is used to measure the strength of an external magnetic field. The device is oriented in the way described in Fig. 5.1. It is made from a cube of p-type silicon with hole concentration $5 \cdot 10^{15}$ cm^{-3} where the length of each side of the cube is 1 mm. A current of 3 mA is applied through the device. The voltage measured across the device is 2.4 mV. Find the strength of the external magnetic flux density, $|\vec{B}|$.

5.4. A Hall effect device is used to measure the strength of an external magnetic field. The device is oriented in the way described in Fig. 5.1. It is made from a material of length $l = 3$ mm, width $w = 0.5$ mm, and thickness $d_{thick} = 0.5$ mm. It has a hole concentration of $p = 10^{20}$ m^{-3}. In an experiment, the devices was placed in an external magnetic field of $|\vec{B}| = 2.5 \frac{\text{Wb}}{\text{m}^2}$ and a voltage of 9 mV was measured. What current was used in the experiment?

5.5. Two expressions were given for the Hall resistance:
$R_H = \frac{B_z}{qp} \cdot \frac{w}{l \cdot d_{thick}}$ and $R_H = \frac{h}{q^2 n}$.
Show that both expressions have the units of ohms.

6 Photovoltaics

6.1 Introduction

This chapter discusses solar cells and optical detectors, both of which are devices that convert optical electromagnetic energy to electricity. The next chapter discusses lamps, LEDs, and lasers which convert energy in the opposite direction. The *photovoltaic effect* is the idea that if a light shines on a pure piece of semiconductor, electron-hole pairs form. In the presence of an external electric field, these charges are swept apart, and a voltage develops across the terminals of the semiconductor. It was first demonstrated in 1839 by Edmond Becquerel. In a *photovoltaic device*, also called a *solar cell*, this effect typically occurs at a semiconductor pn junction. This same effect occurs on a smaller scale in photodiodes used to detect light and in optical sensors in digital cameras. To understand the physics behind these devices, we need to further study crystallography in semiconductors. Energy level diagrams, which illustrate the energy needed to remove an electron from a material, are another topic studied in this chapter.

Unlike fossil fuel based power plants, photovoltaic cells produce energy without contributing to pollution. The solar power industry is growing at a fast pace. Worldwide as of April 2017, photovoltaic cells were capable of generating over 303 GW of power, and 75 GW of this total was installed within the past year [67]. This generating capacity was sufficient to satisfy 1.8% of the worldwide demand for electricity [67]. In the United States as of April 2017, photovoltaic cells installed were capable of generating 14.7 GW [67].

6.2 The Wave and Particle Natures of Light

The physics of electromagnetic radiation is described by Maxwell's equations, Eqs. 1.5 - 1.8, and discussed in Sections 1.6.1 and 4.4.1. Optical energy is electromagnetic energy with wavelengths roughly in the range

$$400 \text{ nm} \lesssim \lambda \lesssim 650 \text{ nm}.$$

This wavelength range corresponds to the frequency range

$$4.6 \cdot 10^{14} \text{ Hz} \lesssim f \lesssim 7.5 \cdot 10^{14} \text{ Hz}.$$

We often think of electromagnetic radiation as behaving like a wave. However, it has both wave-like and particle-like behavior.

One way to understand light is to think of it as composed of particles called photons. A *quantum* is a small chunk, and a *photon* is a quantum, small chunk, of light. A related quantity is a *phonon*, which is a quanta, or small chunk, of lattice vibrations. We will discuss phonons in a later section, and they do not relate to light. Although, phonons can perturb light, and that is the basis for acousto-optic devices. The second way to understand light is to think of it as a wave with a wavelength λ measured in nm. White light has a broad bandwidth while the light produced by a laser has a very narrow bandwidth.

These two descriptions of light complement each other. A photon is the smallest unit of light, and it has a particular wavelength. The energy of a photon of light with wavelength λ is given by

$$E = hf = \frac{hc}{\lambda}. \tag{6.1}$$

The quantity h is called the Planck constant, and it has a tiny value, $h = 6.626 \cdot 10^{-34}$ J $\cdot$ s. The quantity c is the speed of light in free space, $c = 2.998 \cdot 10^8 \, \frac{\text{m}}{\text{s}}$.

In SI units, energy is measured in joules. However, other units are sometimes used by optical engineers because the energy of an individual photon is tiny compared to a joule. Another unit that is used is the electronvolt, or eV. The magnitude of the charge of an electron is $q = 1.602 \cdot 10^{-19}$ C. The *electronvolt* is the energy acquired by a charge of this magnitude in the presence of a voltage difference of one volt [68, p. 8]. Energy in joules and energy in eV are related by a factor of q.

$$E_{[J]} = q \cdot E_{[eV]} \tag{6.2}$$

Equations 6.1 and 6.2 can be combined to relate the energy of a photon in eV and the corresponding wavelength in nm.

$$\frac{1240}{\lambda_{[nm]}} = E_{[eV]}. \tag{6.3}$$

Sometimes, energy is specified in the unit of *wave number*, cm^{-1}, which represents the reciprocal of the wavelength of the corresponding photon. Energy in joules and energy in wave number are related by

$$E_{[J]} = \frac{hc}{\lambda} \tag{6.4}$$

$$E_{[J]} = \frac{6.626 \cdot 10^{-34} \text{ J} \cdot \text{s} \cdot 2.998 \cdot 10^8 \, \frac{\text{m}}{\text{s}} \cdot 100 \, \frac{\text{cm}}{\text{m}}}{\lambda_{[cm]}} \tag{6.5}$$

$$E_{[J]} = 1.986 \cdot 10^{-23} E_{[cm^{-1}]}. \tag{6.6}$$

6 PHOTOVOLTAICS

The human eye can sense light from approximately $\lambda = 400$ nm to $\lambda = 650$ nm. Using the expressions above, we can calculate in different units the energy range over which the human eye can respond. An individual red photon with $\lambda = 650$ nm has energy

$$E_{red} = 3.056 \cdot 10^{-19} \text{ J} = 1.908 \text{ eV} = 1.538 \cdot 10^4 \text{ cm}^{-1} \quad (6.7)$$

in the different units. Similarly, an individual blue photon with $\lambda = 400$ nm has energy

$$E_{blue} = 4.966 \cdot 10^{-19} \text{ J} = 3.100 \text{ eV} = 2.500 \cdot 10^4 \text{ cm}^{-1}. \quad (6.8)$$

We can calculate the energy of individual photons of electromagnetic radiation at radio frequencies, at microwave frequencies, or in other frequency ranges too. For example, the radio station WEAX broadcasts with a frequency $f = 88$ MHz. This corresponds to a wavelength of $\lambda = 3.407$ m. An individual photon at this frequency has energy

$$E = 5.831 \cdot 10^{-26} \text{ J} = 3.640 \cdot 10^{-7} \text{ eV}. \quad (6.9)$$

As another example, wi-fi operates at frequencies near $f = 2.4$ GHz which corresponds to the wavelength $\lambda = 0.125$ m. Each photon at this frequency has energy

$$E = 1.590 \cdot 10^{-24} \text{ J} = 9.927 \cdot 10^{-6} \text{ eV}. \quad (6.10)$$

Ultraviolet light has a wavelength slightly shorter than blue light. A photon of ultraviolet light with wavelength $\lambda = 350$ nm, which corresponds to frequency $f = 8.57 \cdot 10^{14}$ Hz, has energy

$$E = 5.676 \cdot 10^{-19} \text{ J} = 3.543 \text{ eV}. \quad (6.11)$$

X-rays operate at wavelengths near $\lambda = 10^{-10}$ m. An x-ray photon with wavelength $\lambda = 10^{-10}$ m has energy

$$E = 1.986 \cdot 10^{-15} \text{ J} = 1.240 \cdot 10^4 \text{ eV}. \quad (6.12)$$

Why do we talk about radio waves but not radio particles while we treat light as both wave-like and particle-like? A person is around 1.5 to 2 m tall. The wavelength of the radio station broadcast in the example above was $\lambda_{RF} \approx 3.4$ m while the wavelength of blue light was $\lambda_{blue\ light} \approx 400$ nm. Both radio frequency and optical signals are electromagnetic radiation. Both are well described by Maxwell's equations. Both have wave-like and particle-like properties. Humans typically talk about the wave-like nature of radio waves because they are on a scale we can measure with a meter stick. However, with the correct tools, we can observe both the wave-like and particle-like behavior of light.

Why is UV light more dangerous than visible light? Why are x-rays so dangerous? Each photon of x-ray radiation has around a thousand times more energy than a photon of green light. This type of radiation is called *ionizing radiation* because each photon has enough energy to rip an electron from skin or muscles. UV radiation also has enough energy per photon to rip an electron off while red light and blue light do not have enough energy. Photons of radio frequency and microwave electromagnetic radiation contain nowhere near enough energy per photon to do this damage. These types of radiation can still pose a safety hazard if enough photons land on your skin. Microwave ovens are used to cook food. However, they do not pose the hazards of ionizing radiation.

6.3 Semiconductors and Energy Level Diagrams

6.3.1 Semiconductor Definitions

Some semiconductors are made up of atoms of a single type like pure Si or pure Ge. Others contain a combination of elements in column 13 and column 15 of the periodic table. Semiconductors of this type include AlAs, AlSb, GaAs, and InP. Other semiconductors contain a combination of elements in columns 12 and 16 of the periodic table. Examples of this type include ZnTe, CdSe, and ZnS [9]. Most semiconductors involve elements located somewhere near silicon on the periodic table, but more complicated compositions and structures are also possible. Materials made from three different elements of the periodic table are called *ternary* compounds, and materials made from four elements are called *quaternary* compounds.

To understand the operation of devices like solar cells, photodetectors, and LEDs, we need to study the flow of charges in semiconductors. Electrical properties in semiconductors are determined by the flow of both valence electrons and holes. *Valence electrons*, as opposed to inner shell electrons, are the electrons most easily ripped off an atom. A *hole* is an absence of an electron. Valence electrons and holes are known as *charge carriers* because they are charged and they move through the semiconductor when an external voltage is applied. At a finite temperature, electrons are continuously in motion, and some electron-hole pairs may form an *exciton*. These electron-hole pairs naturally combine, also called decay, within a short time. However, at any time, some charge carriers are present in semiconductors at temperatures above absolute zero due to the motion of charges.

Crystalline semiconductors can be classified as intrinsic or extrinsic [9, p. 65]. An *intrinsic* semiconductor crystal is a crystal with no lattice defects or impurities. At absolute zero, $T = 0$ K, an intrinsic semiconductor has no free electrons or holes. All valence electrons are involved in chemical bonds, and there are no holes. At finite temperature, some charge carriers are present due to the motion of electrons at finite temperature. The concentration of these charge carriers is measured in units $\frac{\text{electrons}}{\text{m}^3}$, $\frac{\text{holes}}{\text{m}^3}$, $\frac{\text{electrons}}{\text{cm}^3}$ or $\frac{\text{holes}}{\text{cm}^3}$. The *intrinsic carrier concentration* is the density of electrons in a pure semiconductor, and it is a function of the temperature T. At higher temperatures, more charge carriers will be present even if there are no impurities or defects in the crystalline semiconductor due to more motion of charges. If we apply a voltage across an intrinsic semiconductor at $T = 0$ K, no charges flow. When the equilibrium concentration of electrons n or holes p is different from the intrinsic carrier concentration n_i then we say that the semiconductor is *extrinsic*. If either impurities or crystal defects are present, the material will be extrinsic. If a voltage is applied across an extrinsic semiconductor at $T = 0$ K, charges will flow. If a voltage is applied across either an extrinsic or intrinsic semiconductor at temperatures above absolute zero, charge carriers will be present and will flow.

The process of introducing more electrons or holes into a semiconductor is called *doping*. A semiconductor with an excess of electrons compared to an intrinsic semiconductor is called *n-type*. A semiconductor with an excess of holes is called *p-type*. Silicon typically has four valence electrons which are involved in bonding. Phosphorous has five valence electrons, and aluminum has three. When a phosphorous atom replaces a silicon atom in a silicon crystal, it is called a *donor* because it donates an electron. When an aluminum atom replaces a silicon atom, it is called an *acceptor*. Column 15 elements are donors to silicon and column 13 elements are acceptors. If silicon is an impurity in AlP, it may act as a donor or acceptor. If it replaces an aluminum atom, it acts as a donor. If it replaces a phosphorous atom, it acts as an acceptor.

How can we dope a piece of silicon? More specifically, how can we dope a semiconductor with boron? Boron is sold at some hardware stores. It is sometimes used as an ingredient in soap. Start with a silicon wafer, and remove any oxide which has formed on the surface. Each silicon atoms forms bonds with four nearest neighbors. At the surface though, there is no fourth neighbor, so silicon atoms bond with oxygen from the air. Smear some boron onto the wafer, or place a chunk of boron on top of the wafer. Place it in a furnace at slightly less than silicon's melting temperature, around 1000 °C. Some boron will diffuse in and replace silicon atoms. Remove the excess boron. The same procedure can be used to dope with other donors or acceptors. What is the most dangerous part of the process? Etching the oxide off the silicon because hydrofluoric acid HF, a dangerous acid, is used [69].

Sometimes it is possible to grow one layer of a semiconductor material on top of a layer of a different type of material. A stack of different semiconductors on top of each other is called a *heterostructure*. Not all materials can be made into heterostructures. GaAs and AlAs have almost the same atomic spacings, so heterostructures of these materials can be formed. The spacing between atoms, also called *lattice constant*, in AlAs is 0.546 nm, and the spacing between atoms in GaAs is 0.545 nm [9]. If the atomic spacing in the two materials is too different, mechanical strain in the resulting material will pull it apart. Even moderate mechanical strain can negatively impact optical properties of a device because defects may be introduced at the interface between the materials. These defects can introduce additional energy levels which can trap charge carriers.

6.3.2 Energy Levels in Isolated Atoms and in Semiconductors

In a solar cell, light shining on a semiconductor causes electrons to flow which allows the device to convert light to electricity. How much energy does it take to cause an electron in a semiconductor to flow? To answer this question, we will look at energy levels of:

- An isolated Al atom at $T = 0$ K

- An isolated P atom at $T = 0$ K

- Isolated Al atom and P atoms at $T > 0$ K

- An AlP crystal at $T = 0$ K

- An AlP crystal at $T > 0$ K

Aluminum has an electron configuration of $1s^2 2s^2 2p^6 3s^2 3p^1$. It has 13 total electrons, and it has 3 valence electrons. More specifically, it has two valence electrons in the 3s subshell and one in the 3p subshell. Phosphorous has an electron configuration of $1s^2 2s^2 2p^6 3s^2 3p^3$, so it has 5 valence electrons. Ideas in this section apply to materials regardless of whether they are crystalline, amorphous, or polycrystalline.

6 PHOTOVOLTAICS

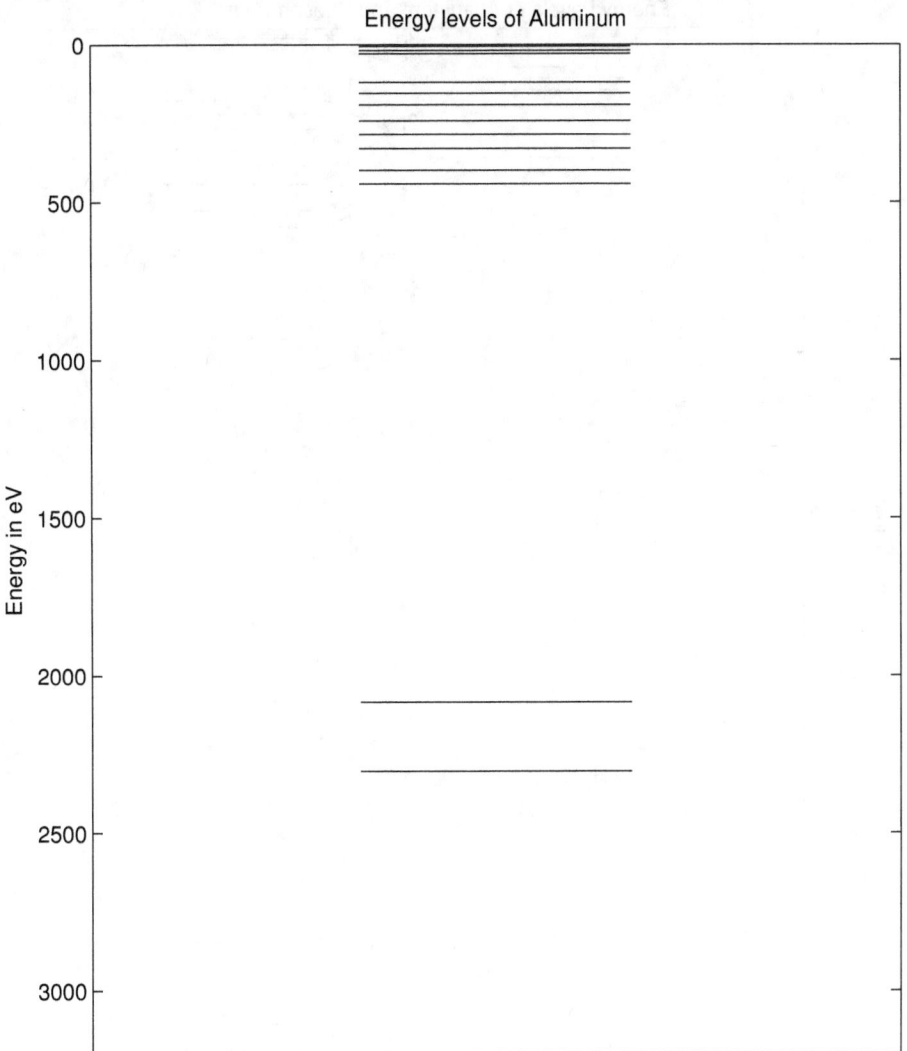

Figure 6.1: Energy level diagram of an isolated aluminum atom at $T = 0$ K plotted using data from [70].

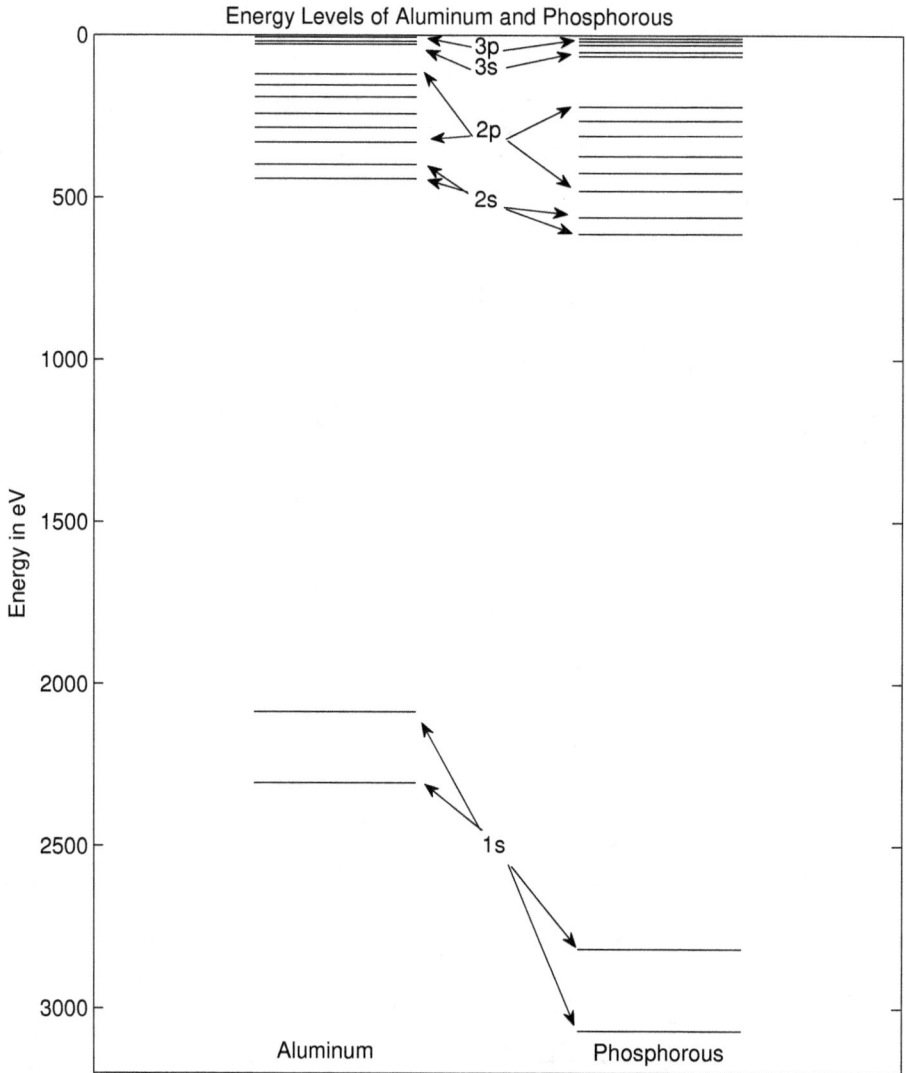

Figure 6.2: Energy level diagram of isolated aluminum and phosphorous atoms at $T = 0$ K plotted using data from [70].

6 PHOTOVOLTAICS

Energy Levels of Electrons of Isolated Al and Isolated P Atoms at $T = 0$ K

To understand the interaction of light and a semiconductor, start by considering an isolated Al atom and an isolated P atom at absolute zero, $T = 0$ K. How much energy does it take to rip off electrons of Al? It takes significantly less energy to rip off a valence electron than an electron from an inner shell. In fact, when we say an electron is a valence electron, or an electron is in a valence shell, we mean that the electron is in the shell for which it takes the least energy to rip off an electron. We do not mean that the electron is further from the nucleus, although often it is. When we say an electron is in an inner shell, we mean the electron is in a shell for which it takes more energy to rip off an electron. This text focuses on energy conversion devices which operate at moderate energies, so all of the devices discussed involve interactions of only valence electrons. Inner shell electrons will not be involved. It is also possible to excite, but not rip off, an electron. When an electron is excited, its internal momentum changes and its quantum numbers change. The terms valence electron and quantum number were both defined in Sec. 1.5.2. Less energy is required to excite than rip off an electron. The energy required to excite or rip off electrons can be supplied by thermal energy, an external voltage, an external optical field, or other external sources.

Figure 6.1 is a plot of the energy required to excite or remove electrons from an isolated neutral Al atom at $T = 0$ K. The figure was plotted using data from [70]. While energy levels are drawn using actual data, the thickness of the lines is not drawn to scale. Energy is on the vertical axis. Allowed energy levels are shown by horizontal lines. Each electron can only have energy corresponding to one of these discrete possible energy levels. At $T = 0$ K, electrons occupy the lowest possible energy levels. One electron can occupy each line, so the lowest 13 energy levels are occupied by electrons. While not shown due to the resolution of the figure, the density of allowed energy levels increases as energy approaches zero at the top of the figure. Since we are considering the case of absolute zero temperature, these upper energy levels are not occupied by electrons.

The left side of Fig. 6.2 replots the allowed energy levels of the electrons in an isolated Al atom at $T = 0$ K. The energy levels are also labeled. The right side of the figure plots the allowed energy levels of electrons in an isolated P atom also at $T = 0$ K. Data on phosphorous energy levels also comes from [70]. As with the Al atom, the electrons of the P atom can only occupy certain specific discrete energy levels. Since the atoms are at absolute zero, the electrons occupy the lowest energy levels possible. Figure

6.3 contains the same information, but is zoomed in vertically to show the valence electron levels more clearly.

The P atom has two more electrons than the Al atom. Phosphorous atoms have more protons, so the electrons are a bit more tightly bound to the nucleus. For this reason, it takes a bit more energy to rip the electrons off, and the allowed energy levels are a bit different than for Al.

The amount of energy required to rip a 3p electron off the atom is the vertical distance from the 3p level to the ground line at the top of the figure. The amount of energy required to rip a 2p electron off is the vertical distance from the 2p level to the ground line. As expected, these figures show that it requires more energy to rip off the inner shell 2p electron than the valence shell 3p electron. If enough energy is supplied, an electron will be ripped off, and the electron will flow freely through the material. If some energy is supplied but not enough to rip off the electron, the electron can get excited to a higher energy level. The energy required to excite an electron is given by the vertical distance in the figure from an occupied to an unoccupied energy level. In either case, we say that an electron-hole pair forms. If the amount of energy supplied is too small to excite an electron from a filled to unfilled state, the external energy will not be absorbed.

Energy Levels of Electrons of Isolated Al and Isolated P Atoms at $T > 0$ K

How do the energy levels change when the Al and P atoms are at temperatures above absolute zero, where electrons are continuously vibrating and moving? First, the energy levels broaden. The electrons can still only take certain energy levels, but there is a wider range to the allowed energy levels. Second, occasionally, electrons spontaneously get excited into higher states. For example, a 3p electron may get excited into the 4s state temporarily. If it does, it will quickly return to the ground state.

Energy Levels of AlP at $T = 0$ K

How much energy does it take to rip an electron off an AlP crystal at $T = 0$ K? The three valence electrons of each Al atom and the five valence electrons of each P atom form chemical bonds. The energy required to rip off these electrons is slightly different than the energy required to rip off the equivalent electrons of isolated Al and isolated P atoms. Figure 6.4 illustrates the energy levels of the valence electrons of AlP. Unlike in the previous figures, these energy levels do not come from actual data. Instead, they are meant as a rough illustration of the effect. The amount of energy

6 PHOTOVOLTAICS

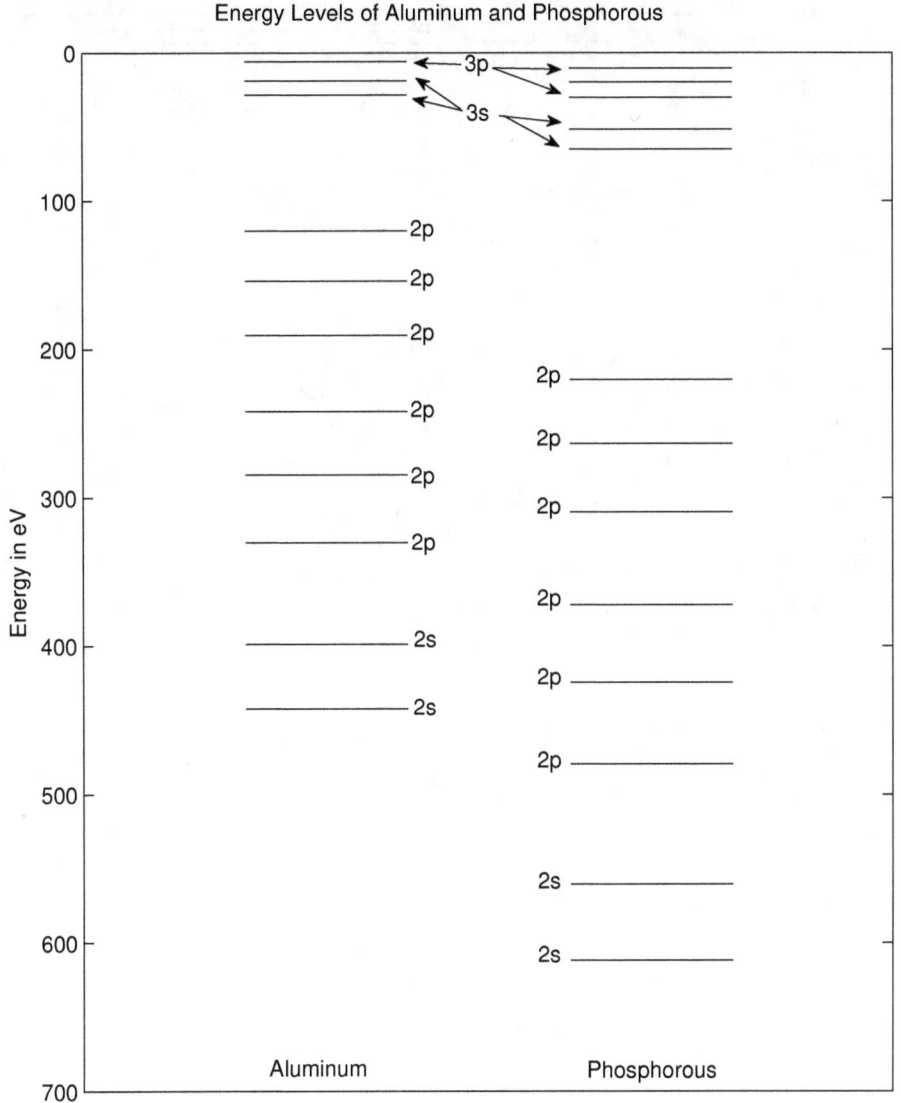

Figure 6.3: Zoomed in version of the energy level diagram of isolated aluminum and phosphorous atoms at $T = 0$ K plotted using data from [70].

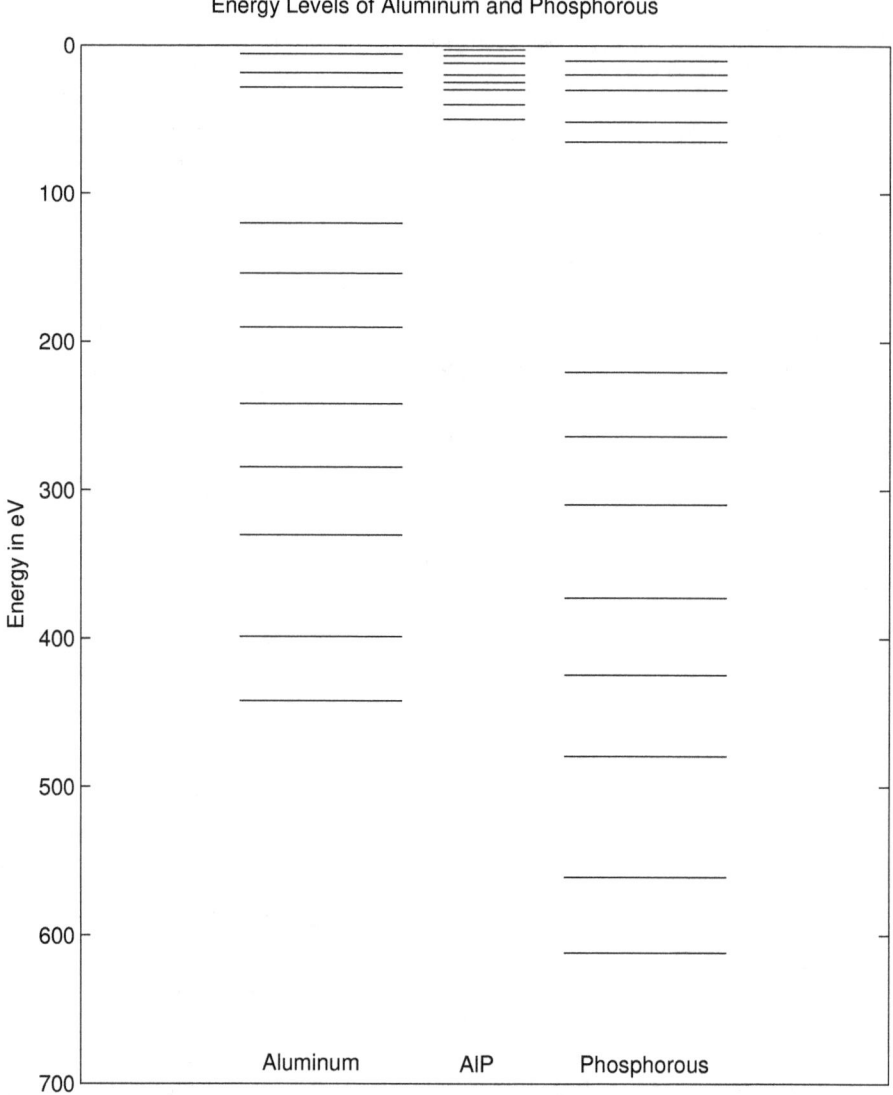

Figure 6.4: Energy level diagram at $T = 0$ K of an isolated aluminum atom, AlP crystal, and isolated phosphorous atom. Energy levels for the isolated atoms are from [70]. Energy levels for AlP are a rough illustration and not from actual data.

6 PHOTOVOLTAICS

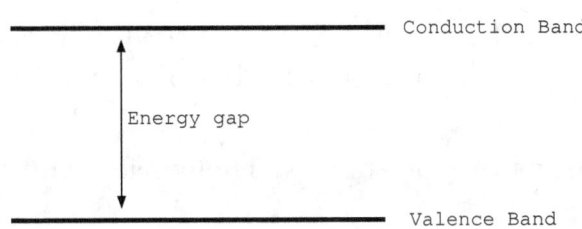

Figure 6.5: Energy level diagram of a semiconductor zoomed in to show only the conduction and valence band.

required to rip off an electron is represented on the energy level diagram by the vertical distance from that level to the ground level at the top of the diagram. The energies needed to remove inner shell electrons do not significantly change from the energy levels of isolated atoms.

Energy levels due to electrons shared amongst atoms in a solid semiconductor are called energy *bands*. The filled energy level closest to the top of an energy level diagram for a semiconductor is called the *valence band*. The energy level above it is called the *conduction band*. The *energy gap* E_g, also called the *bandgap*, is the energy difference from the top of the valence band to the bottom of the conduction band. The term *valence electron* refers to an outer shell electron while the term *valence band* refers to a possible energy level it may occupy. At $T = 0$ K, the valence band is typically filled, and the conduction band may be empty or partially empty. We often are only interested in the valence and conduction bands because we are interested in energy conversion processes involving small amounts of energy. For this reason, we often plot energy level diagrams zoomed in vertically to just show these two energy levels as shown in Fig. 6.5.

If the AlP crystal has defects or impurities, the energy levels broaden a bit because the electrical potential (in volts) seen by each Al and each P atom is slightly different from the potential seen by other Al and P atoms in the crystal. Thus, it takes slightly different amounts of energy to rip off each electron. For this reason, energy levels in amorphous materials are quite a bit broader than energy levels in crystals of the same composition [10]. If the AlP crystal has defects or impurities, additional allowed energy levels may be present. Some of these energy levels may even fall within the energy gap.

Energy Levels of AlP at $T > 0$ K

As with isolated atoms, there are two differences between energy levels for crystals such as AlP at $T > 0$ K compared to at $T = 0$ K. First, energy levels broaden. Second, some electrons get excited to higher energy levels and quickly, perhaps in a few microseconds, decay back down.

6.3.3 Definitions of Conductors, Dielectrics, and Semiconductors

Conductors, dielectrics, and semiconductors were defined in section 1.5.1. Now that we have seen example energy level diagrams, we should revisit these definitions as well as define the term semimetal. In the presence of an applied external voltage, electric field, optical field, or other energy source, valence electrons flow easily in a *conductor* [10, p. 429] [11, ch. 4]. In a conductor, the conduction band is partially filled with electrons, so there are many available energy states for electrons remaining in the conduction band. With just a little bit of external energy, possibly even from vibrations that naturally occur at $T > 0$ K, valence electrons flow easily. Inner shell electrons can be ripped off their atoms and flow, but significantly more energy is needed to rip off inner shell than valence electrons.

In the presence of an applied external voltage, electric field, optical field, or other energy source, electrons do not flow easily in an *insulator* [10, p. 429] [11, ch. 4]. The valence band is filled and the conduction band is empty. The energy gap between valence band and conduction band in an insulator is typically above 3 eV. A little heat or energy from vibrations is not enough to excite an electron from one allowed energy state to another. If a large enough external source of energy is applied, though, an electron can be excited or ripped off of an insulator.

In Sec. 3.3, electro-optic materials were discussed. Some insulators are electro-optic which means that in the presence of an external electric or optical field, the spatial distribution of electrons changes slightly which cause a material polarization to build up. Photons of the external electric or optical field in this case do not have enough energy to excite electrons in the insulator, so the internal momentum of electrons in the material does not change. The electro-optic effect occurs in insulators and involves external energies too small to excite electrons from one allowed energy state to another while the affects discussed in Sec. 6.3 involve semiconductors and external energies large enough to excite electrons from one energy level to another.

At $T = 0$ K in a *semiconductor*, the valence band is full, and the conduction band is empty. The energy gap of a semiconductor is small, in the range 0.5 eV $\lesssim E_g \lesssim$ 3 eV. In the presence of a small applied voltage, electric field, or optical field, a semiconductor acts as an insulator. In the presence of a large applied voltage or other energy source, a semiconductor acts as a conductor, and electrons flow. Photodiodes and solar cells are made from semiconductors. If enough energy is supplied to a photodiode, for example from an optical beam, the valence electrons will flow. More specifically, the photons of the external optical beam must have more energy than the energy gap of the semiconductor for the valence electrons to flow.

The term *semimetal* is used to describe conductors with low electron concentration. Similar to conductors, in a semimetal at $T = 0$ K, there is no energy gap because the conduction band is partially filled with electrons, and there are plenty of available energy states. The concentration of electrons for semimetals, however, is in the range $n < 10^{22} \frac{\text{electrons}}{\text{cm}^3}$ while n is greater for conductors [26, p. 304].

6.3.4 Why Are Solar Cells and Photodetectors Made from Semiconductors?

Energy level diagrams for AlP were illustrated above. The energy gap of AlP is $E_g = 2.45$ eV, so it is a semiconductor [9] [10, p. 432,543]. If a beam of light with photons of energy $E < 2.45$ eV is applied to a piece of AlP, the photons will not be absorbed, and no electrons will be excited. If a beam of light with photons of energy $E \geq 2.45$ eV is applied to a piece of AlP, some of those photons may be absorbed. When a photon is absorbed, an electron will be excited from the valence band to the conduction band. A blue photon with energy $E = 3.1$ eV will be absorbed by AlP, for example, but a red photon with energy $E = 1.9$ eV will not. When the electron is excited, the internal momentum of the electron necessarily changes. The excited electron quickly spontaneously decays back to its lowest energy state, and it may emit a photon or a phonon in the process. If a beam of light with photons of significantly higher energy is applied to a piece of AlP, it is possible to rip off electrons entirely from their atom.

Why are solar cells and optical photodetectors made from semiconductors instead of insulators? Sunlight is composed of light at multiple wavelengths, and it is most intense at wavelengths that correspond to yellow and green light. Green photons have energies near $E \approx 2.2$ eV, and visible photons have energies in the range 1.9 eV $< E < 3.1$ eV. Solar cells are made from materials with an energy gap less than the energy of most of the photons from sunlight. Semiconductors are used because the energy of each photon is large enough to excite the electrons in the material. Insulators are not used because most of the photons of visible light do not have enough energy to excite electrons in the material. The material should not have an energy gap that is too large otherwise photons will not be absorbed.

Material	Gap in eV
AlP	2.45
GaP	2.26
InP	1.35

Material	Gap in eV
ZnS	3.6
ZnSe	2.7
ZnTe	2.25

Material	Gap in eV
GaP	2.26
GaAs	1.43
GaSb	0.70

Table 6.1: Energy gap of various semiconductors.

Why are solar cells and optical photodetectors made from semiconductors instead of conductors? When light shines on a solar cell or photodetector, photons of light are absorbed by the material. If the photon absorbed has energy greater than the energy gap of the material, the electron quickly decays to the top of the conduction band. With some more time, it decays back to the lowest energy state. In a solar cell or photodetector, a pn junction is used to cause the electrons to flow before decaying back to the ground state. The amount of energy converted to electricity per excited electron depends on the energy gap of the material, not the energy of the incoming photon. Only energy E_g per photon absorbed is converted to electricity regardless of the original energy of the photon. Thus, the energy gap of the material used to make a solar cell or photodetector should be large so that as much energy per excited electron is converted to electricity as possible. The material should not have an energy gap that is too small otherwise very little of the energy will be converted to electricity. The electron and hole will release the excess energy, $hf - E_g$, quickly in the form of heat or lattice vibrations called phonons.

Each semiconductor has a different energy gap E_g. Many solar cells and photodetectors are made from silicon, which is a semiconductor with $E_g = 1.1$ eV. Predicting the energy gap of a material is quite difficult. However, all else equal, if an element of a semiconductor is replaced with one below it in the periodic table, the energy gap tends to get smaller. This trend is illustrated in Table 6.1. Data for the table comes from [9]. This trend is also illustrated in Fig. 6.6, which plots the energy gap and lattice constant for various semiconductors. Figure 6.6 is taken from reference [71]. The horizontal axis represents the interatomic spacing in units of angstroms, where one angstrom equals 10^{10} meters. The vertical axis represents the energy gap in eV. This figure illustrates energy gaps and lattice constants for materials of a wide range of compositions. For example, the energy gap for aluminum phosphide can be found from the point labeled AlP, and the energy gap of aluminum arsenide can be found from the point labeled AlAs. Energy gap for semiconductors of composition $AlAs_xP_{1-x}$ can be found from the line between these points.

6 PHOTOVOLTAICS

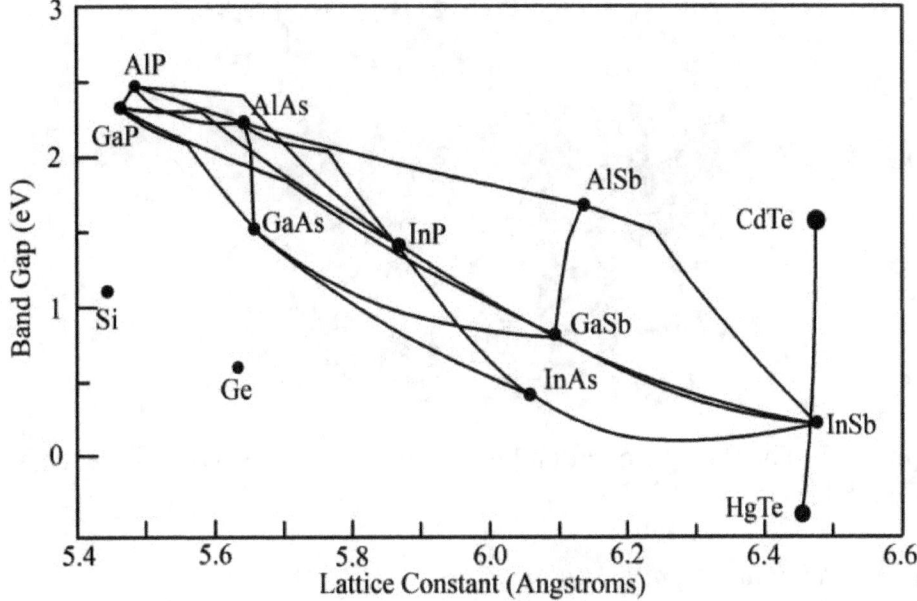

Figure 6.6: Energy gap versus interatomic spacing for multiple semiconductors. Used with permission from [71].

Some solar cells are made from layered material with the largest energy gap material on the top. For example, a solar cell could be made from a top layer of ZnS, a middle layer of ZnSe, and a bottom layer of ZnTe. Photons with energy $E > 3.6$ eV would be absorbed in the ZnS layer. Photons with energy 2.7 eV$< E <$3.6 eV would be absorbed by the ZnSe layer, and photons with energy 2.25 eV$< E <$2.7 eV would be absorbed by the ZnTe layer. Each photon of energy absorbed by the ZnS layer and converted to electricity would have more energy than each photon absorbed by the ZnSe layer. Solar cells made from layers in this way can be more efficient at converting energy from optical energy to electricity than equivalent solar cells made of a single material.

The photo in Fig. 6.7 shows naturally occurring zinc sulfide, also called sphalerite, collected near Sheffler's Rock shop near Alexandria, Missouri. The dark mineral embedded in the middle of the rock is the sphalerite.

6.3.5 Electron Energy Distribution

The *Fermi energy level* of a semiconductor, denoted E_f, represents the energy level at which the probability of finding an electron is one half [9] [10, p. 432,543]. The Fermi level depends on temperature, and it depends on the impurities in the semiconductor. Chemists sometime call the Fermi

Figure 6.7: The dark mineral embedded in the rock is naturally occurring zinc sulfide.

level by the name *chemical potential*, μ_{chem}.

In a pure semiconductor at $T = 0$ K, all electrons occupy the lowest possible states. The valence band is completely filled, and the conduction band is completely empty. The Fermi level, E_f, is the energy level at the middle of the energy gap. No electrons are found at energy E_f because no electrons can have an energy inside the energy gap. However, the Fermi level is a useful measure to describe the material.

In a pure semiconductor at $T > 0$ K, some electrons are excited into higher energy levels. As the temperature increases, more electrons are likely to be found at higher energy levels more often. The probability that an electron is in energy level E varies with temperature as $e^{-E/k_B T}$ [9] [10]. The quantity k_B is the *Boltzmann constant*.

$$k_B = 1.381 \cdot 10^{-23} \frac{\text{J}}{\text{K}} = 8.617 \cdot 10^{-5} \frac{\text{eV}}{\text{K}} \qquad (6.13)$$

The Fermi level for a material with $T > 0$ K is slightly higher than the Fermi level for a material with $T = 0$ K because more electrons are likely to be excited.

The probability of finding an electron at energy level E at temperature T is

$$F(E, T) = \frac{1}{1 + e^{(E-E_f)/k_B T}}. \qquad (6.14)$$

Equation 6.14 is called the *Fermi Dirac distribution*, and like any probability, it ranges $0 \leq F \leq 1$. For energy levels far above the conduction band, $(E - E_f)$ is large and positive, so electrons are quite unlikely to be found, $F \approx 0$. For energy levels far below the valence band, $(E - E_f)$ is large and negative, so electrons are quite likely to be found, $F \approx 1$.

The concentration and type of impurities influence the energy of the Fermi level. A p-type material has a lack of electrons. For this reason in a p-type material, E_f is closer to the valence band than the middle of the energy gap. An n-type material has an excess of electrons. For this reason in a n-type material, E_f is closer to the conduction band.

6.4 Crystallography Revisited

6.4.1 Real Space and Reciprocal Space

Physicists and chemists are often interested in where electrons or nucleons of atoms are likely to be found with respect to position in real space. Ideas of a lattice, basis, and crystal structure were discussed in Sec. 2.3.2. To review, a *lattice* describes the arrangements of points. The *basis* describes how atoms are arranged at each lattice point. The lattice and basis together form the *crystal structure*. A 3D lattice is described by three *lattice vectors* $\vec{a_1}$, $\vec{a_2}$, and $\vec{a_3}$. If they are chosen as short as possible, they are called *primitive lattice vectors*. The magnitude of a primitive lattice vector may be around 0.1 nm. The primitive lattice vectors define a cell called a *primitive cell*. Since a lattice is periodic, if we know how to describe one primitive cell, we can describe the entire lattice.

For each lattice, there is a corresponding *reciprocal lattice* defined by a set of vectors. Both contain the same information in different forms. For a 3D lattice with primitive vectors $\vec{a_1}$, $\vec{a_2}$, and $\vec{a_3}$, the vectors of the reciprocal lattice are labeled by the vectors $\vec{b_1}$, $\vec{b_2}$, and $\vec{b_3}$.

$$\vec{b_1} = \frac{2\pi \vec{a_2} \times \vec{a_3}}{\vec{a_1} \cdot \vec{a_2} \times \vec{a_3}} \quad (6.15)$$

$$\vec{b_2} = \frac{2\pi \vec{a_3} \times \vec{a_1}}{\vec{a_1} \cdot \vec{a_2} \times \vec{a_3}} \quad (6.16)$$

$$\vec{b_3} = \frac{2\pi \vec{a_1} \times \vec{a_2}}{\vec{a_1} \cdot \vec{a_2} \times \vec{a_3}} \quad (6.17)$$

Notice that $\vec{b_1}$ is perpendicular to $\vec{a_2}$ and $\vec{a_3}$. Also, $\vec{b_1}$ is parallel to $\vec{a_1}$. More specifically, $|\vec{b_1}| \cdot |\vec{a_1}| = 2\pi$. (Factors of 2π show up due to choice of units, $\frac{\text{cycles}}{\text{m}}$ vs $\frac{\text{rad}}{\text{m}}$.) Thus if vector $\vec{a_1}$ is long, $\vec{b_1}$ will be short. Just as we can get from one lattice point to another by traveling integer multiples of the $\vec{a_n}$ lattice vectors, we can get from any one point to the next of the reciprocal lattice by traveling integer multiples of the $\vec{b_n}$ lattice vector.

Lattice vectors in real space have units of length, m. Lattice vectors in reciprocal space have units m^{-1}.

The reciprocal lattice gives information about the *spatial frequency* of atoms. If the planes of atoms in a crystal are closely spaced in one direction, $|\vec{a_1}|$ is relatively small. The corresponding reciprocal vector $|\vec{b_1}|$ is relatively large. The reciprocal lattice represents the spatial frequency of the atom in units m^{-1}. If the planes of atoms in a crystal are far apart, $|\vec{a_1}|$ is large and $|\vec{b_1}|$ is small.

If a beam of light shines on a crystal where the wavelength of light is close to the crystal spacing, light will be diffracted, and the diffraction pattern is related to the reciprocal lattice. The *Brillouin zone* is a primitive cell for a reciprocal lattice. The volume of a unit cell in reciprocal space over a unit cell in real space is given by

$$\frac{\text{vol. Brillouin zone}}{\text{vol. primitive cell in real space}} = \frac{\vec{b_1} \cdot \vec{b_2} \times \vec{b_3}}{\vec{a_1} \cdot \vec{a_2} \times \vec{a_3}} = (2\pi)^3. \qquad (6.18)$$

As for the real space lattice, to understand the reciprocal space lattice, we need to only understand one cell because the reciprocal space lattice is periodic.

6.4.2 E versus k Diagrams

The energy level diagrams, discussed in Section 6.3, plot allowed energies of electrons where the vertical axis represented energy. No variation is shown on the horizontal axis. The most useful energy level diagrams for semiconductors are zoomed in so that only the valence and conduction band are shown. In many cases, it is useful to plot energy level diagrams versus position in real space. For such a diagram the vertical axis represents energy, and the horizontal axis represents position. It is also useful to plot energy level diagrams versus position in reciprocal space.

Kinetic energy is given by

$$E_{kinetic} = \frac{1}{2}m|\vec{v}|^2 = \frac{1}{2m}|\vec{M}|^2 \qquad (6.19)$$

where $\vec{v}$ represents velocity in $\frac{m}{s}$ and m represents mass in kg. Momentum is given by $\vec{M} = m\vec{v}$ in units $\frac{\text{kg} \cdot \text{m}}{\text{s}} = \frac{\text{J} \cdot \text{s}}{\text{m}}$. Electrons in crystals at $T > 0$ K vibrate, and certain vibrations are resonant in the crystal. The *crystal momentum* $\vec{M}_{crystal}$ represents the internal momentum of due to vibrations. It can be expressed as

$$\vec{M}_{crystal} = \hbar \vec{k} \qquad (6.20)$$

6 PHOTOVOLTAICS

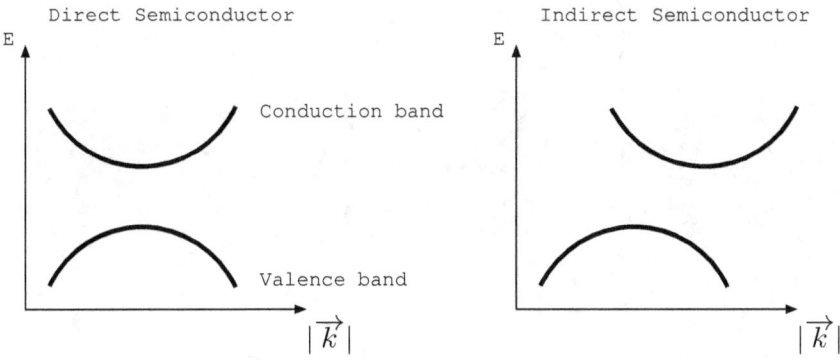

Figure 6.8: Energy plotted vs. $|\vec{k}|$ for a direct and indirect semiconductor.

and has units of momentum $\frac{\text{kg} \cdot \text{m}}{\text{s}}$. The quantity $\vec{k}$ is called the *wave vector*, and it has units m^{-1}. It represents change in spatial frequency, a distance in reciprocal space. The constant

$$\hbar = \frac{h}{2\pi} \qquad (6.21)$$

is called *h-bar* and is the Planck constant divided by 2π. Kinetic energy can be written in terms of the wave vector.

$$E_{kinetic} = \frac{\hbar^2 |\vec{k}|^2}{2m} \qquad (6.22)$$

Equation 6.22 describes how energy of an electron varies with wave vector $|\vec{k}|$ which incorporates information about lattice vibrations. The energy is quadratic in wave vector, so plots of energy versus $|\vec{k}|$ are parabolic. Equation 6.22 is just a model, and it applies best near the top of the valence band and bottom of the conduction band.

Energy versus $|\vec{k}|$ diagrams plot allowed energy levels. Think of the $|\vec{k}|$ axis as change in position in reciprocal space. If the top of the valence band and bottom of the conduction band occur at the same $|\vec{k}|$ value in a semiconductor, we say that it is *direct*. If the top of valence band and bottom of conduction band occur at different $|\vec{k}|$ values, we say that the semiconductor is *indirect*. The left part of Fig. 6.8 shows an energy versus $|\vec{k}|$ diagram for a direct semiconductor, and the right part of Fig. 6.8 shows one for an indirect semiconductor. GaAs, InP, and ZnTe are direct semiconductors. Si, Ge, AlAs, and GaP are indirect semiconductors. Along different crystal axes, the band structure changes somewhat. The horizontal axis of an energy versus $|\vec{k}|$ diagram may be specified along a particular axis in reciprocal space.

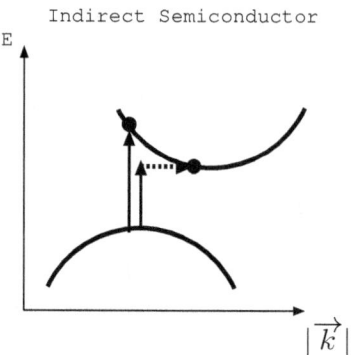

Figure 6.9: Two possible mechanisms of photon absorption in an indirect gap semiconductor.

What happens when we shine light on a direct semiconductor? A photon of sufficient energy can excite an electron from the valence band to the conduction band to create an electron-hole pair. What happens in an indirect semiconductor? Figure 6.9 illustrates two possibilities. As illustrated by the longer arrow, an electron can be excited directly from the valence to conduction band. However, this requires a photon of more energy than the vertical distance between the top of the valence band and the bottom of the conduction band [25, p. 200]. Alternatively, as illustrated by the other two arrows, excitation from the top of the valence band to the bottom of the conduction band may involve a photon and a phonon. Both energy and momentum must be conserved, so a change in crystal momentum is needed to excite an electron in this case. Solar cells and photodetectors may be made from either direct or indirect semiconductors.

6.5 Pn Junctions

Many devices, including photovoltaic devices, LEDs, photodiodes, semiconductor lasers, and thermoelectric devices are essentially made from pn junctions. To understand photovoltaic devices and these other energy conversion devices, we need to understand pn junctions. Consider a semiconductor crystal composed of an n-type material (with excess electrons) on one side and a p-type material (lacking electrons, in other words, with an excess of holes) on the other side. The junction of the p-type and n-type materials is called a *pn junction*. Assume the junction is abrupt and is at thermal equilibrium.

Some pn junctions are made from elemental semiconductors like Si, and other pn junctions are made from compound semiconductors like GaAs.

6 PHOTOVOLTAICS

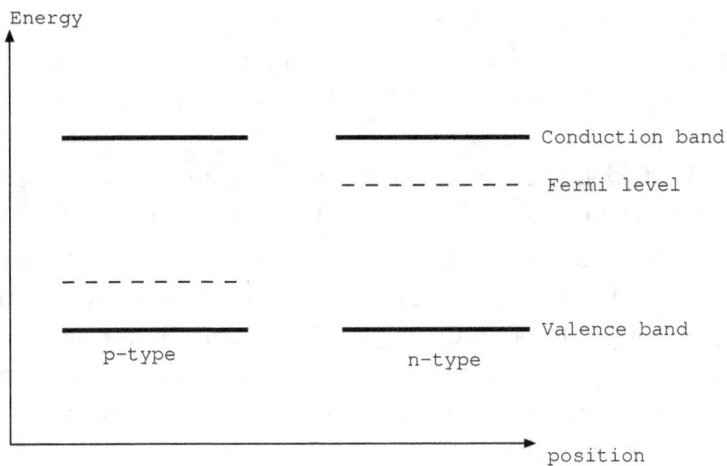

Figure 6.10: Energy level diagram of p-type and n-type semiconductors.

Some pn junctions have the same material on both sides while other pn junctions have different materials on either side. For example, a pn junction can be made from an n-type layer of GaAs and a p-type layer of GaAs. It can also be made from an n-type layer of GaAs and a p-type layer of AlAs.

What happens when we put a p-type material and an n-type material together to form a pn junction? Valence electrons and holes move. Nuclei and inner shell electrons do not. Some excess electrons from the n-type region go towards the p-type region. Some excess holes from the p-type region go towards the n-type region. These charge carriers diffuse, are swept away from, a region near the junction. This region near the junction which is lacking charge carriers is called the *depletion layer* [10, p. 564]. As shown in Fig. 6.10, the Fermi level E_f is near the valence band for p-type materials. P-type material lacks electrons, so the energy where it is equally likely to find an electron state occupied and unoccupied is closer to the valence band. For a similar reason, the Fermi level E_f is near the conduction band for n-type materials. Figure 6.11 shows the energy level diagram versus position for the pn junction, and Fermi levels of the two materials are lined up in this figure.

Consider a junction where the n-type material is silicon doped with phosphorous atoms and the p-type material is silicon doped with aluminum atoms. The n-type side of the pn junction has an excess of positive charges because some phosphorous atoms replace Si atoms in the material. Phosphorous atoms have one more proton than silicon atoms. They also have one more electron, but the valence electron is a charge carrier which diffuses away from the junction. Similarly, the p-type side of the junction has an excess of negative charges because some aluminum atoms replace silicon atoms. Aluminum atoms have one less proton than Si atoms. They also have one less electron, but the hole is a charge carrier which also diffuses away from the junction.

An electric field forms across the junction due to the net charge distribution near the junction. Electric field intensity is the force per unit charge, and it has the units $\frac{V}{m}$. There is also necessarily a voltage drop across a pn junction in equilibrium, and this voltage is called the *contact potential* V_0 in the units of volts. While the contact potential is a voltage, it cannot be measured by placing a voltmeter across a pn junction because additional junctions would be formed at each lead of the voltmeter with additional voltages introduced [9, p. 141].

Figure 6.11 illustrates the energy level diagram of a pn junction. The horizontal axis represents position, and the vertical axis represents energy. It is related to the figures in Section 6.3. However, Fig. 6.11 is zoomed in vertically, and it is plotted versus position near the junction. It also shows the relationship between the energy level diagram and the circuit symbol for a diode, and the depletion layer is labeled. The vertical distance qV_0, also labeled in Fig. 6.11, represents the amount of energy required to move an electron across the junction [9, p. 141].

Figure 6.12 shows the energy level diagram for a forward biased pn junction. In a *forward biased* pn junction, current flows from the p-type to n-type side of the junction. More specifically, holes flow from the p-type to n-type region, and some of these holes neutralize excess charges in the depletion layer. The depletion layer becomes narrower. The electric field preventing the flow of charges gets smaller, and the voltage drop across the junction gets smaller. The energy $q(V_0 - V_x)$ is labeled in Fig. 6.12 for a forward biased pn junction where the voltage V_x is the voltage supplied. This energy represents the energy needed to get charges to flow across the junction, and it is smaller than the corresponding energy in the case of the unbiased junction. Charges flow more easily in the case of a forward biased pn junction, and the diode acts as a wire.

Figure 6.13 shows the energy level diagram for a reversed biased pn junction. For a *reverse biased* pn junction, the voltage across the junction $V_0 + V_x$ is larger than for an unbiased junction, and the energy needed for charges to flow $q(V_0 + V_x)$ is larger than for an unbiased junction. Reversed biased pn junctions act as open circuits, and charges do not flow due to this amount of energy required.

A *light emitting diode* (LED) is a device that converts electricity to optical electromagnetic energy, and it is made from a semiconductor pn junction. In use, a forward bias is put across the LED as shown in Fig.

6 PHOTOVOLTAICS

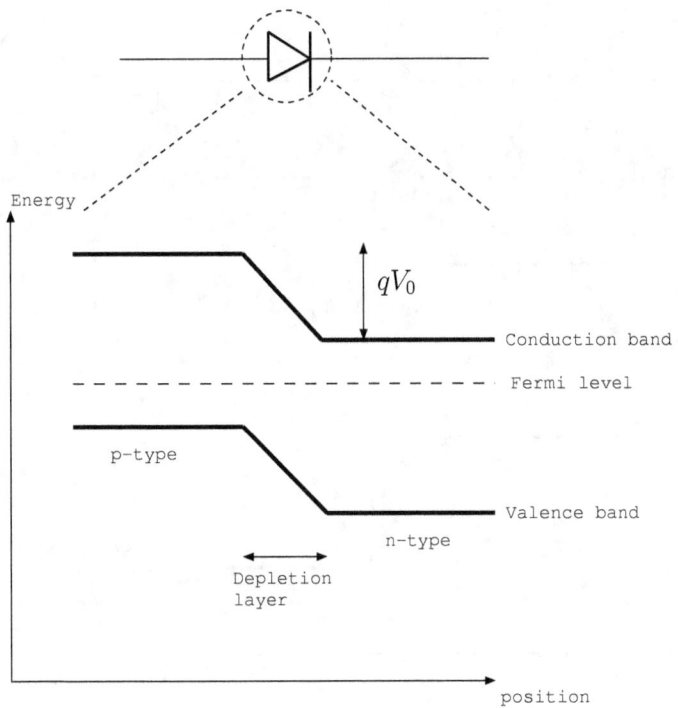

Figure 6.11: Energy level diagram of an unbiased pn junction.

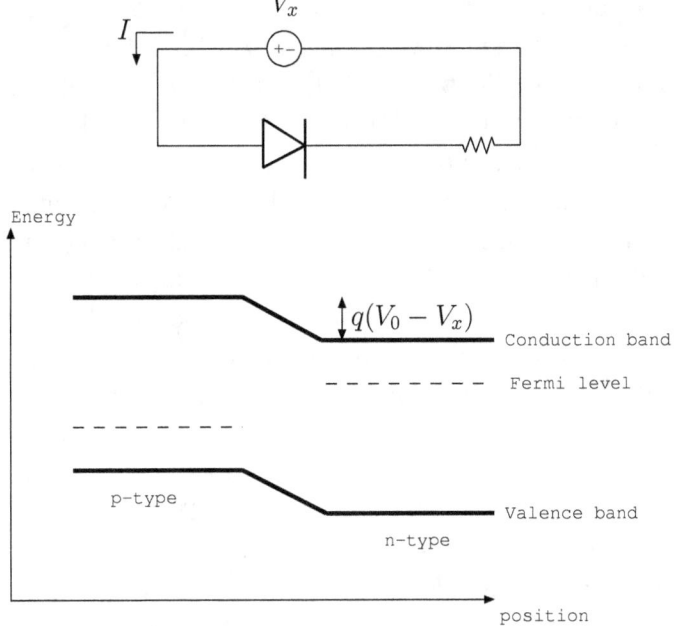

Figure 6.12: Energy level diagram of a forward biased pn junction.

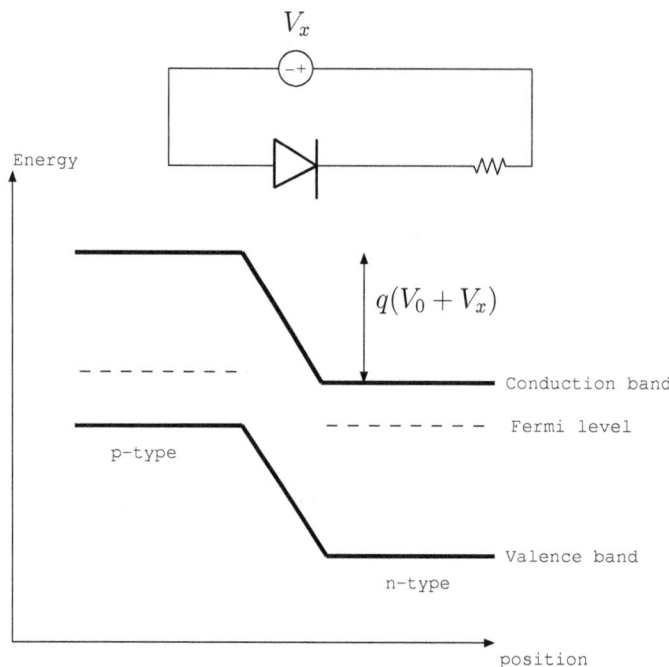

Figure 6.13: Energy level diagram of a reversed biased pn junction.

6.12. Holes flow from the p-type to n-type region. Some of these holes combine with electrons in the depletion layer. In an LED, photons are emitted in this process. The energy of the emitted photon corresponds to the energy of the energy gap. Some LEDs have an additional intrinsic, undoped, layer at the junction, between the p-type and n-type layers to improve the efficiency of the device.

A solar cell and an optical photodetector are also essentially pn junctions. Both of these devices convert optical electromagnetic energy to electricity. When light shines on these devices, electron-hole pairs are created at the junction. Due to the charge distribution across the junction, many of the electrons and holes created are swept away from the junction before they can recombine [9]. This flow of charges is a current, so the optical electromagnetic energy is converted into electricity. When light shines on a photovoltaic device, a voltage can be measured across the junction, and this effect is called the photovoltaic effect [9, p. 212].

The vertical distance between the conduction band and the valence band on an energy level diagram is the energy gap E_g. The energy gap of the material used to make a solar cell or photodetector determines the properties of the device. Photons with energy greater than the energy gap have enough energy to form electron-hole pairs while photons with less

6 PHOTOVOLTAICS

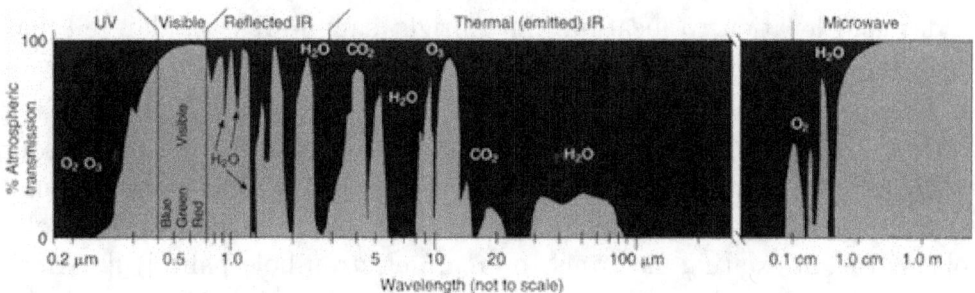

Figure 6.14: Diagram of atmospheric windows– wavelengths at which electromagnetic radiation will penetrate the Earth's atmosphere. Chemical Notation (CO_2, O_3) indicates the gas responsible for blocking sunlight at a particular wavelength. This figure is used with permission [72].

energy cannot.

If a temperature gradient is applied across a pn junction, charges flow. When one side of the device is heated, charges move more rapidly and these energetic charges diffuse to the cooler side. This effect, called the Seebeck thermoelectric effect, is discussed in Chapter 8.

6.6 Solar Cells

6.6.1 Solar Cell Efficiency

Energy conversion devices are never 100% efficient. Efficiency is defined as the output power over the input power. Efficiency of a solar cell is often defined as the ratio of electrical power out to optical power in to the device.

$$\eta_{eff} = \frac{P_{electrical\ out}}{P_{optical\ in}} \qquad (6.23)$$

Not all sunlight reaches a solar cell because some of it is absorbed by the earth's atmosphere. This atmospheric absorption is strongly dependent on wavelength. Figure 6.14 is a plot of the transmissivity of the atmosphere as a function of wavelength. It plots the percent of light which passes through the atmosphere without getting absorbed. Some gases in the atmosphere, such as water vapor and CO_2, absorb a significant amount of energy at particular wavelengths. The figure indicates which gas is responsible for atmospheric absorption at some particular wavelengths. For example, ozone O_3 absorbs ultraviolet light. Ozone in the atmosphere offers benefits because ultraviolet light can damage eyes and skin. The intensity of the optical power from the sun that is hits a solar cell varies from day to day and location to location. In a bright sunny area, a solar cell may receive around 0.1 $\frac{W}{cm^2}$ [73, p. 7].

Even if energy from the sunlight reaches a solar cell, the energy is not converted to electricity with perfect efficiency. There are multiple reasons for this inefficiency, and some of these reasons relate to the fact that not all light that hits a solar cell is absorbed. Light may heat up the solar cell instead of exciting electrons to create electron-hole pairs [74]. Alternatively, light may be reflected off the solar cell surface [74]. Many solar cells have an antireflection coating to reduce reflections, but they are not eliminated. The surface of other solar cells are manufactured to be rough instead of smooth to reduce reflections. Furthermore, if a photon hits an electron that is already excited, the photon will not be absorbed. Additionally, solar cells have wires throughout the surface to capture the produced electricity. These wires are often thin and in a finger-like configuration. Light that hits these wires does not reach the semiconductor portion of the solar cell and is not efficiently converted to electricity. To reduce this issue, wires of some solar cells are made from materials that are partially transparent conductors, such as indium tin oxide or tin oxide SnO_2 [74]. Indium tin oxide is a transparent conductor with a moderately high electrical conductivity of $\sigma = 10^6 \frac{1}{\Omega \cdot m}$ [75].

Other reasons that solar cells are not perfectly efficient have to do with what happens after a photon excites an electron. An electron may be excited, but it may decay before it gets swept from the junction [74]. A photon may excite an electron to a level above the conduction band, but the electron may quickly decay to the top of the conduction band losing some energy to heat. Internal resistance in the bulk n-type or p-type regions may convert electricity to heat. There may also be internal resistance of wiring in the system. Also unmatched loads make solar cells less efficient than matched loads [74].

The voltage across and the current produced by an illuminated solar cell are both functions of temperature. Reference [76] demonstrates, both theoretically and experimentally, that efficiency of a solar cell decreases as temperature increases. A number of mechanisms occurring in a solar cell are dependent on temperature. First, as the temperature increases, the allowed energy levels broaden. For this reason, the energy gap E_g, which is proportional to the voltage produced by the solar cell, is smaller at higher temperatures. As temperature increases, this voltage produced by the solar cell decreases roughly linearly [76]. Second, the current due to recombination of electron-hole pairs at the junction is a function of temperature. At higher temperatures, more electron-hole pairs recombine at the junction, so the overall current produced by the solar cell is less. For this reason, as temperature increases, the overall current produced by the solar cell decreases roughly exponentially [76]. This effect on the current is the main reason that solar cell efficiency depends on temperature. Other mechanisms are temperature dependent, but are less significant [76].

6.6.2 Solar Cell Technologies

There are four major solar cell technologies being developed: crystalline, thin film, multijunction cells, and emerging photovoltaic technologies [77]. However, these categories are not distinct because some solar cells fit into multiple categories simultaneously. Figure 6.15, from [77], compares solar cells of these technologies. More specifically, it shows record efficiencies for each of these types of solar cells as well as the year the records were achieved.

The first category is crystalline, and these cells may be made from single crystals or from polycrystalline material [78]. The first generation of solar cells was made with this technology. For a simple recipe for how to produce a crystalline solar cell, see [69]. Most solar cells produced today, around 80% of the market, are silicon cells in this category. Typical efficiency of a crystalline solar cell available today may be around 20% [78]. Polycrystalline solar cells are often cheaper and a bit less efficient than single crystalline cells.

The second category is thin film. To make these solar cells, thin films of semiconductors are deposited on a substrate such as glass or steel. The substrate may be rigid or flexible. The solar cell itself may be made of layers of material only a few microns thick. Thin film solar cells may be cheaper than other types of solar cells [78]. Often they are less efficient than crystalline cells, but they have other advantages [78]. One material used to make thin film solar cells is amorphous silicon. Another material in use is CdTe, which has a energy gap 1.45 eV. Cadmium and tellurium are both toxic, but they may be easier to deposit in thin films than silicon.

The third category is multijunction, also called compound, solar cells. These solar cells are made of a dozen or more layers of semiconductor stacked on top of each other [78]. These layers form multiple pn junctions. Larger gap semiconductors are on the upper layers, and smaller gap semiconductors are closer to the substrate. These solar cells can be quite efficient. Cells with efficiency up to 46% have been demonstrated in labs [77].

The last category is emerging technology solar cells. Multiple creative strategies are being used to develop solar cells. Nanotechnology strategies include using solar cells made from carbon nanotubes and from quantum dot based materials [78]. Organic solar cells also fall into this category. The active part of these solar cells is a thin, often 100-200 nm, layer of an organic material [79]. One advantage of organic solar cells is that their processing may not require as high of temperatures as the processing of solar cells made from pn junctions of inorganic semiconductors [79].

6.6.3 Solar Cell Systems

Solar cells are used in a wide range of devices. Inexpensive lawn ornaments with solar cells are available at hardware stores for less than a dollar. Small photovoltaic devices used as optical sensors are equally inexpensive. On the other extreme, solar cells power the NASA Mars rovers Spirit and Opportunity as well as satellites orbiting the earth. Also, large arrays of solar cells are used to generate electricity.

A typical solar cell produces around a watt of electrical power while a typical house may require around 4 kW of power [73]. To produce the necessary power, individual solar cells are connected together into modules, and the modules are connected together into solar panels. In a typical installation on the roof of a house, a panel may be composed of around 40 solar cells, and 10 or 20 panels may be mounted roof [73]. A typical solar panel installation on the roof of a building has a number of components in addition to the solar panel arrays. The additional components are often referred to as the *balance of the system*, and they consist of batteries, mounting or tracking hardware, solar concentrators, and power conditioners. These components are illustrated in Fig. 6.16.

The mounting system is composed of the foundation, mechanical supports, brackets, and wiring needed to physically mount and connect the solar panel. Some solar panels are mounted in a fixed position. Other solar panels are mounted on systems that angle the panels towards the sun. Some tracking systems rotate the panel around a single east-west axis. Others have two axes. Two axis tracking systems are often used with solar concentrators. A concentrator is a mirror or lens system designed to capture more of the sun's light onto the panels.

Solar panel systems require batteries or some other energy storage mechanism to provide electrical power at night, on cloudy days, and other times when inadequate sunlight falls on the solar panels. Solar panels can last 30 years or more with only about 1% or 2% degradation per year. Also, solar panels rarely need maintenance, and they cannot easily be repaired. If a solar panel fails, the entire panel is replaced. However, batteries have a typical lifetime of three to nine years, and they are often the first part of a solar panel system that needs replacement [73].

6 PHOTOVOLTAICS

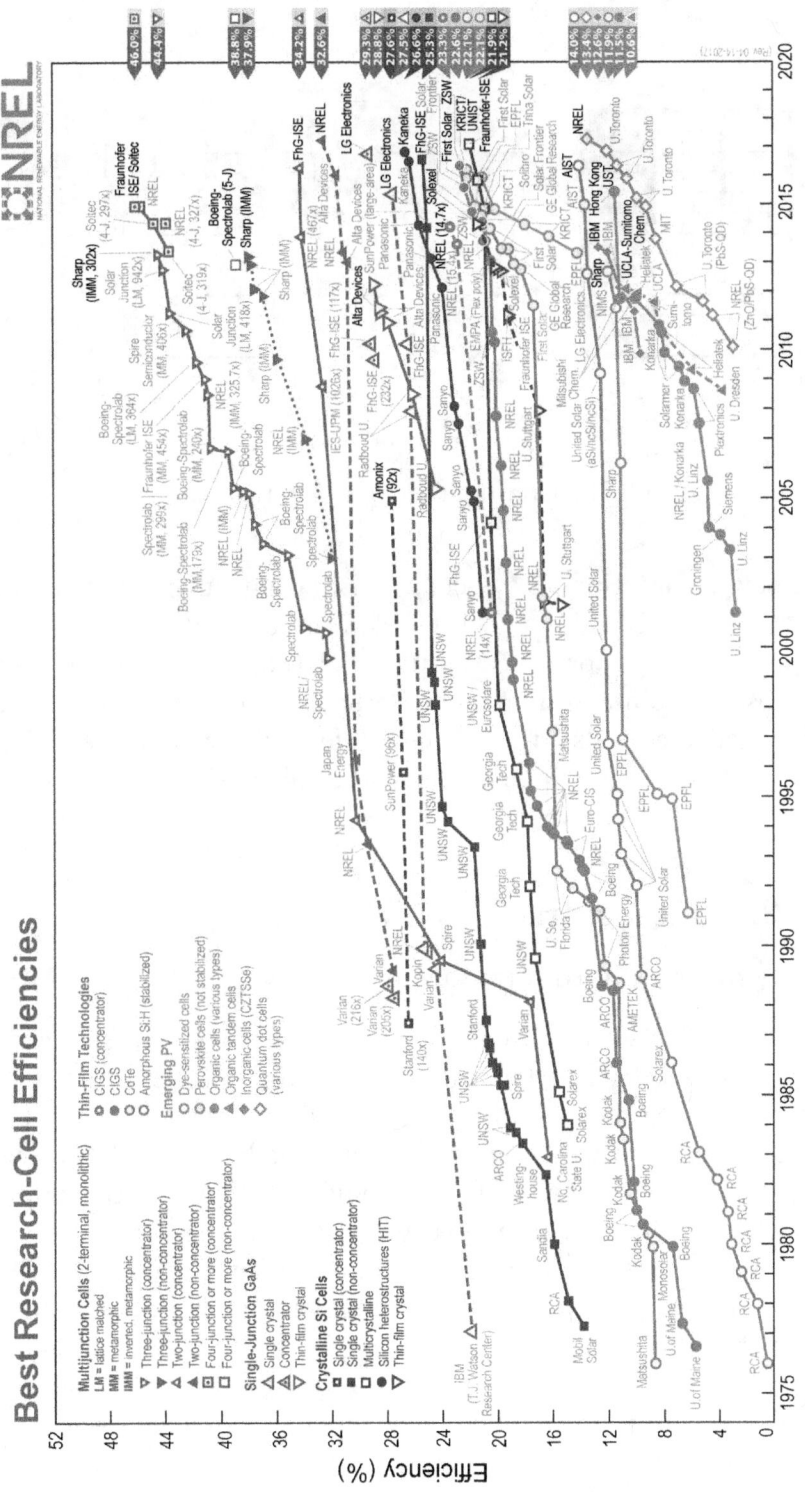

Figure 6.15: Best efficiency of various types of solar cells. This plot is courtesy of the National Renewable Energy Laboratory, Golden, CO [77].

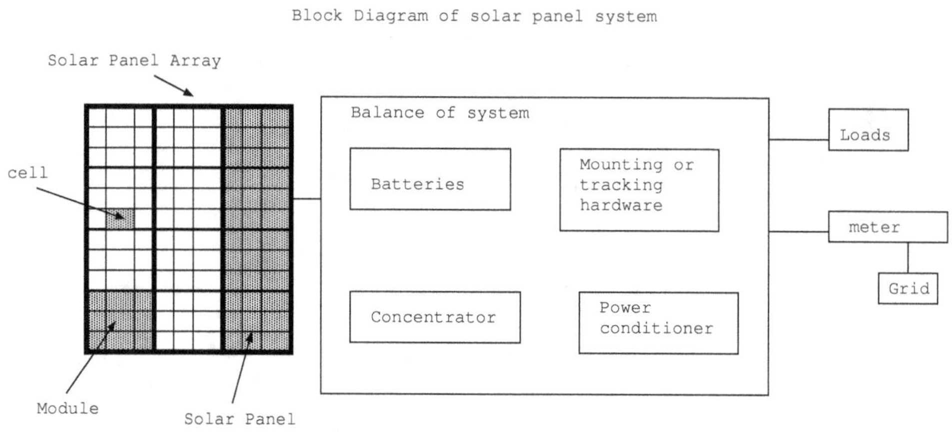

Figure 6.16: Components of a solar panel system.

The power conditioning system consists of an inverter which converts DC electricity to AC and, for grid tied systems, a system to match the phase of the produced AC power to the phase of the grid. Power conditioning systems also contain a system to limit the current or voltage to maximize the power delivered. Also, they include safeguards such as fuses to prevent injury or damage to equipment. The typical lifetime for the electronics may be around 10-15 years [73].

6.7 Photodetectors

6.7.1 Types of Photodetectors

Photodetectors are sensors used to convert light, at optical or other nearby frequencies, to electricity. One way to classify photodetectors is by their type of active material, which may be a solid or a gas. The first type of detectors are *semiconductor photodetectors* made from solid semiconductor pn junctions. The choice of semiconductor influences the wavelengths of light which can be absorbed because only photons with energy greater than or equal to the energy gap of the semiconductor can be absorbed. For example, silicon has an energy gap of 1.11 eV, so it is able to absorb the photons in both the visible range $1.9 \text{ eV} < E < 3.1 \text{ eV}$ as well as photons in the near infrared range $1.1 \text{ eV} < E < 1.9 \text{ eV}$. In some semiconductor photodetectors, a thin intrinsic (undoped) layer is added between the p-type material and the n-type material at the junction. In these *semiconductor p-i-n junction photodetectors*, the added layer widens the depletion layer. It also decreases the internal capacitance of the junction thereby increasing the detector response time [10, p. 660]. The second type of detectors are made from gas filled vacuum tubes, and these detectors are called *phototubes* [10, p. 646]. A voltage is placed across electrodes in the tubes. When light shines on the phototube, energy from a photon of light can rip off an electron from a gas atom. The electron and ion flow towards the electrodes, thereby producing electricity. The most common type of phototube is the *photomultiplier tube*. This device has multiple electrodes, and when an electron hits one of these electrodes, additional electrons are emitted. These electrons can hit additional electrodes to produce even more electrons. Because each incoming photon produces a cascade of electrons, photomultiplier tubes have high internal amplification.

Another way to classify photodetectors depends on whether incoming photons have enough energy to rip off electrons or just excite them. The first type of detectors are called *photoelectric detectors*, and they operate based on a process called *photoelectric emission* [10, p. 645] [27, p. 171]. In these detectors, incoming light has energy greater than or equal to the energy from the valence band to the ground level at the top of an energy level diagram. These detectors convert light to electricity because incoming photons of light rip electrons off their atoms, and the flow of the resulting electrons is a current. The second type of detectors are called *photoconductive detectors* or sometimes *photovoltaic detectors*, and they operate based on a process called *photoconductivity* [10, p. 647]. In these detectors, incoming light has energy equal to the difference between the valence and conduction bands, not enough to rip off electrons. These detectors convert light to electricity because incoming photons excite electrons, and the conductivity of the detector is higher when light shines on it. Solid semiconductor photodetectors can operate based on either photoelectric emission or photoconductivity, but most operate based on photoconductivity. Phototubes typically operate based on photoelectric emission.

Some photodetectors have a single element while others are made from an array of elements. A digital camera may contain millions of individual photodetectors. These elements are integrated with a *charge-coupled device* (CCD), which is circuitry to sequentially transfer the electrical output of each photodetector of the array [9, p. 359]. The CCD was invented in 1969 by Willard S. Boyle and George E. Smith. For this invention, they shared the 2009 Physics Nobel Prize with Charles K. Kao, who was awarded the prize for his work on optical fibers [80].

Eyes in animals are photodetectors. The retina of the human eye is an array composed of around 120 million rod cells and 6 to 7 million cone cells [81]. These cells convert light to electrical impulses which are sent to the brain.

6.7.2 Measures of Photodetectors

The frequency response is one of the most important measures of a photodetector. Often it is plotted versus wavelength or photon energy instead of frequency. A photodetector is only sensitive within a particular wavelength range, and the frequency response is often not flat.

As with all types of sensors, signal to noise ratio is another important measure. While photodetectors have many sources of noise, one major source is thermal noise due to the random motion of charges as they flow through a solid [9, p. 220]. To mitigate thermal noise in photodetectors used to detect very weak signals, the detectors are cooled with thermoelectric devices or using liquid nitrogen. A measure related to signal to noise ratio is the noise equivalent power. It is defined as the optical power in watts that produces a signal to noise ratio of one [82].

Another measure of a photodetector is the *detectivity*, denoted D*, in units $\frac{cm \cdot (Hz^{1/2})}{W}$. It is a measure of the strength of the output assuming a one watt optical input. By definition, it is equal to the square root of the area of the sensor times the bandwidth under consideration divided by the noise equivalent power [82] [83, p. 654].

$$D* = \frac{\sqrt{\text{Area} \cdot \text{Bandwidth}}}{\text{Noise Equivalent Power}}$$

Figure 6.17 shows detectivity versus wavelength for optical detectors made of various semiconductors.

Photodetectors are also characterized by their response times. Response time is defined as the time needed for a photodetector to respond to a step-like optical input [82]. Typical response times can range from picoseconds to milliseconds [83, p. 656]. There may be a tradeoff between response time and sensitivity, so some detectors are designed for fast operation while others are design for higher sensitivity [9, p. 220].

6 PHOTOVOLTAICS

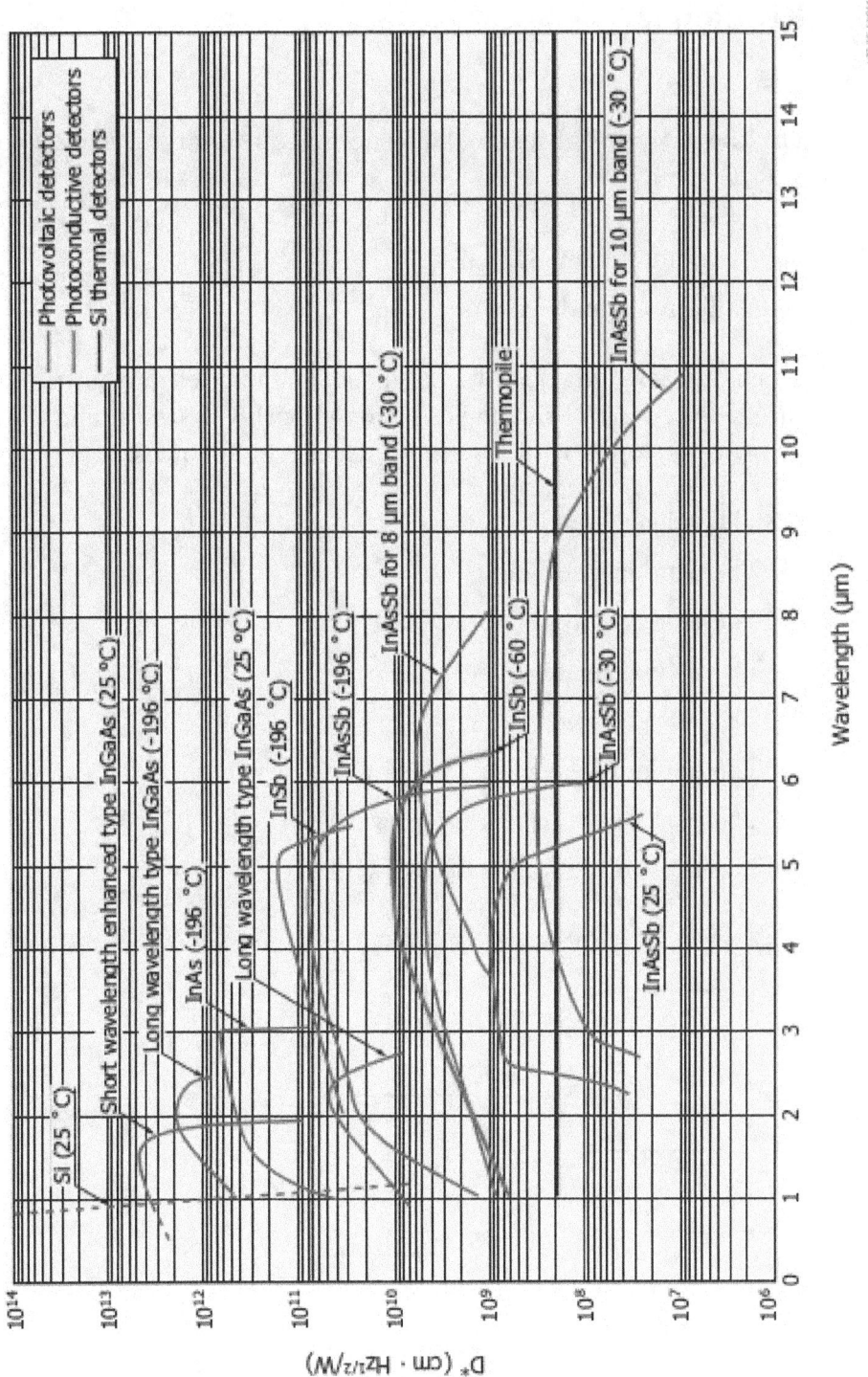

Figure 6.17: Spectral response of a variety of photodetectors. This figure is used with permission from Hamamatsu [82].

6.8 Problems

6.1. Rank the materials from smallest energy gap to largest energy gap:

- Indium arsenide, InAs
- Aluminum arsenide, AlAs
- Gallium arsenide, GaAs

6.2. The energy level diagram for a silicon pn junction is shown in the figure below. Part of the device is doped with Ga atoms, and part of the device is doped with As atoms. Label the following:

- The valence band
- The conduction band
- The energy gap
- The n-type region
- The p-type region
- The depletion layer
- The part of the device doped with Ga
- The part of the device doped with As

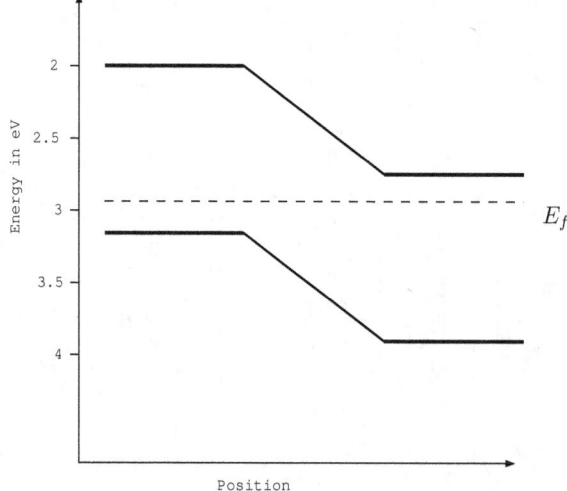

6.3. The figure in the previous problem shows the energy level diagram for a semiconductor pn junction.

   (a) If this pn junction is used in an LED, what will be the wavelength in nm of the light emitted by the LED?

   (b) If this pn junction is used as a solar cell, what range of wavelengths of light will be absorbed by the solar cell?

6.4. A semiconductor is used to make an LED that emits red light at $\lambda = 630$ nm.

   (a) Find the energy gap in eV of the semiconductor.

   (b) Find the energy in joules of a photon emitted.

   (c) Find the energy in joules for Avogadro constant number of these photons.

6.5. The figure below shows the energy level diagram for a gallium arsenide LED.

   (a) Find the energy gap.

   (b) Find the energy of a photon emitted by the LED.

   (c) Find the frequency in Hz of a photon emitted by the LED.

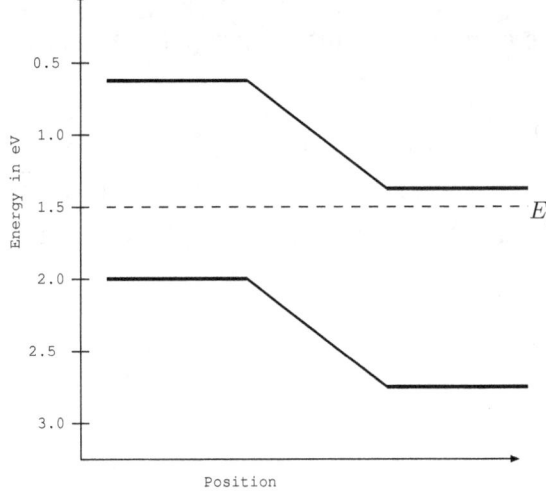

6.6. Use Fig. 6.6 to answer this question.

   (a) Suppose you would like to make an LED that emits red light with a wavelength of 650 nm. Suggest three possible semiconductor materials that could be used.

   (b) Suppose you would like to make a layered solar cell using layers of the following materials: InP, $In_{0.5}Ga_{0.5}As$, and $AlAs_{0.5}Sb_{0.5}$. Which layer would be on top, in the middle, and on the bottom of the device, and why?

6.7. Use Fig. 6.6 to answer this question.

   (a) Find the energy gap of $InP_{0.1}As_{0.9}$ in the units of joules.

   (b) If $InP_{0.1}As_{0.9}$ is used to make an LED, find the expected frequency, in Hz, of the photons emitted.

   (c) Would it be better to make a solar cell out of gallium phosphide or indium phosphide? Why?

6.8. A solar panel produces an average power of 800 W. The panel is in a location which receives an average of 0.07 $\frac{W}{cm^2}$ of optical energy from the sun. Assume the panel has an efficiency of 9%.

   (a) Calculate the surface area of the solar panel in units m^2.

   (b) Calculate the average amount of energy (in eV) produced in one week.

6.9. A solar panel has an area of 50 m^2, and it produces an average of 4 kW of power. The panel is in a location which receives an average of 0.085 $\frac{W}{cm^2}$ of optical energy from the sun. Calculate the efficiency of the panel.

7 Lamps, LEDs, and Lasers

7.1 Introduction

Chapter 6 discussed devices that convert light to electricity. In this chapter, we discuss devices that convert electricity to light. These devices vary widely in size and shape from tiny Light Emitting Diodes (LEDs) and semiconductor lasers to large high power gas lasers. In addition to LEDs and lasers, lamps and optical amplifiers are also discussed.

We take lamps for granted now because they are present in practically all buildings. However, their invention dramatically improved human productivity because lamps allowed people to constructively use indoor spaces at night. Similarly, lasers have improved productivity in many activities. We encounter them almost daily in our use of communications networks, DVD players, medical devices, and in other applications.

7.2 Absorption, Spontaneous Emission, Stimulated Emission

Absorption, spontaneous emission, and stimulated emission are three related energy conversion processes. Chapter 6 discussed devices based on absorption including solar cells and photodetectors. Devices which operate based on spontaneous emission include LEDs and lamps. Optical amplifiers and lasers operate based on stimulated emission.

7.2.1 Absorption

Absorption is the process in which optical energy is converted to internal energy of electrons, atoms, or molecules. When a photon is absorbed, the energy may cause an electron in an atom to go from a lower to a higher energy level, thereby changing the internal momentum of the electron and the electron's internal quantum numbers. This process was illustrated in Chapter 6 by energy level diagrams. Energy in a solar cell or photodetector is then converted to electricity because the excited charge carriers can travel more freely through the material. The electrons absorbing the energy may be part of atoms which make up solids, liquids, gases, or plasmas. They may be around isolated neutral atoms, ionic compounds, or complicated organic molecules. Furthermore the electrons absorbing the energy may be part of conductive, insulating, or semiconducting materials. The photons absorbed may be optical photons, with individual energies in the range 1.9 to 3.1 eV that can be detected by human eyes. Alternatively, they may have energies that are multiple orders of magnitude larger or smaller than the energy of a visible photon. For example, in isolated neutral neon atoms in the ground state, electrons occupy the 2p energy level but not the 3s energy level. These energy levels are separated by an energy gap of $E_g = 1.96$ eV which corresponds with energy of red photons of wavelength 632.8 nm [31]. If a photon of this energy impinges upon neon gas, the photon may be absorbed, and an electron of a neon atom would be excited to the higher energy level. Photons of smaller energy would not be absorbed. Photons of larger energy may be absorbed depending on allowed energy levels. As another example, the energy gap of the semiconductor gallium phosphide, GaP, is 2.2 eV which corresponds with the energy of a green photon of wavelength 549 nm. If a photon of this energy impinges on a piece of gallium phosphide, it may be absorbed.

7.2.2 Spontaneous Emission

Spontaneous emission is an energy conversion process in which an excited electron or molecule decays to an available lower energy level and in the process gives off a photon. This process occurs naturally and does not involve interaction of other photons. The average time for decay by spontaneous emission is called the *spontaneous emission lifetime*. For some excited energy levels this spontaneous decay occurs on average within nanoseconds while in other materials it occurs within a few seconds [10, p. 480]. As with absorption, this process can occur in isolated atoms, ionic compounds, molecules, and other types of materials, and it can occur in solids, liquids, and gases. Energy is conserved when the electron decays to the lower level, and that energy must go somewhere. The energy may be converted to heat, mechanical vibrations, or electromagnetic photons. If it is converted to photons, the process is called spontaneous emission, and the energy of the photon produced is equal to the energy difference between the electron energy levels involved. The emitted photon may have any direction, phase, and electromagnetic polarization.

There are many ways in which an electron can be excited to a higher energy level [10, p. 455]. Spontaneous emission processes may be classified based on the source of energy which excites the electrons, and these classes are listed in Table 7.1. If the initial source of energy for spontaneous emission is supplied optically, the process is called *photoluminescence*. Glow in the dark materials emit light by this process. If the initial form of energy is supplied by a chemical reaction, the process is called *chemiluminescence*. Glow sticks produce spontaneous emission by chemiluminescence. If the initial form of energy is supplied by a voltage, the process is called *electro-*

	Spontaneous emission energy source
Photoluminescence	Optical electromagnetic waves
Chemiluminescence	Chemical reactions
Electroluminescence	Applied voltages
Sonoluminescence	Sound waves
Bioluminescence	Biological processes

Table 7.1: Spontaneous emission is classified based on the source of energy [10, p. 455].

luminescence. LEDs emit light by electroluminescence. If the initial form of energy is caused by sound waves, the process is called *sonoluminescence*. If the initial form of energy is due to accelerated electrons hitting a target, this process is called *cathodoluminescence*. If spontaneous emission occurs in a living organism, such a firefly, the process is called *bioluminescence*.

At temperatures above absolute zero, some electrons in atoms are thermally excited to energy levels above the ground state. These electrons decay and emit a photon by spontaneous emission. Any object at a temperature above absolute zero naturally emits photons by spontaneous emission, and this process is called *blackbody radiation*. In 1900, Max Planck derived a formula for the energy density per unit bandwidth of a blackbody radiator by making the assumption that only discrete energies are allowed [10, p. 453]. His work agreed with known experimental data, and it is one of the fundamental ideas of quantum mechanics. More specifically, the spectral energy density per unit bandwidth, u in units $\frac{\text{J} \cdot \text{s}}{\text{m}^3}$, is given by

$$u = \frac{8\pi f^2}{c^3} \cdot \frac{hf}{e^{(hf/k_B T)} - 1}. \tag{7.1}$$

Equation 7.1 includes a number of constants including c the speed of light in free space, h the Planck constant, and k_B the Boltzmann constant. Additionally, f is frequency in Hz, and T is temperature in kelvins. For a nice derivation, see [84, p. 186]. The first term represents the number of modes per unit frequency per unit volume while the second term represents the average energy per mode. The expression can be written as a function of wavelength instead of frequency with the substitution $f = \frac{c}{\lambda}$.

Photons emitted by a blackbody radiator have a relatively wide range of wavelengths, and this bandwidth depends on temperature. Figure 7.1 plots the energy density per unit bandwidth for blackbody radiators as a function of wavelength at temperatures 3000, 4000, and 5000 K. Room temperature corresponds to around 300 K. Visible photons have wavelengths between

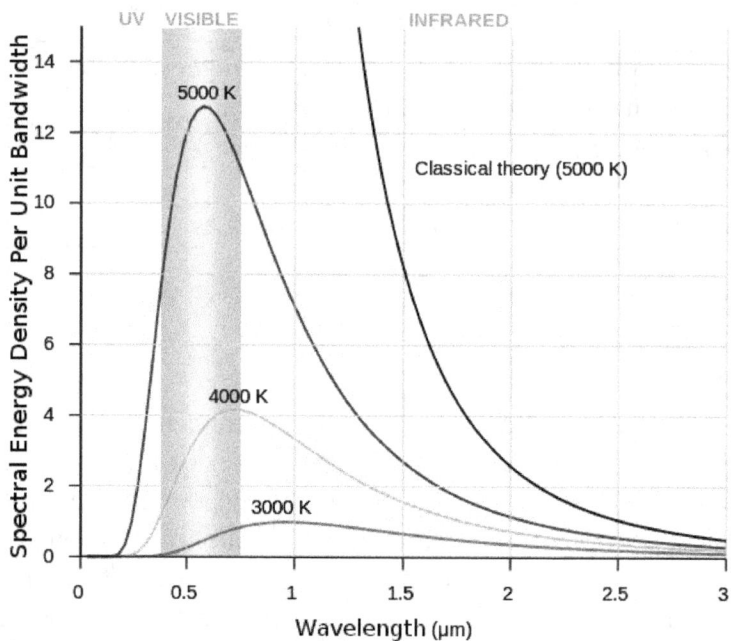

Figure 7.1: Spectral energy density of a blackbody radiator. This figure is in the public domain [85].

400 nm $< \lambda <$ 650 nm. From the figure, we can see that black body radiators at higher temperatures emit both more photons and have a larger fraction of photons emitted fall in the visible range.

7.2.3 Stimulated Emission

Stimulated emission is the process in which an excited electron or molecule interacts with a photon, decays to an available lower energy level, and in the process gives off a photon. As with the other processes, this process can occur in isolated atoms, ionic compounds, organic molecules, and other types of materials, and it can occur in solids, liquids, and gases. If an incoming photon, with energy equal to the difference between allowed energy levels, interacts with an electron in an excited state, stimulated emission can occur. The energy of the excited electron will be converted to the energy of a photon. The stimulated photon will have the same frequency, direction, phase, and electromagnetic polarization as the incoming photon which initiated the process [10, p. 436].

7 LAMPS, LEDS, AND LASERS

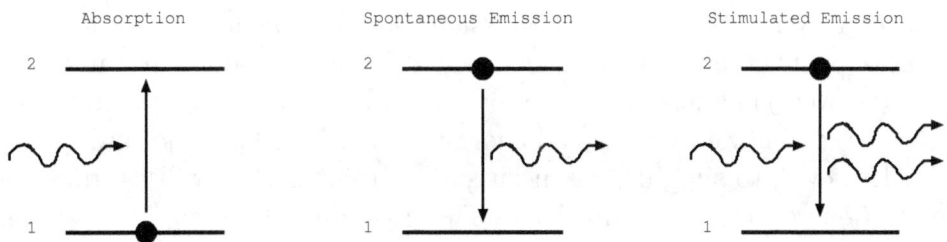

Figure 7.2: Energy level diagrams illustrating absorption, spontaneous emission, and stimulated emission.

7.2.4 Rate Equations and Einstein Coefficients

The processes of absorption, spontaneous emission, and stimulated emission are illustrated by energy level diagrams in Fig. 7.2. Energy is on the vertical axis, and nothing is plotted on the horizontal axis. Only two energy levels are shown, so this diagram illustrates only a small fraction of possible energy levels of a material. The lower energy level is labeled 1. It may represent, for example, the highest occupied energy level of an electron in an isolated atom, or it may represent the valence band of a semiconductor. The higher energy level is labeled 2, and it may represent the lowest unoccupied energy level of an electron in an isolated atom or the conduction band of a semiconductor. The dot represents an electron occupying the energy level at the start of the process. The squiggly arrows represent a photon absorbed or emitted by the process. The vertical arrow shows how the internal energy of the electron changes in the process. During absorption, an electron takes energy from an incoming photon, and the internal energy of the electron increases. During spontaneous emission, the internal energy of an electron decreases, and a photon is emitted. Stimulated emission occurs when a photon, with energy equal to the energy gap of the levels, interacts with the electron. In the process, the electron decays to the lower energy level, and a photon is produced with the same frequency, direction, phase, and electromagnetic polarization as the original photon. The figures do not illustrate a change in position of the electrons. Instead, they illustrate a change in energy and internal momentum.

The descriptions of the processes above involve changes in energy levels of an electron. However, absorption, spontaneous emission, and stimulated emission can instead involve vibrational energy states of molecules. For example, a photon may be absorbed by a molecule, and the energy may cause the molecule to go from one allowed vibrational state to another with higher internal energy. Similarly, this molecule may spontaneously decay from the higher energy state to a lower energy state emitting a photon by spontaneous emission or by stimulated emission. An example involving molecular vibration states is a carbon dioxide laser. This laser produces infrared light by stimulated emission at $\lambda = 10.6$ μm, and the stimulated emission occurs between allowed vibrational energy levels of the CO_2 molecule [31, p. 217]. However, to simplify the discussion in this text, we will assume that electron energy levels are involved. This assumption is true in most, but not all, energy conversion devices.

What factors determine the rate of these processes? Assume only two energy levels are involved. The number of electrons per unit volume in the lower state will be denoted n_1, and the number of electrons per unit volume in the upper state will be denoted n_2. The rate of absorption will be denoted $\frac{dn_2}{dt}\big|_{abs}$, the rate of spontaneous emission will be denoted $\frac{dn_2}{dt}\big|_{spont}$, and the rate of stimulated emission will be denoted $\frac{dn_2}{dt}\big|_{stim}$. Since only two energy levels are involved in this system, we can describe the rates of the processes either in terms of the upper or lower energy levels. For example, we can write the rate of absorption either as the change in population density with respect to time of the upper state or the change in population density with respect to time of the lower state.

$$\frac{dn_2}{dt}\bigg|_{abs} = -\frac{dn_1}{dt}\bigg|_{abs} \qquad (7.2)$$

Absorption can only occur if there is an electron present in the lower energy level. Furthermore, the rate of absorption is proportional to the number of electrons in the lower state. Additionally, the rate of absorption depends on the number of incoming photons. As in Eq. 7.1, u represent the spectral energy density per unit bandwidth in units $\frac{J \cdot s}{m^3}$. We can model the rate of absorption in terms of these factors [84, ch. 6] [86, ch. 7].

$$\frac{dn_2}{dt}\bigg|_{abs} = -\frac{dn_1}{dt}\bigg|_{abs} = B_{12} n_1 u \qquad (7.3)$$

The constant of proportionality B_{12} is called an *Einstein B coefficient*, and it has units $\frac{m^3}{J \cdot s^2}$.

Spontaneous emission depends on the number of electrons in the upper energy level. We can model the rate of spontaneous emission as

$$\frac{dn_2}{dt}\bigg|_{spont} = -\frac{dn_1}{dt}\bigg|_{spont} = -A_{21} n_2 \qquad (7.4)$$

The constant of proportionality A_{21} is called the *Einstein A coefficient*, and it has units $\frac{1}{s}$ [84, ch. 6] [86, ch. 7]. No photons are needed to initiate spontaneous emission.

7 LAMPS, LEDS, AND LASERS

We can model the rate of stimulated emission as

$$\left.\frac{dn_2}{dt}\right|_{stim} = -\left.\frac{dn_1}{dt}\right|_{stim} = -B_{21}n_2 u. \tag{7.5}$$

The constant of proportionality B_{21} is known as another *Einstein B coefficient*, and it also has units $\frac{m^3}{J \cdot s^2}$ [84, ch. 6] [86, ch. 7]. The rate of stimulated emission is dependent on the number of electrons in the upper energy level. Stimulated emission requires an incoming photon, so the rate also depends on the spectral energy density per unit bandwidth u.

By considering the factors that affect the rate of absorption, spontaneous emission, and stimulated emission, we can see some similarities and differences in the processes. As absorption occurs, the population of electrons in the upper energy level increases, and the population of the lower energy level decreases. As both spontaneous and stimulated emission occurs, the population of the upper energy level decreases, and the population of the lower energy level increases. Both the rate of absorption and the rate of stimulated emission depend on both the population of electrons in an energy level and the energy of incoming photons while the rate of spontaneous emission does not depend on the energy of incoming photons. This similarity between absorption and stimulated emission is reflected in the rate equations, Eqs. 7.3 and 7.5.

Einstein showed that if one of the coefficients describing the absorption, spontaneous emission, or stimulated emission is known, the other coefficients can be calculated from it. We can combine the terms above to find the overall upper state population rate.

$$\frac{dn_2}{dt} = -A_{21}n_2 + B_{12}n_1 u - B_{21}n_2 u \tag{7.6}$$

At equilibrium, where photons are absorbed and emitted at the same rate, this population rate is zero.

$$\left.\frac{dn_2}{dt}\right|_{equilibrium} = 0 = -A_{21}n_2 + B_{12}n_1 u - B_{21}n_2 u \tag{7.7}$$

We can solve for the energy density per unit bandwidth, u.

$$B_{12}n_1 u - B_{21}n_2 u = A_{21}n_2 \tag{7.8}$$

$$u = \frac{A_{21}}{\frac{n_1}{n_2}B_{12} - B_{21}} \tag{7.9}$$

In the expression above, $\frac{n_1}{n_2}$ represents the electron density in the lower energy state divided by the electron density in the upper state in equilibrium. This quantity is a function of temperature. Assuming many allowed energy states, the number of occupied states decreases exponentially with temperature, an idea known as Boltzmann statistics.

$$\frac{n_2}{n_1} = \frac{g_2}{g_1} e^{\frac{-hf}{k_B T}} \qquad (7.10)$$

The quantity $\frac{g_2}{g_1}$ represents the *degeneracy level* which is the number of allowed electrons in the upper state over the number of allowed electrons in the lower state [84, p. 186]. In this expression, g_1 and g_2 are unitless measures of the number of ways electrons can occupy an energy states. Equations 7.10 and 7.11 can be combined.

$$u = \frac{A_{21}}{\left(\frac{g_1}{g_2} e^{\frac{hf}{k_B T}}\right) B_{12} - B_{21}} \qquad (7.11)$$

$$u = \frac{\frac{A_{21}}{B_{21}}}{\frac{g_1 B_{12}}{g_2 B_{21}} e^{\frac{hf}{k_B T}} - 1} \qquad (7.12)$$

Consider a blackbody radiator, a conducting wire which is continually supplied with heat so that it remains at temperature T in equilibrium.

$$\left.\frac{dn_2}{dt}\right|_{equilibrium} = 0 \qquad (7.13)$$

One expression for the energy density per unit bandwidth of this system is given by Eq. 7.1. Equation 7.12 gives a second expression for the energy density per unit bandwidth, and it was found by considering the relative rates of absorption, spontaneous emission, and stimulated emission. These equations can be combined to relate the rates of the different processes.

$$\frac{8\pi h f^3}{c^3} \cdot \frac{1}{e^{(hf/k_B T)} - 1} = \frac{\frac{A_{21}}{B_{21}}}{\frac{g_1 B_{12}}{g_2 B_{21}} e^{\frac{hf}{k_B T}} - 1} \qquad (7.14)$$

The above equation is true for the conditions

$$\frac{A_{21}}{B_{21}} = \frac{8\pi h f^3}{c^3} \qquad (7.15)$$

and

$$\frac{g_1 B_{12}}{g_2 B_{21}} = 1. \qquad (7.16)$$

If we know one of the Einstein coefficients, we can quickly calculate the other two Einstein coefficients from Eqs. 7.15 and 7.16.

These equations provide further insight into the operation of lasers and other devices based on stimulated emission. The overall nonequilibrium upper state population rate is given by

$$\frac{dn_2}{dt} = -A_{21}n_2 + B_{21}\frac{g_2}{g_1}n_1 u - B_{21}n_2 u \qquad (7.17)$$

which can be simplified with some algebra.

$$\frac{dn_2}{dt} = -A_{21}n_2 - uB_{21}\left(n_2 - \frac{g_2}{g_1}n_1\right) \qquad (7.18)$$

The term in parenthesis is the net upper state population. Optical amplification and lasing can only occur when the term in parenthesis is positive. The condition

$$n_2 - \frac{g_2}{g_1}n_1 > 0 \qquad (7.19)$$

is called a *population inversion* [86, p. 189]. It only occurs when enough energy is being supplied to the system, by optical, electrical, or thermal means, so that there are more electrons in the upper energy level than the lower energy level. Population inversion has nothing to do with inversion symmetry discussed in Sec. 2.3.2. See Appendix C for a discussion of inversion and other overloaded terms.

7.3 Devices Involving Spontaneous Emission

Spontaneous emission occurs in many commercially available consumer products. This section discusses three categories of devices that convert electricity to light by spontaneous emission: incandescent lamps, gas discharge lamps, and LEDs.

7.3.1 Incandescent Lamps

An *incandescent lamp* is a device that converts electricity to light by blackbody radiation. These devices are typically constructed from a solid metal filament inside a glass walled vacuum tube. A current passes through the filament which heats it to a temperature of thousands of degrees. High temperatures are used because the visible spectral response of daylight is close to the visible spectral response of a blackbody radiator at a temperature of 6500 K [87]. The main limitation of incandescent lamps is their efficiency. Much of the electromagnetic radiation emitted by a blackbody radiator falls outside the visible range.

The main advantage of incandescent lamps over other technologies is their simplicity. For this reason, incandescent lamps were some of the earliest lamps developed. Humphry Davy demonstrated that blackbody radiation could be used to produce visible light in 1802, and practical incandescent lamps date to the 1850s [88]. In order to develop these practical incandescent lamps, vacuum pumping technology had to be developed, and technology to purify the metal used to make lamp filaments was required [88].

In some ways, an incandescent lamp is similar to an antenna. In both cases, the input takes the form of electricity, and this electrical energy is converted to electromagnetic energy by passing through a conducting wire. In an antenna, the input is time varying to encode information, and the output is at radio or microwave frequencies. However, in an incandescent lamp, the input is typically AC and does not contain information. The desired output of an incandescent lamp is visible light, but it also produces heat and electromagnetic radiation at infrared frequencies and at other non-visible frequencies. Additionally, antennas are typically designed to operate at a wavelength close to the length of the antenna, and such antennas can produce waves with specific electromagnetic polarization and radiation patterns. Spontaneous emission in incandescent lamps, however, is necessarily unpolarized and incoherent.

7.3.2 Gas Discharge Lamps

A *gas discharge* occurs when a conducting path forms through a plasma, an ionized gas [89]. Gas discharge devices convert electricity to light by spontaneous emission when this type of conducting path forms. In 1802 in addition to demonstrating blackbody radiation and proposing the idea of a fuel cell, Humphry Davy demonstrated a gas discharge device [3, p. 222] [88]. W. Petrov demonstrated a gas discharge around the same time [88]. One of the first practical gas discharge lamps, a carbon arc lamp, was built by Leon Foucoult in 1850, and it was used for theater lighting [88]. Development of gas discharge lamps required the ability to purify gases in addition to the development of vacuum pumping technology [88]. Examples of gas discharge devices in use today include include sodium vapor lamps, mercury arc lamps, fluorescent lamps, and neon advertising signs [89].

A gas discharge lamp is made from a sealed tube containing two electrodes and filled by a gas. The glass tube contains the gas, maintains the gas pressure, and keeps away impurities. The pressure of the gas inside the tube can range from 10^{-4} Pa to 10^5 Pa for different lamps [87, p. 206]. Typical electrode spacing is on the order of centimeters [87]. Some neon bulbs have an electrode spacing of 1 mm while many fluorescent tubes have an electrode spacing over 1 m. Hundreds to millions of volts are applied across the electrodes [89]. Transformers are used to achieve these high voltage levels. The voltage between the electrodes ionizes the gas inside the tube and provides a supply of free electrons which travel along the conducting path between the electrodes [89]. The gas may be ionized, and electrons supplied, by other methods such as chemical reactions, a static electric field, or an optical field instead [87, Ch. 5]. Electrons may also be supplied to the gas by *thermionic emission*, boiling electrons off the cathode.

The optical properties of the lamp are determined by the gas inside the tube. Energy supplied by the electric field across the electrodes, or other means, excites electrons of the gas atoms to higher energy levels. Spontaneous emission occurs between distinct allowed energy levels only, so the emission occurs over relatively narrow wavelength ranges. Gases are chosen to have allowed energy level transitions in the desired wavelength range. Typical gases used include helium, neon, sodium, and mercury [87, p. 514].

Gas discharge lamps are classified as either glow discharge devices or arc discharge devices. Figure 7.3 shows an example plot of the current between electrodes as a function of voltage. As shown in the figure, the current-voltage characteristic of a gas discharge tube is quite nonlinear. However, it can be broken up into three general regions, denoted the dark region, the glow region, and the arc region. The regions are distinguished by a change in slope of the current-voltage plot. This figure is used with permission from [89] which provides more details on the physics of gas discharges.

The *dark region* of operation corresponds to low currents and voltages, and devices operating in this region are said to have a *dark* or *Townsend discharge*. Optical emission from devices operating in this region are not self sustaining. While atoms of the gas may ionize and collide with other atoms, no chain reaction of ionization occurs. The transition between the dark and glow discharges is called the *spark* [87, p. 160]. In Fig. 7.3, V_S is the sparking voltage. The second region, corresponding to higher currents, is called the *glow region*, and this region is called self sustaining because ions collide and ionize additional gas atoms producing more free electrons in an avalanche process. Significant spontaneous emission occurs in the glow discharge region [87] [89]. The third region, corresponding to even higher current, is called the *arc region*. Arc discharges are also self sustaining [87, p. 290], and spontaneous emission is produced. Once the

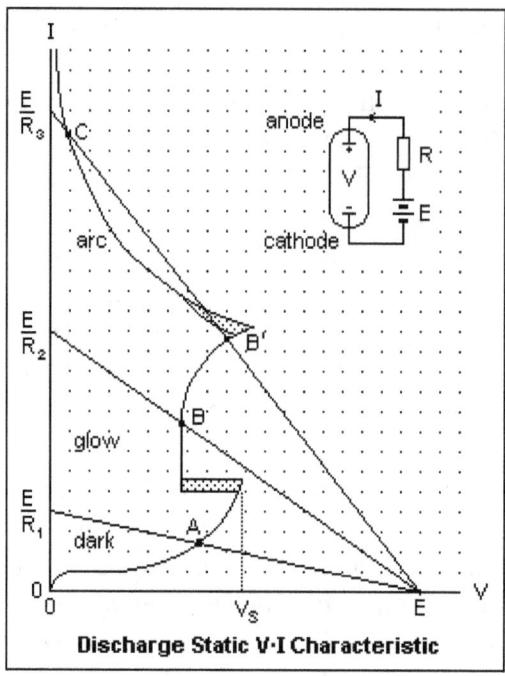

Figure 7.3: Example current-voltage characteristics of a gas discharge lamp. Figure used with permission from [89].

arc discharge is established, relatively low voltages are required to maintain it compared to the voltages needed to maintain the glow discharge.

Fluorescent lamps are a type of gas discharge device that involves the use of chemicals with desired optical properties, called *phosphors* [87, p. 542]. The gas and electrode voltage used in fluorescent lamps is chosen to so that the spontaneous emission produced is at ultraviolet frequencies. These UV photons may be produced by either an arc or glow discharge. The UV photons produced are absorbed by the phosphor molecules, and the phosphor molecules emit light at lower frequencies. Examples of phosphors used include zinc silicate, calcium tungstate, and zinc sulfide [87, p. 542].

7.3.3 LEDs

LEDs are devices that convert electricity to light by spontaneous emission. They are made from pn junctions in semiconductors. Pn junctions were discussed in Section 6.5. When a forward bias is applied across a pn junction, electrons and holes are injected into the junction. The energy from the power supply excites electrons from the valence to the conduction bands. These excited electrons can flow through the material much more easily than unexcited electrons. Some of the electrons and holes near the junction combine and spontaneously emit photons in the process. Some LEDs have a thin intrinsic, undoped, layer between the p-type and n-type layers at the junction to improve efficiency.

LEDs emit light over a relatively narrow range of frequencies. The frequency of light emitted is determined by the energy gap of the semiconductor. Semiconductors are used because the energy gap of semiconductors corresponds to the energy of near ultraviolet, visible, or infrared photons. While light emitted by an LED has a narrow range of frequencies, lasers emit light with a much narrower range of frequencies. LEDs emit light within a narrow frequency range, but applications, such as residential lighting, require white light with a broader bandwidth. One strategy used to produce white light from an LED is to use phosphors. In such a device, an LED converts electricity to near UV or blue light. The phosphors absorb the blue light and emit light at lower energies, at wavelengths in the visible range. For this reason, blue LEDs were particularly important for generating white light. It took decades from the invention of red LEDs in the 1960s until reliable blue LEDs were developed in the 1980s and 1990s. In 2014, Isamu Akasaki, Hiroshi Amano and Shuji Nakamura were awarded the Nobel Prize in physics for their work developing blue LEDs. This effort required the development of deposition technology for new materials like gallium nitride, and it required being able to deposit these materials in very pure layers without mechanical strain tearing the materials apart [90].

A related device which emits light by spontaneous emission is an organic light emitting diode, OLED. In an OLED, a voltage excites electrons in a thin layer, 100-200 nm, of an organic material, and the type of organic material used determines the wavelength of light emitted [91]. Some flat panel displays are made from arrays of OLEDs. White light in these displays is achieved from a combination of red, green, and blue OLEDs near to each other [91].

LEDs are small devices that can often fit into a cubic millimeter. For this reason, they can be integrated into electronics more easily than devices like incandescent lamps and gas discharge lamps which require vacuum tubes. LEDs require low voltages electricity to operate. Since they require a small amount of input electrical power, they produce a small amount of output optical power. Incandescent lamps and gas discharge lamps have advantages in high power applications, but arrays of LEDs can also be used in these applications. Another advantage of LEDs is that they have a longer useful lifetime. In gas discharge lamps, the electrodes sputter, depositing material onto the surface of the tube, limiting the lifetime of the device.

7.4 Devices Involving Stimulated Emission

7.4.1 Introduction

Lasers are devices that produce optical energy through stimulated emission and involve optical feedback. The word *laser* is an acronym for Light Amplification by Stimulated Emission of Radiation. Lasers come in a wide range of sizes and shapes. Some lasers produce continuous output power, denoted cw for continuous wave, and other lasers operate pulsed. One advantage of pulsed operation is that the peak intensity of the light produced can be extremely high even with moderate average input power. Some lasers are designed to operate at room temperature while other lasers require external cooling.

The development of many energy conversion devices required technological breakthroughs. The development of lasers, however, was preceded by breakthroughs in understanding of energy conversion processes in atoms and molecules. The idea of amplification by stimulated emission was first developed in the mid 1950s, [31, p. 183] [83, p. 687]. A *maser*, which operated at microwave frequencies, was demonstrated only a few years later by Gordon, Zeiger, and Townes in around 1955 [83, p. 687]. In 1960, a ruby laser with visible output at $\lambda = 694$ nm was demonstrated by Maiman, [83, p. 687]. Lasing in semiconductors was predicted in 1961 [92] and demonstrated within a year in gallium arsenide [93]. The development of semiconductor lasers required both the theoretical prediction as well as development in the ability to deposit pure thin semiconductor layers. Thin crystalline layers grown on top of a substrate are called *epitaxial layers*. Early semiconductor lasers were made by growing epitaxial layers from a liquid melt, through a process called liquid phase epitaxy [94]. In subsequent years, other methods which allowed more control and precision were developed including molecular beam epitaxy [95] and metal organic chemical vapor deposition [96].

7 LAMPS, LEDS, AND LASERS

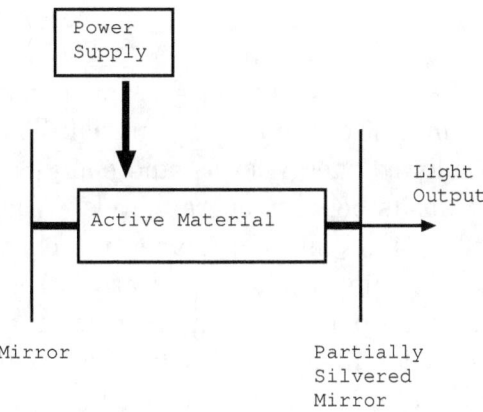

Figure 7.4: Components of a laser.

7.4.2 Laser Components

Lasers have three main components: a power supply also called a pump, an active material, and a cavity. These components are illustrated in Fig. 7.4 where mirrors form the cavity. Input energy from the power supply excites electrons or molecules in the active material. A photon interacts with the excited electrons or molecules of the active material stimulating the emission of a photon at the same frequency, phase, direction, and electromagnetic polarization. The cavity reflects the photon back to the active material so that it can stimulate another photon, and this process continues to occur as these photons stimulate additional identical photons.

Pumps

Laser power supplies are called *pumps*. Energy may be supplied to lasers in different ways. For many lasers, energy is supplied electrically. For example, the pump of a semiconductor laser is typically a battery which supplies a DC current. These lasers are energy conversion devices which convert the input electricity to light. For other lasers, energy is supplied optically, so the pump is a lamp or another laser. These lasers are energy conversion devices which convert light with large energy per photon to light with smaller energy per photon. The power supply of early ruby lasers were flashlamps [86, p. 351]. As another example, argon ion lasers are used to pump titanium doped sapphire lasers. Argon ion lasers can be tuned to emit photons with energy 2.54 eV ($\lambda = 488$ nm). These photons excite electrons in titanium doped sapphire. Titanium doped sapphire lasers are tunable solid state lasers which emit near infrared light [86, p. 392].

Active Materials

Active materials can be solids, liquids, or gases, and lasers can be classified based on the state of matter of the active material. The *active material* of a laser has multiple allowed energy levels, and energy conversion occurs as the active material transits between energy levels. When an electron transits between energy levels, its internal momentum changes, not its spatial position. Typically, the pump excites an electron from a lower to higher allowed energy level, and a photon is emitted when the electron goes from a higher to lower energy level. In some lasers such as carbon dioxide lasers, however, molecular vibration states are involved instead of electron energy states.

Optical amplification and lasing can only occur when there is a population inversion in the active material. The term *population inversion* means that more electrons are in the upper energy level than the lower energy level. The condition for a population inversion was defined by Eq. 7.19. A photon begins the process of stimulated emission, and another photon is produced in the process. Only in the case of a population inversion can the resulting photon be more likely to stimulate another photon than decay by spontaneous emission, by emitting phonons, or by other means.

In some lasers, called two level lasers, the pump excites an electron from a lower energy level to a higher energy level, and lasing occurs as the electron transits back and forth between the same two levels. In other lasers, more energy levels must be considered. Figure 7.5 illustrates possible electron transitions in two, three, and four level lasers, but other three and four level schemes are possible too. In the three level system illustrated in the figure, the pump excites electrons from level one to level three. The electrons quickly decay to level two, possibly emitting heat, and lasing occurs as electrons transit from level two to level one. In the four level scheme illustrated, the pump excites electrons from level one to four. The electrons quickly decay from level four to three, emitting heat in the process. Lasing occurs between energy levels three and two. The electrons then decay between levels two and one, again emitting heat, vibration, or some other form of energy. Some four level systems lase more easily than two level systems because a population inversion may be easier to achieve in four than two level systems. Lasing requires a population inversion, and level two may be less likely to be occupied than level one.

7 LAMPS, LEDS, AND LASERS

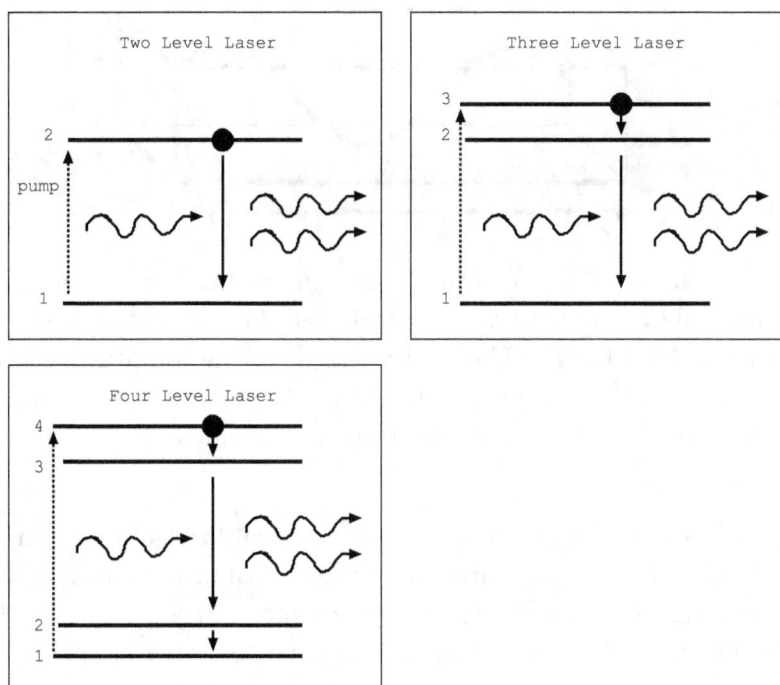

Figure 7.5: Example energy level diagram for two, three, and four level lasers.

Cavities

Laser *cavities* have two main functions. They confine photons to the active material and they act as optical filters. The simplest optical cavity is made from two mirrors as shown in Fig. 7.4. This type of cavity is called a *Fabry Perot cavity*. More complicated cavities have multiple mirrors, lenses, and other optical components to focus the desired photons within the active material and reject photons at frequencies other than the desired frequency. Semiconductor lasers do not use separate mirrors to form the cavity. In some semiconductor lasers, the edges of the semiconductors act as mirrors because the index of refraction of the semiconductor is larger than that of the surrounding air thereby reflecting a portion of the light back inside the semiconductor. The edges of these lasers are formed by cleaving along crystal planes to produce extremely flat surfaces. In other semiconductor lasers, multiple thin layers of material act as mirrors.

Even without an active material present, an optical cavity acts as an optical filter that selectively passes or rejects light of different wavelengths. To understand this idea, consider the rectangular cavity shown in Fig. 7.6. Assume that the cavity has partial mirrors on the left and right side so that

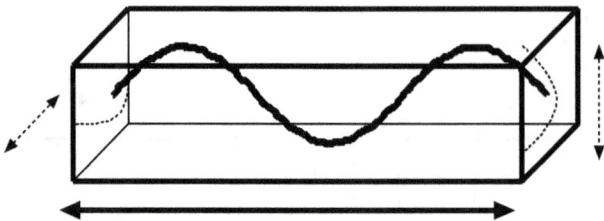

Figure 7.6: The solid arrow shows the longitudinal direction while the dotted arrows show the transverse directions. The solid sinusoid shows an allowed longitudinal mode. The cavity length in the longitudinal direction is equal to $\frac{3\lambda}{2}$. The dotted sinusoids show allowed transverse modes. The cavity lengths in the transverse directions are equal to $\frac{\lambda}{2}$.

some light can enter the cavity on the left side and some light can exit the cavity on the right. The direction along the length of the cavity, illustrated by the solid arrow, is called the *longitudinal direction*. The other two directions, illustrated by dotted arrows, are called *transverse directions*. If the longitudinal length of the cavity is exactly equal to an integer number of half wavelengths of the light, the wave will constructively interfere with itself. However, if the longitudinal length of the cavity is not equal to an integer number of half wavelengths, it will destructively interfere. The same ideas apply in the transverse directions. In the figure, the longitudinal length of the cavity is equal to three half wavelengths shown by the solid sinusoid. The transverse lengths are both equal to one half wavelength shown by the dotted sinusoids. Because of this constructive or destructive interference, cavities selectively allow certain wavelengths of light to pass through while they attenuate other wavelengths of light. In a typical laser cavity, the ratio of the longitudinal length to the transverse lengths is much larger than is shown in Fig. 7.6. Figure 7.6 illustrates a rectangular cavity while many lasers have cylindrical cavities instead. The same ideas apply, so only certain allowed longitudinal and transverse modes propagate in cylindrical cavities too [86, p. 133,145].

If there is a pump and an active material in a cavity, this filtering effect encourages lasing to occur at specific wavelengths due to the feedback the cavity provides. As discussed above, stimulated emission occurs when a photon interacts with an excited electron. The result is another photon of the same frequency, electromagnetic polarization, phase, and direction as the original photon. When the pump first turns on, electrons are excited, but no photons are present. Very soon, some photons are produced by spontaneous emission. Some of these photons stimulate the emission of additional photons. Since the cavity selectively attenuates some wavelengths but not others, photons produced by stimulated emission are more likely to occur at certain wavelengths corresponding to modes in the longitudinal direction. For these modes, the cavity length is equal to an integer multiple of half wavelengths. Due to the feedback of the laser cavity, these photons go on to stimulate additional identical photons. For this reason, the output of a laser necessarily has a very narrow wavelength range.

7.4.3 Laser Efficiency

The overall efficiency of a laser is the ratio of the output optical power over the input power. Many lasers are electrically pumped, and the overall efficiency, also known as the *wall plug efficiency*, for these lasers is the ratio of the output optical power over the input electrical power [10, p. 604].

$$\eta_{eff} = \frac{P_{optical\ out}}{P_{electrical\ in}} \quad (7.20)$$

The pump, active material, and cavity all affect a laser's efficiency. The overall efficiency is the product of a component due to the pump η_{pump}, a component due to the active material $\eta_{quantum}$, and a component due to the cavity η_{cavity} [86].

$$\eta_{eff} = \eta_{pump} \cdot \eta_{quantum} \cdot \eta_{cavity} \quad (7.21)$$

These factors vary widely from one type of laser to another.

In an optically pumped laser, a lamp or another laser excites the electrons of the active material. In this case, some of the pump light may get reflected from the surface or transmitted through instead of absorbed by the active material. Also, some of the pump energy may be converted directly to heat. Additionally, especially in the case of lamps which emit light over a wide range of frequencies, the pump light may have too little energy per photon to excite the electrons, or the light may have too much energy per photon thereby exciting electrons to a different upper energy level. Also, some of the pump light may interact with electrons that are already in excited energy states. In an electrically pumped laser, electricity excites the electrons of the active material. Some of the electrical energy may be converted to heat instead of exciting the electrons. All of these factors involving the pump contribute to η_{pump} and the overall laser efficiency η_{eff}.

The contribution to the overall laser efficiency due to the active material $\eta_{quantum}$ is more commonly known as the *internal quantum efficiency*. Some fraction of excited electrons decay to a lower energy level and emit a photon by spontaneous or stimulated emission. Alternatively, other excited electrons decay to a lower energy level while emitting heat or lattice vibrations instead. The *internal quantum efficiency* is the ratio of the rate with which excited electrons decay and produce a photon over the rate at which all excited electrons decay [10, p. 562]. It depends on temperature, the concentration of impurities or crystalline defects, and other factors [10, p. 596].

Efficiency is also determined by the laser cavity. A laser cavity reflects photons towards the active material. However, the laser cavity must let some light exit. In many lasers, the cavity is formed by mirrors. While these mirrors reflect most of the light, some light is absorbed and some light is transmitted through the mirrors as laser output. Many lasers which use mirrors include lenses, prisms, and other optical components in the cavity to focus or filter light to the active material. These components may also reflect or absorb some light and thereby decrease the laser efficiency. As mentioned above, the cavity of many semiconductor lasers is formed by the interface between the active material and the surrounding air. While external mirrors can reflect over 99% of photons [86, p. 159], mirrors formed by semiconductor air interfaces are much less efficient. The amount of light reflected depends on the index of refraction of the material. In gallium arsenide, for example, the index of refraction is 3.52 which corresponds to only 31% of light reflected at each interface [97].

The influence on efficiency of internal absorption and mirror reflectivity can be summarized in a single relationship [98].

$$\eta_{eff} = \eta_{eff-other} \frac{\ln\left(\frac{1}{R}\right)}{\alpha l + \ln\left(\frac{1}{R}\right)} \quad (7.22)$$

In this equation, R is the unitless mirror reflectivity, α is the absorption coefficient of the active material in units m^{-1}, and l is the length of the active material in m. The term $\eta_{eff-other}$ represents the efficiency due to all other factors besides absorption and mirror reflectivity, and η_{eff} is the overall efficiency. Equation 7.22 can be rewritten with some algebra.

$$\eta_{eff} = \eta_{eff-other} \left(1 - \frac{1}{1 + \frac{1}{\alpha l}\ln\left(\frac{1}{R}\right)}\right) \quad (7.23)$$

These efficiency concepts generalize to other energy conversion devices which produce light. Equation 7.20 also describes the overall efficiency of LEDs and lamps in addition to electrically pumped lasers. The concepts of efficiency due to the pump and internal quantum efficiency also apply to LEDs and lamps. However, η_{cavity} is not useful in describing these devices because LEDs and lamps do not contain a cavity.

7.4.4 Laser Bandwidth

Compared to LEDs and gas discharge lamps, incandescent lamps emit light over a much wider range of wavelengths. Compared to these devices, lasers emit light over a much narrower range of wavelengths. One reason that lasers emit over such a narrow wavelength range is that photons generated by stimulated emission have the same wavelength as the stimulating photon. As explained above, another reason is that only light at integer half multiples of the length of an optical cavity constructively interfere.

This narrow bandwidth of lasers compared to other sources of light is a major advantage in many applications. For example, lasers generate communication signals sent down optical fibers. Multiple signals can simultaneously be sent down a single fiber with each signal produced by a laser at a slightly different frequency. Due to the narrow bandwidth, these signals can be separately detected at the receiver.

Bandwidth of devices which emit light is typically specified by the *full width half maximum bandwidth* (FWHM). More specifically, intensity of light emitted is plotted as a function of wavelength where optical intensity is proportional to the square of the electric field. To find the FWHM, identify the wavelength of maximum intensity, and identify the wavelengths corresponding to half this intensity. The wavelength difference between these points of half intensity is called the FWHM, and this quantity is specified in meters or more likely nanometers. Sometimes FWHM is specified in units of Hz instead. A frequency response plot, showing intensity of light emitted versus frequency, is used to find FWHM in Hz. Again two points at half maximum intensity are identified on the plot. The frequency difference between these points of half intensity is the FWHM in Hz. A related measure is called the *quality factor*, and lasers with narrow bandwidth have high quality factor. It is defined as the ratio of the wavelength in nm of peak intensity emitted over the FWHM in nm. Alternatively, it is defined as the ratio of the frequency of peak intensity over the FWHM in Hz.

$$\text{Quality factor} = \frac{\lambda_{\text{peak intensity in nm}}}{\text{FWHM}_{[nm]}} = \frac{f_{\text{peak intensity in Hz}}}{\text{FWHM}_{[Hz]}} \quad (7.24)$$

As an example, consider Figs. 7.7 and 7.8 which are from [99]. Figure 7.7 relates to a dye laser where the active material is a liquid solution of

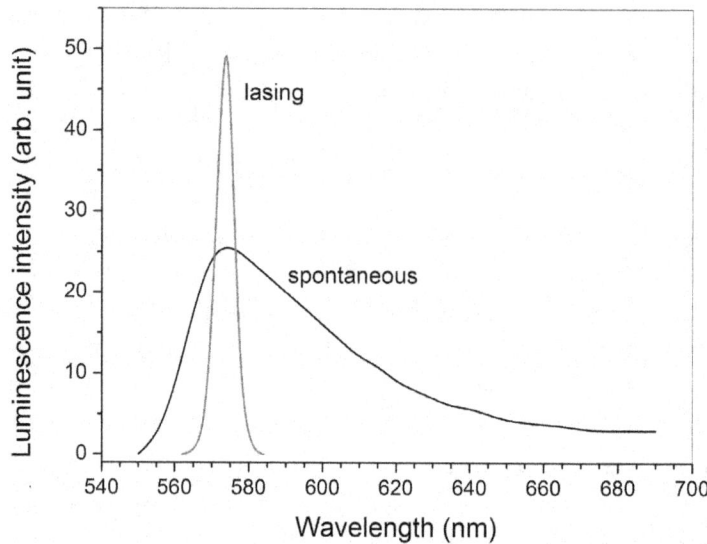

Figure 7.7: Optical intensity versus wavelength for a dye laser with an active material of rhodamine 6G mixed with silver nanoparticles. The curves correspond to two different pump energies, one above the lasing threshold and the other below the lasing threshold. This figure is used with permission from [99].

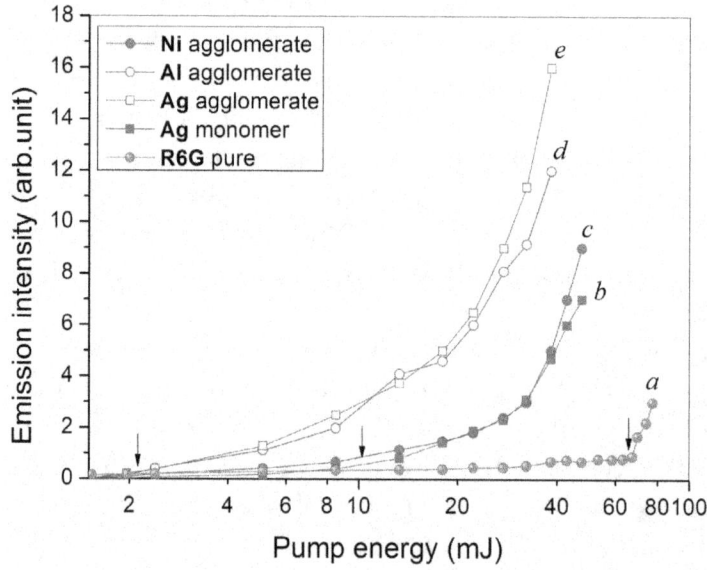

Figure 7.8: Optical intensity versus pump energy for dye lasers. Curve a describes a laser with rhodamine 6G as the dye while the other curves describe a laser with rhodamine 6G mixed with various nanoparticles. This figure is used with permission from [99].

7 LAMPS, LEDS, AND LASERS

the organic dye rhodamine 6G mixed with silver nanoparticles. Curve a of Fig. 7.8 relates to a dye laser with rhodamine 6G as the active material. The other curves of Fig. 7.8 relate to dye lasers with active materials made from rhodamine 6G doped with various nanoparticles. Typically, lasing will only occur if the active material is pumped strongly enough. If less energy is supplied, spontaneous emission occurs. Above a threshold, some spontaneous emission still occurs, but stimulated emission dominates. Figure 7.7 plots the intensity of light emitted at two different pumping levels, above and below the threshold for lasing. From this figure, we can see that the bandwidth of light emitted when the device is producing only spontaneous emission is much broader than the bandwidth of light emitted when the device is lasing. The FWHM and quality factor in each case can be approximated from this figure. For the spontaneous emission curve, $\text{FWHM}_{[nm]} \approx 45$ nm and quality factor ≈ 13. For the stimulated emission curve, $\text{FWHM}_{[nm]} \approx 5$ nm and quality factor ≈ 115. While these values are for dye lasers, other types of lasers, especially gas lasers, can have FWHM values which are orders of magnitude smaller, and values of 0.01 nm are achievable [83, p. 625]. Figure 7.8 illustrates another feature typical of lasers. Consider curve a which shows the intensity of the output versus pump energy supplied. The arrow in the figure near 65 mJ indicates the lasing threshold. Once lasing occurs, the intensity of the light emitted increases due to the optical feedback, so a discontinuity in the slope of plots of this type can be seen at the lasing threshold.

7.4.5 Laser Types

Engineers have developed many types of lasers utilizing a wide range of active materials. Lasers can be classified based on the type of active material as gas lasers, dye lasers, solid state lasers, or semiconductor lasers. Most lasers fit into one of these four categories, but there are exceptions such as free electron lasers where lasing occurs between energy levels of unbound electrons [31, p. 277] [86, p. 417].

Gas lasers

In a gas laser, the active material is a gas, and lasing occurs between energy levels of a neutral or ionized atom. Gas lasers are constructed from a gas filled glass tube. Electrodes inside the tube supply power to excite electrons of the gas atoms, and external mirrors form the cavity. One of the more common gas lasers is the helium neon laser, which typically operates at 632.8 nm [31, ch. 10]. However, the laser cavity may be designed so that lasing occurs at 3.39 μm and at other wavelengths too [31, ch. 10]. Another example of a common gas laser is the argon ion laser in which lasing occurs between energy levels of ionized argon. One advantage of gas lasers compared to other types of lasers is that they can be electrically pumped. Another advantage is that gas lasers can be designed to have high output powers. For this reason, gas lasers are used in applications requiring high power such as cutting, welding, and weaponry [86, p. 405]. Carbon dioxide lasers can produce hundreds of kilowatts of power when operating continuous wave and terawatts of power when operating pulsed [86, p. 405]. However, gas lasers are often physically large in size and not as portable as semiconductor lasers. High power gas lasers typically also require water cooling or another form of cooling.

Dye Lasers

In dye lasers, the active material is a solute in a liquid, and dye lasers are often optically pumped by other lasers [86, p. 386]. Lasing may occur between molecular vibration energy levels as opposed to electron energy levels [31, p. 225] [86, p. 386]. An advantage of dye lasers is that they may be tunable over a wide range of wavelengths. However, dye lasers require regular maintenance because the dyes have a finite useful lifetime [86, p. 391]. One example of a dye used is the organic molecule rhodamine 6G, and lasers using this dye are tunable from $570 < \lambda < 610$ nm [31, p. 228] [86, p. 387]. Figures 7.7 and 7.8 illustrate the behavior of a dye laser of this type.

Solid state lasers

The active material of a solid state laser is a solid insulating material, often a high purity crystal, doped with some element. Lasing occurs between electron energy levels of the dopant embedded in the solid. External mirrors are used to form the cavity. Solid state lasers are typically optically pumped by lamps or other lasers. A ruby laser is a solid state laser with an active material made from a crystal of sapphire, Al_2O_3, doped with around 0.05% by weight of chromium Cr^{3+} ions [31, ch. 10]. Ruby lasers are three level lasers [10, p. 476]. Another common solid state laser is a neodymium yttrium aluminum garnet laser, often denoted Nd:YAG, which is a four level laser. The active material of this laser is yttrium aluminum garnet $Y_3Al_5O_{12}$ doped with around 1% of neodymium Nd^{3+} ions, and this laser produces infrared light at $\lambda = 1.0641$ μm [10, p. 478] [31, p. 208] [86, p. 539]. Another common laser is the titanium doped sapphire laser, denoted

7 LAMPS, LEDS, AND LASERS

Ti:Sapph. The active material of this laser is sapphire Al_2O_3 doped with about one percent of titanium ions Ti^{3+}. This laser is tunable in the range $700 < \lambda < 1020$ nm [86, p. 392]. Tuning is achieved through an adjustable prism inside the laser cavity and through coatings on the mirrors of the cavity. Due to the tunability, these lasers are used for spectroscopy and materials research.

Semiconductor lasers

The active material of a semiconductor laser is a solid semiconductor pn junctions. An intrinsic, undoped, layer may be added between the p-type layer and the n-type layer at the junction to increase the width of the depletion region and improve overall efficiency [10, p. 567]. As with diodes and LEDs, the entire device typically fits inside a cubic millimeter. The wavelength emitted depends on the energy gap of the semiconductor. The first semiconductor lasers were made from gallium arsenide and produced infrared light [93]. Since then, semiconductor lasers emitting at all visible frequencies have been produced. It took over thirty years from the time the first semiconductors were produced to the time reliable blue lasers were produced [90] [100]. The first blue semiconductor lasers were produced using ZnMgSSe, and more commonly now GaN is used. Developing this technology required the ability to deposit very pure layers of the semiconductors without developing mechanical strain in the layers.

Almost all semiconductor lasers are made from direct semiconductors. It is for this reason that the first semiconductor lasers were made from GaAs even though silicon processing technology was more developed at the time [93]. Direct semiconductors were defined in Section 6.4 and illustrated in Fig. 6.8. In a direct semiconductor, the top of the valence band and the bottom of the conduction band line up in a plot of energy levels versus wave vector $|\vec{k}|$.

Figure 7.9 is a sketch of energy levels versus wave vector for a direct semiconductor and an indirect semiconductor. In both cases, an electron is excited to the conduction band. In both cases, the electron can decay by spontaneous emission from the conduction band to the valence band. In both cases, both energy and momentum must be conserved. In the direct semiconductor case, the electron can decay by emitting a photon. The electron does not need to change momentum in the process. While it is not shown in the figure, the electron can also decay by stimulated emission. In the indirect case, spontaneous emission can occur, but this process necessarily requires a change in momentum of the electron too. While it is possible that spontaneous emission can occur and produce a

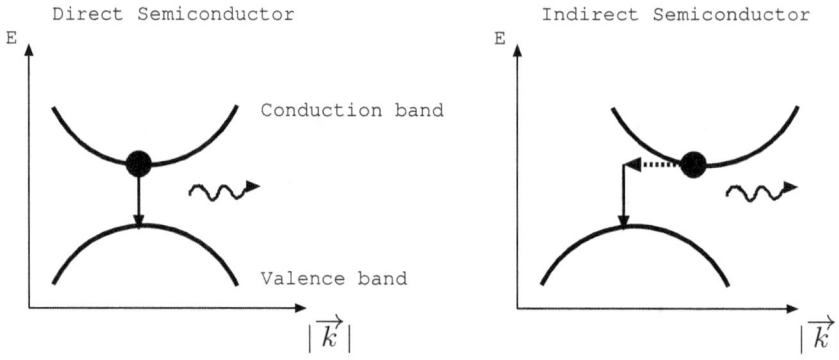

Figure 7.9: Energy level diagram vs. wave vector illustrating spontaneous emission in a direct and indirect semiconductor.

photon, often the electron decays by producing heat or vibrations instead of a photon of light [86, p. 444]. For this reason, stimulated emission is significantly less likely to occur in indirect than direct semiconductors.

As discussed above, semiconductor lasers do not have external mirrors. Semiconductor lasers can be broadly classified into two categories, *edge emitters* and *vertical cavity surface emitting lasers* (VCSELs) [101] depending on whether the optical emission is from the edge or the surface of the device. In edge emitting lasers, the cavity is often formed by the edges of the semiconductor. In other edge emitting lasers called distributed feedback semiconductor lasers, a grating, which acts as an optical filter, is etched into the semiconductor. In vertical cavity surface emitting lasers, multiple epitaxial layers of different materials form mirrors above and below the active material.

A main advantage of semiconductor lasers over other types of lasers is their small size. They can be integrated into both consumer devices like laser pointers and DVD players as well as industrial equipment and communication networks. Another large advantage is that they are electrically pumped. They also often do not need external cooling due to their relatively high overall efficiency. Another advantage is that the output wavelength can be designed by selecting the composition. For example, semiconductor lasers of composition $In_{1-x}Ga_xAs_{1-y}P_y$ produce infrared light in the range 1.1 μm $< \lambda <$ 1.6 μm. This frequency range is particularly useful for optical communication networks. Fiber optic cables are made from SiO_2 glass, a material with very low but nonzero absorption. Absorption is a function of wavelength, and the absorption minimum of silica glass is near 1.55 μm [10, p. 882]. These fibers also have low, but nonzero dispersion. Dispersion refers to the spread of pulses as they propagate through the fiber. The dispersion minimum in silica glass is around 1.3 μm. [10, p. 879]. Semiconductor lasers producing light in this range can be used to transmit signals down optical fibers, and these signals will have very low absorption and dispersion. A limitation is power output. While a semiconductor laser can produce over a watt of power, gas lasers can produce orders of magnitude more power.

7.4.6 Optical Amplifiers

Optical amplifiers are quite similar to lasers, and they can be made from all types of active materials used to make lasers including gases, solid state materials, semiconductors, and dyes [10, p. 477]. An *optical amplifier* consists of a pump and active material, but it does not have a cavity. The pump excites electrons of the active material to an upper energy level. Photons of an incoming optical signal cause additional photons to be generated by stimulated emission. *Amplification* occurs because these incoming photons generate additional photons, but *lasing* does not occur without the optical feedback provided by the cavity.

Erbium doped fiber amplifiers are one of the most useful types of optical amplifiers because of their use in optical communication networks [10, p. 882]. These devices can amplify optical signals without the need to convert them to or from electrical signals. They are solid state devices where stimulated emission occurs between energy levels of erbium, a dopant, in silica glass fibers. Energy from a semiconductor laser acts as the pump which excites electrons of the erbium atoms. Erbium doped fiber amplifiers are very useful because they can amplify optical signals near the fiber absorption minimum at $1.55 \mu m$.

7.5 Relationship Between Devices

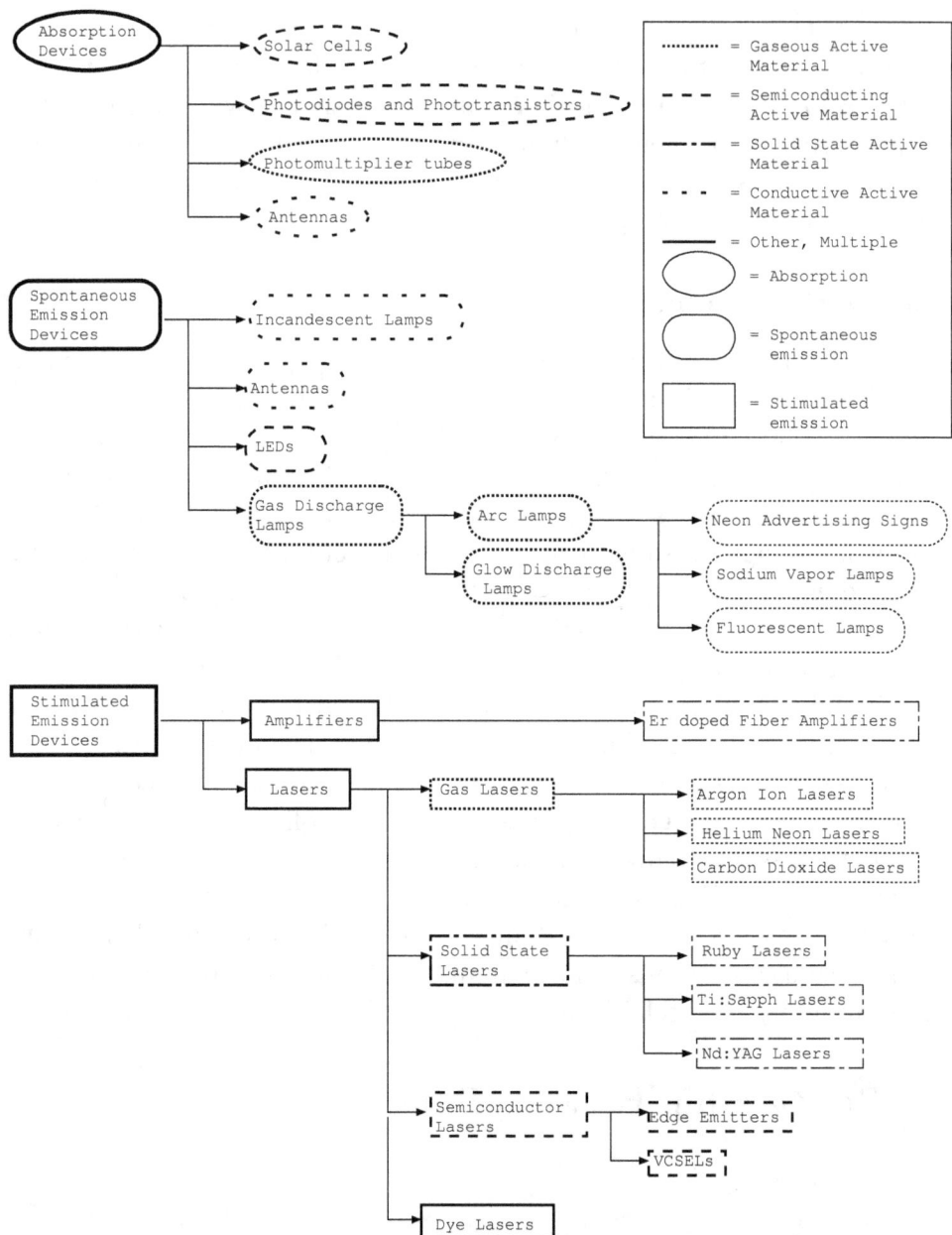

Figure 7.10: Devices which convert between electricity and light can be classified based on whether they involve absorption, spontaneous emission, or stimulated emission. Thick borders indicate categories of devices while thin borders indicate example types of devices.

7 LAMPS, LEDS, AND LASERS

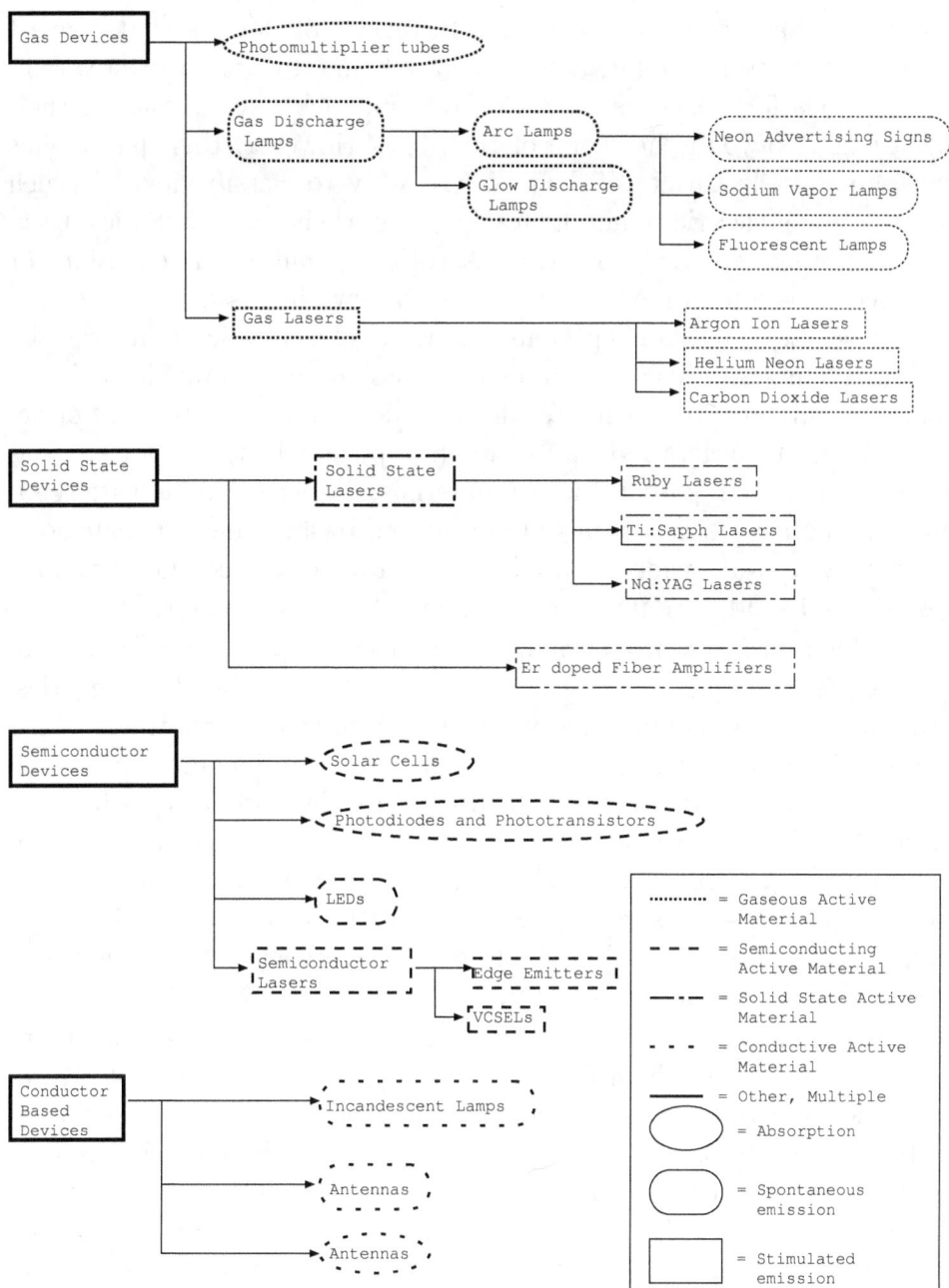

Figure 7.11: Devices which convert between electricity and light can be classified based on the type of active material. Thick borders indicate categories of devices while thin borders indicate example types of devices.

Scientists have come up with a wide variety of devices that convert between electricity and light, and they have come up with a wide variety of applications for these devices. Covering all such devices and all their applications is beyond the scope of this text. However, this chapter has shown some of the variety of devices. One way to classify devices which convert between electricity and light is to group them into categories based on whether they primarily involve absorption, spontaneous emission, or stimulated emission. Figure 7.10 illustrates how to classify many of the devices discussed in this chapter in this way. Ovals indicate absorption, rounded rectangles indicate spontaneous emission, and rectangles indicate stimulated emission. In the figure, thick borders indicate categories of devices while thin borders indicate example types of devices. Dotted lines indicate devices with gaseous active materials, dashed lines indicate semiconducting active materials, mixed dotted and dashed lines indicate solid state active materials, widely spaced dotted lines indicate conductive active materials, and solid lines indicate mixed or other active materials.

This diagram is far from complete because many other categories of devices exist, and these categories may be broken into further subcategories. Furthermore, only a handful of example devices are shown relating to some of the specific devices discussed above. This diagram includes absorption based devices discussed in Chapter 6. It also includes antennas discussed in Chapter 4. Light is a form of electromagnetic radiation with frequencies in the visible range. Light can be absorbed by a solar cell and spontaneously emitted by an LED, for example. Similarly, electromagnetic waves at longer wavelengths can be absorbed or spontaneously emitted by antennas which are devices with conductive active materials.

A different way of classifying devices which convert between electricity and light is to classify them based on the type of active material. All of these devices involve the interaction of light and atoms. The atoms involved may be part of a gas, may be dopants inside an insulating solid, may be part of a bulk semiconductor material, or may be part of a conductive solid. This way of classifying devices is illustrated in the Fig. 7.11. Dotted lines indicate devices with gaseous active materials, dashed lines indicate semiconducting active materials, mixed dotted and dashed lines indicate solid state active materials, widely spaced dotted lines indicate conductive active materials, and solid lines indicate mixed or other active materials. As in the previous figure, ovals indicate absorption, rounded rectangles indicate spontaneous emission, and rectangles indicate stimulated emission. Antennas are shown twice in the conductor based devices category because receiving antennas involve absorption while transmitting antennas involve spontaneous emission. Also as in the previous figure, thick borders indicate categories of devices while thin borders indicate example types of devices. Again, this figure does not show a complete list of all possible devices or device categories, but it does illustrate relationships between some devices discussed in this chapter.

Devices are usually designed to involve only one of these processes of absorption, spontaneous emission, or stimulated emission. However, it is possible for multiple of these processes to occur in a single device depending on how it is operated. For example, a semiconductor laser converts electricity to light by stimulated emission when current above the lasing threshold is supplied. If a weaker current is supplied, the device will act as an LED which converts electricity to light by spontaneous emission. If light shines on the device and the voltage across the device is measured, the same device acts as a photodetector which converts light to electricity by absorption. Similarly, photomultiplier tubes, gas discharge lamps, and gas lasers all involve tubes of gas with electrodes to supply or measure electricity. Like many energy conversion devices, these devices may convert electricity to light when operated in one direction and convert light to electricity when operated in reverse.

7.6 Problems

7.1. Identify whether the devices below operate based on spontaneous emission, stimulated emission, or absorption.

- Light emitting diode
- Gas discharge lamp
- Argon ion laser
- Solar panel
- Semiconductor laser

7.2. Consider a blackbody radiator at a temperature of 6500 K. Use Matlab, or similar software, to answer this question.

(a) Find the frequency which corresponds to peak spectral energy density per unit bandwidth.

(b) Find the wavelength which corresponds to peak spectral energy density per unit bandwidth.

(c) Find the value of the spectral energy density per unit bandwidth in $\frac{\text{J}\cdot\text{s}}{\text{m}^3}$ at the frequency found in part a.

7.3. A semiconductor laser which emits $\lambda = 500$ nm light has a length of 800 μm. The width is 12 μm, and the thickness is 5 μm. How many wavelengths long is the device in the longitudinal direction? How many wavelengths long is the device in each transverse direction?

7.4. Assume a semiconductor laser has a length of 800 μm. Laser emission can occur when the cavity length is equal to an integer number of half wavelengths. What wavelengths in the range 650 nm $< \lambda <$ 652 nm can this laser emit, and in each case, list the cavity length in wavelengths.

7.5. Assume two energy levels of a gas laser are separated by 1.4 eV, and assume that they are equally degenerate ($g_1 = g_2$). The spontaneous emission Einstein coefficient for transitions between these energy levels is given by $A_{12} = 3 \cdot 10^6$ s^{-1}. Find the other two Einstein coefficients, B_{12} and B_{21}.

7.6. The energy gap of AlAs is 2.3 eV, and the energy gap of AlSb is 1.7 eV [9, p. 19]. Energy gaps of materials of composition AlAs$_x$Sb$_{1-x}$ with $0 \leq x \leq 1$ vary approximately linearly between these values [9, p. 19]. Suppose you would like to make a semiconductor laser from a material of composition AlAs$_x$Sb$_{1-x}$. Find the value of x that specifies the composition of a material which emits light at wavelength $\lambda = 640$ nm.

7.7. Laser spectra are often modeled by Lorentzian functions. A Lorentzian function centered at the origin with area under the curve of unity has equation
$$y(x) = \frac{1}{\pi} \cdot \frac{0.5 \cdot \text{FWHM}}{x^2 + (0.5 \cdot \text{FWHM})^2}$$
where FWHM is the full width at half maximum. The maximum value of this function is $\frac{2}{\pi \cdot \text{FWHM}}$. The laser spectrum of Fig. 7.7 is centered near $\lambda = 570$ nm, has a FWHM of 5 nm, and it has a maximum luminescence intensity of 49. Find a Lorentzian equation that can model this particular spectrum.

7.8. As discussed in the previous problem, laser spectra are often modeled by Lorentzian functions. To better understand Lorentzian functions, use Matlab or similar software for this problem.

(a) Plot a Lorentzian function centered at the origin with FWHM 5 and maximum amplitude of unity. On the same axis, plot a Gaussian function also centered at the origin with FWHM 5 and maximum amplitude of unity.

(b) Repeat part a, but put the vertical axis of your plots on a log scale.

7.9. The figure illustrates a laser spectrum. Approximately find:

(a) The wavelength of peak intensity
(b) The FWHM
(c) The quality factor

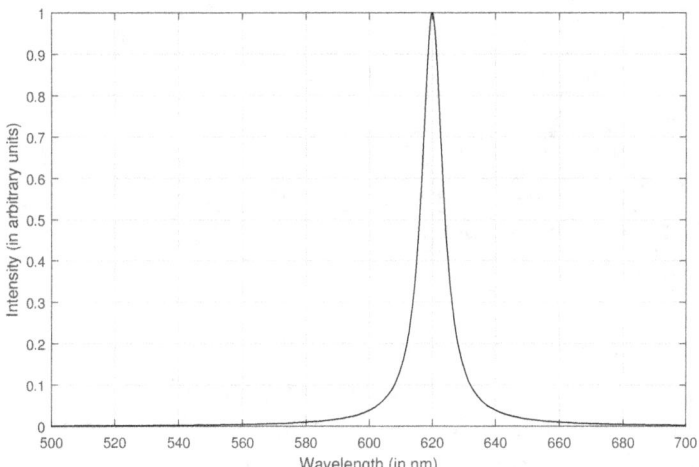

7.10. Three main components of a laser are the pump, active material, and cavity. Four main types of lasers are gas lasers, semiconductor lasers, dye lasers, and solid state lasers. Match the example component with the best description of the type of component and type of laser it is found in specified. (Each answer will be used once.)

Example Component
1. Edges of a AlGaAs crystal
2. Rhodamine 6G liquid solution
3. External mirror made of SiO_2 glass coated with aluminum
4. Battery of a laser pointer
5. SiO_2 glass doped with 1% Er atoms
6. CO_2 gas in an enclosed tube
7. Pn junction made from InGaAs
8. Argon ion laser used to supply energy to excite electrons of a Ti doped Sapphire

Description
A. Cavity of a semiconductor laser
B. Cavity of a gas laser
C. Active material of a semiconductor laser
D. Active material of a gas laser
E. Active material of a dye laser
F. Active material of a solid state laser
G. Pump of a semiconductor laser
H. Pump of a solid state laser

7.11. The intensity from sunlight on a bright sunny day is around 0.1 $\frac{W}{cm^2}$. Laser power can be confined to a very small spot size. Assume a laser produces a beam with spot size 1 mm^2. For what laser power in watts will the intensity of the beam be equivalent to the intensity from sunlight on sunny day? Staring at the sun can damage an eye, so staring at a laser beam of this intensity is dangerous for the same reason.

8 Thermoelectrics

8.1 Introduction

A *thermoelectric device* is a device which converts a temperature differential to electricity, or vice versa, and it is made from a junction of two different conductors or semiconductors. To understand thermoelectric devices, we need to understand the fundamentals of heat transfer and thermodynamics. This chapter begins by discussing these fundamental ideas. Next, thermoelectric effects and thermoelectric devices are discussed.

Many common processes heat an object. Rubbing blocks together, for example, heats them by friction. Burning a log converts the chemical energy in the wood to thermal energy, and applying a current to a resistor also heats it up. How can we cool an object? If we supply electricity to a thermoelectric device, one side heats up and the other cools down. We can place the object we want to cool near the cooler side of the thermoelectric device.

Thermoelectric devices, pyroelectric devices, and thermionic devices all convert energy between a temperature difference and electricity. Pyroelectric devices were discussed in Sec. 3.2. They are made from an insulating material instead of from a junction of conductors or semiconductors. Thermionic devices are discussed in Sec. 10.2, and they involve heating a cathode until electrons evaporate off. Thermoelectric devices, discussed in this chapter, are much more common than pyroelectric devices and thermionic devices due to their efficiency and durability.

8.2 Thermodynamic Properties

A container of air of fixed mass confined to a volume stores energy. We can shrink the volume of the air. This process requires energy, and the shrunken volume of air stores more energy. We can increase the gas pressure, for example, by exerting a force on a piston within which air is confined. This process requires energy, and the air under pressure stores more energy. We can take the fixed volume of air and heat it too. It takes energy to heat the air, and the hotter air stores more energy. Similarly, we can shake the container of air. Again, this process requires energy, and the energy from shaking is stored in the internal energy, the random motion, of the air molecules.

To talk about thermodynamic energy conversion, we need to define four fundamental properties of a system: volume, pressure, temperature, and entropy. All of these properties depend on the current state, not the past

Units for Pressure
$1\ \frac{N}{m^2} = 1$ Pa
1 bar $= 10^5$ Pa
1 mmHg$= 133.322$ Pa
1 atm $= 101,325$ Pa
1 psi $= 6.894757 \cdot 10^3$ Pa

Table 8.1: Pressure unit conversion factors [68].

history, of the sample. These properties can be classified as intensive or extensive [2, p. 10]. An *intensive* property is independent of the size or extent of the material. An *extensive* property depends on the size or extent [2, p. 10].

Volume $\mathbb{V}$ is an extensive property measured in m^3 or liters where 1 L $=$ 0.001 m^3. Pressure $\mathbb{P}$ is an intensive property measured in the SI units of pascals where 1 Pa$=$1 $\frac{N}{m^2}$. Pressure is also measured in a wide variety of other, non-SI, units such as bars, millimeters of mercury, or standard atmospheres as listed in Table 8.1. Pressure measures are often specified in comparison to the lowest possible pressure, of a complete vacuum, and such pressure measurements are called *absolute* pressure measurements [102, p. 15-17]. In some cases, values are specified as the difference above the local atmospheric pressure, and these measurements are called *gauge* pressure measurements [102, p. 15-17]. In other cases, values are specified as the difference below the local atmospheric pressure, and these measurements are called *vacuum* pressure measurements [102, p. 15-17]. Unless otherwise specified, the term pressure in this text refers to absolute pressure, not gauge or vacuum pressure.

Symbol	Quantity	Unit	Ext/int
$\mathbb{V}$	Volume	m^3	Extensive
$\mathbb{P}$	Pressure	Pa	Intensive
T	Temperature	K	Intensive
S	Entropy	$\frac{J}{K}$	Extensive

Table 8.2: Thermodynamic properties.

8 THERMOELECTRICS

Symbol	Name	Value and Unit
k_B	Boltzmann constant	$1.381 \cdot 10^{-23} \; \frac{J}{K}$
$\mathbb{R}$	Molar gas constant	$8.314 \; \frac{J}{mol \cdot K}$
N_a	Avogadro constant	$6.022 \cdot 10^{23} \; \frac{1}{mol}$

Table 8.3: Values of the Boltzmann constant, the molar gas constant, and the Avogadro constant.

Temperature T is an intensive property measured in either the SI units of degrees Celsius or kelvins. By definition, we can relate the two units by

$$T_{[°C]} = T_{[K]} - 273.15 \tag{8.1}$$

[68]. We can also measure temperature in the non-SI unit of degrees Fahrenheit. Temperature in degrees Celsius and temperature in degrees Fahrenheit are related by

$$T_{[°C]} = \left(\frac{T_{[°F]} - 32}{1.8} \right). \tag{8.2}$$

As with absolute pressure measurements, temperature in kelvins is said to be measured on an absolute temperature scale because the lowest possible temperature is given by zero kelvin. All temperatures are either absolute zero or have positive values. We use the term *temperature* to describe a *property* of a system. We use the term *heat transfer* to describe the *process* of transferring energy from a hot to a cold object. Entropy S is measured in units $\frac{J}{K}$, and it is an extensive property. Intuitively, entropy is a measure of the lack of order or organization of a material. The atoms in an amorphous material are less ordered than the atoms in a crystal of the same composition, so the amorphous material has more entropy.

Some further definitions will be needed. The symbol $\mathbb{N}$ represents the number of atoms or molecules of a substance. While it is not usually considered a fundamental thermodynamic property, it is a useful property of a sample. Sometimes it is specified in the units of mols instead of by the number of atoms or molecules. The Avogadro constant

$$N_a = 6.022 \cdot 10^{23} \; \frac{1}{mol} \tag{8.3}$$

is a constant which is used to convert a number given to the number per mol. The molar gas constant is

$$\mathbb{R} = 8.314 \; \frac{J}{mol \cdot K}. \tag{8.4}$$

Material	Bulk modulus $\mathbb{B}$ in GPa	Thermal cond. κ in $\frac{W}{m \cdot K}$	Electrical cond. σ in $\frac{1}{\Omega \cdot m}$	Ref.
Diamond	539	300	$1 \cdot 10^{-12} - 1 \cdot 10^{-2}$	[104]
Stainless steel	143	15.5	$1.3 \cdot 10^6 - 1.5 \cdot 10^6$	[105]
Graphite	18.6	195	$1.6 \cdot 10^4 - 2.0 \cdot 10^7$	[106]
Silicone rubber	1.75	1.38	$1 \cdot 10^{-14} - 3.2 \cdot 10^{-12}$	[107]

Table 8.4: Bulk modulus, thermal conductivity, and electrical conductivity of some materials. The references list ranges of values for bulk modulus and thermal conductivity while this table lists their averages.

The Boltzmann constant is

$$k_B = 1.381 \cdot 10^{-23} \, \frac{J}{K}. \tag{8.5}$$

These three constants are related by

$$k_B = \frac{\mathbb{R}}{N_a}. \tag{8.6}$$

8.3 Bulk Modulus and Related Measures

The *bulk modulus* $\mathbb{B}$ describes how a gas, liquid, or solid changes as it is compressed [103]. More specifically, bulk modulus per unit volume is the change in pressure required to get a given compression of volume,

$$\mathbb{B} = -\mathbb{V}\frac{\partial \mathbb{P}}{\partial \mathbb{V}} \tag{8.7}$$

and bulk modulus is specified in the SI units of pascals or $\frac{N}{m^2}$. The bulk modulus is greater than zero ($\mathbb{B} > 0$) even though there is a minus sign in Eq. 8.7 because volume shrinks when pressure is applied. Table 8.4 lists example bulk modulus values.

Assuming constant temperature, the inverse of the bulk modulus $\frac{1}{\mathbb{B}}$, is also called the *isothermal compressibility* [108]. There is a relationship between this compressibility and the permittivity ϵ discussed in Chapter 2. If we take an insulating material and apply an external electric field, a material polarization is established, and energy is stored in this charge accumulation. The permittivity is a measure of the charge accumulation

8 THERMOELECTRICS

per unit volume for a given strength of external electric field, in units of $\frac{F}{m}$. It is the ratio of the displacement flux density $\vec{D}$ to the electric field intensity $\vec{E}$.

$$\epsilon = \frac{|\vec{D}|}{|\vec{E}|} \tag{8.8}$$

If we take a material and apply an external pressure, the material compresses and energy is stored in this compressed volume. The inverse of the bulk modulus per unit volume is a measure of the change in volume for a given external pressure

$$\frac{1}{\left(\frac{\mathbb{B}}{\mathbb{V}}\right)} = -\frac{\partial \mathbb{V}}{\partial \mathbb{P}} \tag{8.9}$$

in units of $\frac{m^3}{Pa}$. Both Eqs. 8.8 and 8.9 can be called *constitutive relationships* because they describe how a material changes when an external influence is applied.

Multiple other measures describe the variation of a gas, liquid, or solid, with respect to variation of a thermodynamic property. The *specific heat* describes the ability of a material to store thermal energy, and it has units $\frac{J}{g \cdot K}$ [109, p. 98]. More specifically, the specific heat over temperature is equal to the change in entropy with respect to change in temperature [108]. It may be given either assuming a constant volume or assuming a constant pressure.

$$\text{Specific heat at constant volume} = C_v = T\frac{\partial S}{\partial T}\bigg|_{\mathbb{V}} \tag{8.10}$$

$$\text{Specific heat at constant pressure} = T\frac{\partial S}{\partial T}\bigg|_{\mathbb{P}} \tag{8.11}$$

The *Joule-Thomson coefficient* is defined as the ratio of change in temperature to change in pressure for a given total energy of the system

$$\text{Joule-Thomson coefficient} = \frac{\partial T}{\partial \mathbb{P}}, \tag{8.12}$$

and it has units $\frac{K}{Pa}$ [102, p. 685]. When a pressure is applied and overall energy is held fixed but entropy is allowed to vary, some materials cool and others heat. So, this coefficient may be positive, negative, or even zero at an inversion point. Additionally, the *volume expansivity* is defined as

$$\text{Volume expansivity} = \frac{1}{\mathbb{V}}\frac{\partial \mathbb{V}}{\partial T}\bigg|_{\mathbb{P}} \tag{8.13}$$

[108].

8.4 Ideal Gas Law

In most materials, if we know three of the four thermodynamic properties, volume, pressure, temperature, and entropy, we can derive the fourth property as well as other thermodynamic measures. Such materials are called *simple compressible systems* [109, 102]. For such materials, the *ideal gas law* relates pressure, volume, and temperature.

$$\mathbb{P}\mathbb{V} = \mathbb{N}RT. \tag{8.14}$$

While this is not a mathematical law, it is a good description of gases, and it can be used as a rough approximation for liquids and solids. Consider a container filled with a gas. If the volume of the container is compressed while the temperature is kept constant, the pressure increases. If the gas is heated and the pressure is kept constant, the volume increases. The energy stored in a gas that undergoes change in volume at constant temperature is given by

$$E = \int \mathbb{P} d\mathbb{V} \tag{8.15}$$

where the change in energy is specified by

$$\Delta E = \mathbb{P} \Delta \mathbb{V}. \tag{8.16}$$

The ideal gas law can be written incorporating entropy as

$$\mathbb{P}\mathbb{V} = ST. \tag{8.17}$$

For example, consider a 10 L tank that holds 5 mol of argon atoms. The argon gas is at a temperature of $T = 15$ °C. Find the pressure in the tank in pascals and in atm. We begin by converting the volume and temperature to more convenient units, $\mathbb{V} = 0.01$ m^3 and $T = 288.15$ K. Next, the ideal gas law provides the pressure in Pa.

$$\mathbb{P} = \frac{\mathbb{N}RT}{\mathbb{V}} = \frac{5 \text{ mol} \cdot 8.314 \frac{\text{J}}{\text{mol·K}} \cdot 288.15 \text{ K}}{0.01 \text{ m}^3} = 1.20 \cdot 10^6 \text{ Pa} \tag{8.18}$$

Finally, we convert the pressure to the desired units.

$$\mathbb{P} = 1.20 \cdot 10^6 \text{ Pa} \cdot \frac{1 \text{ atm}}{101325 \text{ Pa}} = 11.8 \text{ atm} \tag{8.19}$$

As another example, consider a container that holds neon atoms at a temperature of $T = 25$ °C. Assume that the pressure in the container is 10 kPa, and the mass of the neon in the container is 3000 g. Find the volume of the container. The temperature is 298.15 K. From a periodic table, the atomic weight of a neon atom is 20.18 $\frac{g}{mol}$. Thus, the container holds 148.7 mol of neon atoms. Next, we use the ideal gas law to find the volume.

$$\mathbb{V} = \frac{\mathbb{N}\mathbb{R}T}{\mathbb{P}} = \frac{148.7 \text{ mol} \cdot 8.314 \frac{J}{mol \cdot K} \cdot 298.15 \text{ K}}{10^4 \text{ Pa}} = 36.86 \text{ m}^3 \quad (8.20)$$

8.5 First Law of Thermodynamics

The idea of energy conservation was introduced in Sec. 1.3. Most discussions of thermodynamics also begin with the same idea. The first law of thermodynamics is a statement of energy conservation. Energy can be stored in the material polarization of a capacitor, the chemical potential of a battery, and in many other forms. People studying thermodynamics and heat transfer, however, often make some drastic assumptions. They classify all energy conversion processes as heat transfer or other where the primary component of the latter is mechanical work. At the beginning of introductory thermodynamics courses, all forms of energy besides heat transfer and mechanical work are ignored. Charging a capacitor, discharging a battery, and all other energy conversion processes are grouped in with mechanical work when writing the first law of thermodynamics. The first law of thermodynamics is typically written as

$$\text{(change in int. energy)} = \text{(heat in)} - \text{(work and other forms)}. \quad (8.21)$$

$$\Delta \mathbb{U} = \mathbb{Q} - W \quad (8.22)$$

Each term of the Eq. 8.22 has the units of joules. The symbol $\mathbb{Q}$ represents the energy supplied *in* to the system by heating, and $-W$, with the minus symbol, represents the mechanical work *in* to the system as well as all other forms of energy into the system. The quantity $\Delta \mathbb{U}$ represents the change in internal energy of the system. In a closed system the total energy is conserved. In a closed system, energy is either stored in the system (for example as potential energy or another form of internal energy), is transferred in or out as heat, or is transferred in or out as another form such as mechanical work [109, p. 51].

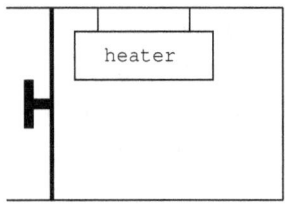 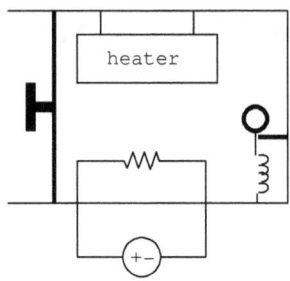

Figure 8.1: Illustration of closed systems containing energy conversion devices.

As an example, consider the closed system shown on the left part of Fig. 8.1 comprised of a cylinder with a piston and a heater. Assume the cylinder contains a fixed volume of gas inside. Suppose the heater is used to transfer 100 J of energy into the piston in an hour while the piston is forced to remain in a fixed position. After the hour, the internal energy of the gas will be 100 J greater than before. Again suppose the heater is used to transfer 100 J of energy into the gas, but this time assume the piston is allowed to move thereby expanding the gas volume. After the hour, the internal energy of the gas will be the original energy of the gas, plus the 100 J supplied by the heater, and minus a factor due to the mechanical work done by the piston.

The first law of thermodynamics says two things. First, energy is conserved. Second, energy can be stored, converted to mechanical work, or converted to heat. We know energy can be converted to other forms too, like electrical or electromagnetic energy. While introductory thermodynamics classes do not usually do so, we can add other devices to the piston as shown on the right part of Fig. 8.1. We can include a battery and put a resistor inside to convert the chemical energy to electrical energy, and the resistor can heat the air in the piston. We can put a mass and a spring in the piston and convert potential energy of a compressed spring to kinetic energy by removing a clip which holds the spring compressed. In a closed system when all energy conversion processes are considered, energy must be conserved.

8.6 Thermoelectric Effects

8.6.1 Three Related Effects

In the 1800s, three effects were experimentally observed. At first, it was not obvious that these experiments were related, but soon they were found

8 THERMOELECTRICS

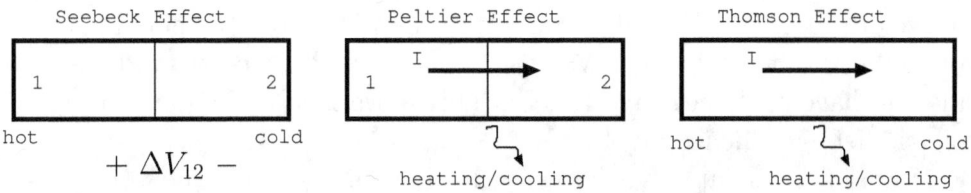

Figure 8.2: The Seebeck effect, Peltier effect, and Thomson effect.

to be three aspects of the same phenomenon [5, p. 113].

The first effect, now called the *Seebeck effect*, was discovered in 1821 by Thomas Seebeck [5, p. 113]. It is observed in a junction of two different metals or semiconductors. As discussed in Section 6.5, junctions are also used to make photovoltaic devices, LEDs, and semiconductor lasers. When the different sides of the junction are held at different temperatures, a voltage develops across the junction. The *Seebeck coefficient* $, in units $\frac{V}{K}$, is defined as the ratio of that voltage to the temperature difference. More specifically, consider a junction where one side is held at a hotter temperature than the other, as shown in the left part of Fig. 8.2. The difference between the Seebeck coefficient in material one $_1$ and the Seebeck coefficient in material two $_2$ is given by the measured voltage across the junction ΔV_{12} divided by the temperature difference between the materials ΔT_{12} [110, p. 24].

$$\$_1 - \$_2 = \frac{\Delta V_{12}}{\Delta T_{12}} \qquad (8.23)$$

The difference between the Seebeck coefficients can be positive or negative because both the temperature difference and the measured voltage can be positive or negative. However, for any given material, the Seebeck coefficient is positive. To find the Seebeck coefficient for an unknown material, form a junction between that material and a material with known Seebeck coefficient, heat one end of the junction hotter than the other, and measure the voltage established. For most materials, the Seebeck coefficient is less than 1 $\frac{\mu V}{K}$. Some of the largest values of the Seebeck coefficient are found in materials containing tellurium. For example, $(Bi_{0.7}Sb_{0.3})_2 Te_3$ has $ \$ \approx 230\ \frac{\mu V}{K}$ and PbTe has $\$ \approx 400\ \frac{\mu V}{K}$ [3].

To understand the physics behind the Seebeck effect, consider the flow of charges across this diode-like device. In metals, valence electrons are the charge carriers, and in semiconductors, both valence electrons and holes are the charge carriers. These charge carriers diffuse from the hot to the cold side of the junction. Consider a junction of two metals with no net charge on either side initially. If an electron moves from the hot side to the cold side, the hot side then will have a net positive charge, and the cold side will have a net negative charge. This movement of charges sets up an electric field and hence a voltage.

If we let the sample reach an equilibrium temperature, no voltage will be measured. A voltage is measured only during the time when charge carriers have diffused from one material to the other but when the material has not reached a uniform temperature. *Thus, for a material to have a large thermoelectric effect, it must have a large electrical conductivity and small thermal conductivity.* Thermoelectric devices are typically made from metals or semimetals because these materials satisfy this condition.

The second effect was discovered by Jean Peltier in 1834 [5, p. 113]. The *Peltier effect* is also observed in a junction of two different metals, semimetals, or semiconductors. It is illustrated in the middle part of Fig. 8.2. When a current, I in amperes, is supplied across a junction, heat is transferred. This effect occurs because charges from the supplied current flow through different materials with different thermal conductivities on the different sides of the junction. The effect is quantified by the *Peltier coefficient* for the junction, Π_{12}, or Peltier coefficients for the materials forming the junction, Π_1 and Π_2. More specifically, the Peltier coefficient is defined as

$$\Pi_{12} = \Pi_1 - \Pi_2 = \frac{\left(\frac{dQ}{dt}\right)}{I} \tag{8.24}$$

in the units of volts [110, p. 24]. The term $\frac{dQ}{dt}$ represents the rate heat is transferred in $\frac{J}{s}$, and it may be positive or negative because the thermal conductivity in the first material may be higher or lower than in the second material. The Seebeck coefficient and the Peltier coefficient are related by

$$\Pi_1 - \Pi_2 = (\$_1 - \$_2) T. \tag{8.25}$$

PbTe is a material with a relatively high Seebeck coefficient. At room temperature, it has coefficients $\$ = 400 \frac{\mu V}{K}$ and

$$\Pi = 400 \frac{\mu V}{K} \cdot 300 K = 0.12 \text{ V}. \tag{8.26}$$

The third effect was first discovered by William Thomson in the 1860s [3]. Thomson also derived the relationship between these three effects. It is illustrated on the right part of Fig. 8.2. When a current passes through a uniform piece of material which has a temperature gradient, heating or cooling will occur, and this result is known as the *Thomson effect* [3, p. 148] [110, p. 24] [5, p. 115]. To observe this effect, apply a temperature gradient across a piece of metal or semiconductor and also apply a current through the length of the material. Heating or cooling can be measured, and this effect is described by another coefficient. The *Thomson coefficient* τ also has units $\frac{V}{K}$. It is defined as the rate of heat generated over the product of the current and temperature difference.

$$\tau = \frac{\frac{dQ}{dt}}{I(T_h - T_c)} \tag{8.27}$$

The Thomson and Seebeck coefficients for a single material are related by

$$\int_0^T \frac{\tau}{T'} dT' = \$ \tag{8.28}$$

where the integral is over temperature [110, p. 24].

These effects work both ways. We can use the Peltier effect, for example, to make either a heating or a cooling device. We can supply a current across a junction to produce a temperature differential, or we can supply a temperature difference to generate a current. All three effects relate to the fact that when the electrical conductivity is larger than the thermal conductivity, energy can be converted between a temperature differential and electricity. As an aside, materials with low electrical conductivity and high thermal conductivity are also used to make energy conversion devices. Components of motors and generators are often made from layers of metal and dielectrics with these properties [111, ch. 8].

8.6.2 Electrical Conductivity

Electrical conductivity σ, in units $\frac{1}{\Omega \cdot m}$, is a measure of the ability of *charges* to flow through a material. Resistivity is the inverse of electrical conductivity, $\rho = \frac{1}{\sigma}$. Example electrical conductivity values are listed in Table 8.4, found in Section 8.3. Few tools are needed to measure these quantities. An ohmmeter can be used to find the resistance R, in ohms, of a sample with known length l and cross sectional area A. The conductivity can be calculated directly,

$$\sigma = \frac{l}{AR}. \tag{8.29}$$

Electrical conductivity is the product of the number of charges flowing and their mobility. For conductors, valence electrons are charge carriers that flow, so conductivity can be expressed as [9, p. 84]

$$\sigma = nq\mu_n. \tag{8.30}$$

In this expression, n is the concentration of valence electrons in units $\frac{\text{electrons}}{\text{m}^3}$, and it was introduced in Sec. 5.2. The magnitude of the charge of an electron is $q = 1.6 \cdot 10^{-19}$ C. *Mobility* of electrons, μ_n, is the ease with which charge carriers drift in a material, and it has units $\frac{\text{m}^2}{\text{V} \cdot \text{s}}$. By definition, mobility is the ratio of the average drift velocity of electrons to the applied electric field in $\frac{\text{V}}{\text{m}}$ [9, p. 84].

$$\mu_n = \frac{-\text{avg drift velocity of electrons}}{\left|\vec{E}\right|} \quad (8.31)$$

For semiconductors, both electrons and holes act as charge carriers, and both contribute to the conductivity,

$$\sigma = q\left(n\mu_n + p\mu_p\right) \quad (8.32)$$

where μ_p is the mobility of the holes, and p is the concentration of holes.

To understand which materials have large electrical conductivities, and hence make good thermoelectric devices, we need to consider the charge concentrations n and p. Conductors and semiconductors have charges that can move through the material while insulators do not. Thus conductors and semiconductors have large electrical conductivity and are used to make thermoelectric devices. Furthermore, a doped semiconductor has more charge carriers than an undoped, also called intrinsic, semiconductor. Thus, doped semiconductors usually have higher electrical conductivity than undoped semiconductors of the same material [110].

Electrical conductivity σ is proportional to the electron and hole mobilities, μ_n and μ_p, and the mobilities are a strong function of temperature [9]. For this reason, the electrical conductivity is a function of temperature. At low temperature, mobilities are limited by impurity scattering while at high temperatures, they are limited by phonon scattering. At some intermediate temperature, mobility and conductivity are maximum, and this peak occurs at different temperatures for different materials. Mobility also depends on whether a material is crystalline or amorphous and on the degree of crystallinity. Mobility and electrical conductivity are both typically higher in crystals than glasses because charges are more likely to get scattered in amorphous materials.

8.6.3 Thermal Conductivity

Thermal conductivity κ is a measure of the ability of *heat* to flow through a material, and it has units $\frac{\text{W}}{\text{m} \cdot \text{K}}$ [109, p. 793]. Example thermal conductivity

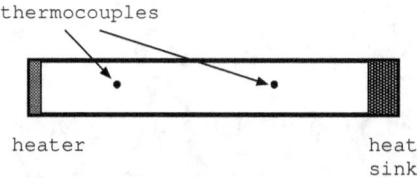

Figure 8.3: Components used to measure thermal conductivity.

values are listed in Table 8.4, found in Section 8.3. Understanding thermal conductivity is complicated because a number of mechanisms are responsible for the conduction of heat. Heat may be transported by phonons, photons, electrons, or other mechanisms, and each mechanism depends on temperature and the properties of the material. Good thermoelectric devices have small thermal conductivity. Often metals have large thermal conductivity and insulators have small thermal conductivity.

The apparatus to measure thermal conductivity consists of a heater, a heat sink, and a number of thermocouples [110] [112]. To measure thermal conductivity experimentally, start with a bar of material with a known cross sectional area A. Heat one end of the bar with respect to the other, wait for a steady thermal state, and measure the temperature at each end of the bar. Next, calculate the temperature gradient $\frac{dT}{dx}$ in units $\frac{K}{m}$ along the length of the bar. Also measure the rate that heat is supplied to the bar, $\frac{dQ}{dt}$, in units $\frac{J}{s}$. By definition, thermal conductivity is the ratio

$$\kappa = \frac{\text{(power dissipated in heater)(distance between thermocouples)}}{\text{(cross sectional area)(change in temp)}} \tag{8.33}$$

[109, p. 49]. The thermal conductivity can be calculated from

$$\kappa = \frac{-\left(\frac{dQ}{dt}\right)}{A\left(\frac{dT}{dx}\right)}. \tag{8.34}$$

This technique works well for low to moderate temperatures and materials with high thermal conductivity. Other methods exist to measure thermal conductivity and are advantageous for different temperature or conductivity ranges.

Another way to understand thermal conductivity is to think of it as the product of the amount of heat transported by some particle times the velocity of that particle. This viewpoint applies whether or not actual particles are involved in the heat transport. More specifically, thermal conductivity is given by

$$\kappa = C_v |\vec{v}| l \tag{8.35}$$

The symbol C_v represents the specific heat at constant volume in $\frac{J}{g \cdot K}$, and the symbol $|\vec{v}|$ represents the magnitude of the transport velocity in $\frac{m}{s}$. The scattering length is represented by l in m.

Regardless of whether electrons, phonons, or something else transports heat through a material, the ability of that heat to get from one end to the other without being scattered or blocked influences the thermal conductivity. Thus, crystals typically have higher thermal conductivity than amorphous materials [113]. The thermal conductivity of a crystal can be lowered by exposing the material to radiation which destroys the crystallinity and increases the likelihood that the heat carrier will be scattered [110]. For glasses, scattering length is roughly the interatomic spacing [112]. Also, thermal conductivity of glasses is less temperature dependent than crystals because high temperatures distort the perfect crystallinity, thereby lowering the thermal conductivity for crystals but not glasses [113].

All the contributing factors, C_v, $|\vec{v}|$, and scattering length l, are temperature dependent, so the thermal conductivity is a function of temperature. The temperature dependence of the factors is discussed in reference [110]. Thermal conductivity, like electrical conductivity, is low at low temperatures then rises to a maximum before decreasing again at higher temperatures [112].

8.6.4 Figure of Merit

The *figure of merit* of a thermoelectric device, Z, is a single measure that summarizes how good a material is for making thermoelectric devices. It is defined as

$$Z = \frac{\$^2 \sigma}{\kappa}, \tag{8.36}$$

and it has units K^{-1}. It depends on the Seebeck coefficient $\$$, electrical conductivity σ, and thermal conductivity κ. A large value of Z indicates that the material is a good choice for use in construction of a thermoelectric device.

The figure of merit depends on temperature because the parameters $\$$, σ, and κ are strong functions of temperature. Thus, the best choice material for a thermoelectric device operating near room temperature may not be the best choice for a device operating at other temperatures. Sometimes ZT is used as a measure instead of just Z to account for the temperature dependence.

The figure of merit does not incorporate all of the temperature related factors to consider in selecting materials for thermoelectric devices. Melting temperature is also important. A thermoelectric device converts more energy when a larger temperature difference is placed across the device. The Seebeck coefficient is inversely proportional to the temperature differential, $\$ = \frac{\Delta V}{\Delta T}$. However, too large of a temperature differential will melt the hot end of the device, and different materials can have very different melting temperatures. For example, lead telluride PbTe melts at 924 °C, and Bi_2Te_3 melts at 580 °C [114, p. 4-52, 4-71].

The figure of merit also depends on doping level because the electrical conductivity is directly proportional to the charge concentrations n and p [110]. Thus, a thermoelectric device made from a doped semiconductor has a higher electrical conductivity and thermoelectric efficiency than a device made from an undoped semiconductor of the same material. The Seebeck coefficient is also dependent on doping level but not as strongly [110]. Thermal conductivity is not a strong function of charge concentrations n and p [110]. Thus, thermoelectric materials are often made from heavily doped semiconductors or from conductors.

The figure of merit also depends on degree of crystallinity. Typically, both the electrical conductivity and thermal conductivity are much higher in crystals than glasses because charge and heat carriers are less likely to get scattered as they travel through crystals than glasses [113]. Since both electrical and thermal conductivity are influenced, the effect of degree of crystallinity on the figure of merit can be complicated.

Thermoelectric devices are typically made from junctions of two different metals or semiconductors. Essentially, a thermoelectric device is a diode. Common materials used include bismuth telluride, lead telluride, and antimony telluride, all of which are semimetals. Bi, Sb, and Pb are all located near each other on the periodic table. Other materials studied for use in thermoelectric devices include [110], BiSeTe, LiMnO, LiFeO, LiCoO, LiNiO, PbS, and ZnSb. These materials are either small gap semiconductors or semimetals. In semiconductor materials with small energy gaps, the ratio of electrical conductivity to thermal conductivity is large. However, this fact must be balanced against the fact that smaller gap semiconductors tend to have lower melting temperatures than larger gap semiconductors [110, ch. 1].

Recently, layered materials and superlattices have been considered as materials for thermoelectric devices [115] [116]. The layers can be tailored to affect the thermal and electrical properties differently and can act like a filter to select out different conduction mechanisms. Understanding of the conductivity mechanisms is a prerequisite to understanding such more

8.7 Thermoelectric Efficiency

8.7.1 Carnot Efficiency

Many devices convert a temperature difference to another form of energy. For example, thermoelectric devices and pyroelectric devices convert a temperature difference to electricity, and Stirling engines and steam turbines convert a temperature difference to mechanical work. There is a fundamental limit to the efficiency of any device that converts a temperature difference into another form of energy. The *Carnot efficiency* is the maximum possible efficiency of such an energy conversion process.

Consider a thermoelectric device made from a junction of two materials that converts a temperature difference to electricity using the Seebeck effect. Assume that one end of the device is connected to a heater, and the other end of the device is connected to a heat sink so that it is at a lower temperature. The temperature of the hot side of the device is denoted T_h, and the temperature of the cold side of the device is denoted T_c. Both temperatures are measured in kelvins, K (or another absolute temperature measure such as Rankine). Assume that the only energy conversion process that occurs converts energy from the temperature difference to electricity. Furthermore, assume that energy is continuously supplied from the heater at a constant rate to maintain the hot end of the device at temperature T_h. The heater is supplying heat to the room it is in. However, assume that the room is so large and the amount of heat from the heater is so small that the temperature of the room remains roughly constant. For this reason, we say that the room is a *thermodynamic reservoir*. Also, assume that we have waited long enough that the temperature of the device has reached a steady state. The temperature is not constant along the length of the device, but it no longer varies with time.

The input to this system is the thermal energy supplied from the heater, E_{in}. The output of this system is the electrical energy extracted out, E_{out}. The device is not used up in the process, so the number of atoms in the device remains constant. As long as energy is supplied from the heater at a constant rate to maintain the hot side at temperature T_h, we can extract electrical energy out of the system at a constant rate. Heat transfer scientists call this type of process a *thermodynamic cycle* or a *heat engine*. A *thermodynamic cycle* is a sequence of energy conversion processes where the device begins and ends in the same state. In a thermodynamic cycle, energy is supplied in one form and is extracted in another form. The device

8 THERMOELECTRICS

or mass involved starts and ends in the same state, so the processes can continue indefinitely as long as the input is continually supplied.

How much energy is supplied in to the system from the heater? The amount of energy required to maintain the hot side at temperature T_h is given by

$$E_{in} = k_B T_h. \tag{8.37}$$

The device is composed of atoms. Each of these atoms has some internal energy. A device at temperature T contains $k_B T$ joules of energy where k_B is the Boltzmann constant. Energy flows from the hot side to the cold side of the device. Above, we assumed that the device was in a room that was so large that the heat from the heater did not raise the temperature of the room. Thus, we must continually supply this energy at a constant rate to keep the hot side of the device at temperature T_h. While the cold side of the device is at a lower temperature T_c, it maintains that temperature regardless of the fact that there is a heater in the room.

How much energy is extracted out of the system as electrical energy? In the Seebeck device, the hot side is held fixed at temperature T_h, and because of the environment it is in, the cold side remains at temperature T_c. Energy is conserved in this system. Thus, the electrical energy extracted from the device is given by

$$E_{out} = k_B T_h - k_B T_c. \tag{8.38}$$

What is the efficiency of this system? Above we assumed that no other energy conversion processes occur, so this is an idealized case. The resulting efficiency that we calculate represents the best possible efficiency of a thermoelectric device operating with sides at temperatures T_h and T_c. Efficiency is defined as

$$\eta_{eff} = \frac{E_{out}}{E_{in}}. \tag{8.39}$$

Using Eqs. 8.37 and 8.38 and some algebra, we can simplify the efficiency expression.

$$\eta_{eff} = \frac{E_{out}}{E_{in}} = \frac{k_B T_h - k_B T_c}{k_B T_h} \tag{8.40}$$

$$\eta_{eff} = \frac{T_h - T_c}{T_h} \tag{8.41}$$

$$\eta_{eff} = 1 - \frac{T_c}{T_h} \tag{8.42}$$

Eq. 8.42 is known as the *Carnot efficiency*. It provides a serious limitation on the efficiency of energy conversion devices which involve converting energy of a temperature difference to another form. The Carnot efficiency applies to thermoelectric devices, steam turbines, coal power plants, pyroelectric devices, and any other energy conversion device that convert a temperature difference into another form of energy. It does not, however, apply to photovoltaic or piezoelectric devices. If the hot side of a device is at the same temperature as the cold side, we cannot extract any energy. If the cold side of a device is at room temperature, then the efficiency cannot be 100%. The Carnot efficiency represents the best possible efficiency, not the actual efficiency of a particular device because it is likely that other energy conversion processes occur too. We can extract more energy from a steam turbine with $T_h = 495$ K than $T_h = 295$ K. However, in both cases, the amount of energy we can extract is limited by the Carnot efficiency. Note that when using Eq. 8.42, T_c and T_h must be specified on an absolute temperature scale, where $T = 0$ is absolute zero. In SI units, we use temperature in kelvins.

As an example, consider a device that converts a temperature difference into kinetic energy. The cold side of the device is at room temperature, $T_c = 300$ K. How hot must the hot side of the device be heated to so that the device achieves 40% efficiency?

$$\eta_{eff} = 1 - \frac{T_c}{T_h} \tag{8.43}$$

$$0.4 = 1 - \frac{300}{T_h} \tag{8.44}$$

According to Eq. 8.42, we find that $T_h = 500$ K.

As another example, suppose we want to convert a temperature differential to electrical energy using a thermoelectric device. Assume that the cold side of the device is at room temperature of $T_c = 72$ °F and the hot side is at human body temperature of $T_h = 96$ °F. What is the best possible efficiency? First the temperatures must be converted from degrees Fahrenheit to kelvins. The resulting temperatures are $T_c = 295$ K and $T_h = 309$ K. Next, using Eq. 8.42, we find the best possible efficiency is only 4.5%.

$$\eta_{eff} = 1 - \frac{295}{309} = 0.045 \tag{8.45}$$

As another example, assume that the temperature outside on a December day is $T_c = 20$ °F and inside room temperature is $T_h = 72$ °C. What is the Carnot efficiency of a thermoelectric device operating at these temperatures? Again we begin by converting the temperatures to kelvins, $T_c = 266$ K and $T_h = 295$ K.

$$\eta_{eff} = 1 - \frac{266}{295} = 0.098 \tag{8.46}$$

8.7.2 Other Factors That Affect Efficiency

The efficiency of practical energy conversion devices is always lower than the Carnot efficiency because it is very unlikely that only a single energy conversion process occurs. All practical materials, even good conductors, have a finite resistance, so energy is converted to thermal energy as charges travel through the bulk of the device and through wires connected to it. Furthermore, heat flows through the device, so if a heater is connected to one side of a device, the other side will be at a higher temperature than the room it is in. For this reason, not all energy supplied by the heater can be converted to electricity.

As an example, consider a material with length $l = 1$ mm $= 10^{-3}$ m and cross sectional area $A = 1$ mm^2 $= 10^{-6}$ m^2. Assume the material has a resistivity of $\rho = 10^{-5}$ $\Omega \cdot$m which is typical for a moderate conductor. Assume a current of $I = 3$ mA flows through the sample. How much power is converted to heat due to resistive heating? The electrical conductivity of the sample is $\sigma = \frac{1}{\rho} = 10^5 \frac{1}{\Omega \cdot m}$. The resistance of the device is given by $R = \frac{\rho l}{A}$. Power is

$$P = I^2 R = I^2 \frac{\rho l}{A} = (3 \cdot 10^{-3})^2 \frac{10^{-5} \cdot 10^{-3}}{10^{-6}} = 9 \cdot 10^{-8} \text{ W}. \tag{8.47}$$

While this amount of power may seem small, it is another factor which diminishes the efficiency of the device. Even if we convert energy from a temperature differential to electricity at the junction of the thermoelectric device, some resistive heating occurs. This heat is wasted in the sense that it isn't converted back to electricity.

The efficiency of most thermoelectric devices is less than 10% [5, p. 140] [117]. As seen by Eq. 8.42, efficiency depends heavily on the temperatures T_c and T_h, and efficiency can be increased by increasing T_h. For many devices, the maximum temperature is limited by material considerations including the melting temperature.

8.8 Applications of Thermoelectrics

Thermoelectric devices are used to cool electronics, food, and people. Computer CPUs, graphics cards, and other types of electronics all generate heat, and these components can be damaged by excessive heat. Small thermoelectric devices can increase the reliability and lifetime of such components. Thermoelectric refrigerators have been used in RVs and submarines [3]. These devices are often less efficient than traditional refrigerators, but they can be small and quiet and require low maintenance. Some butter and cream dispensers in restaurants use thermoelectric devices to keep perishable foods cool [118], and truck-sized thermoelectric refrigerators are used to keep pharmaceuticals cool [118]. Engineers have tried making air conditioning units out of these devices [110]. They are better for the environment than traditional air conditioning units which require freon or other chemicals. However, they are not often used because the efficiencies are a few percent at best [110]. Thermoelectric devices have also been incorporated into military clothing to keep soldiers cool [118].

Thermoelectric devices are used both to make sensors and to control the temperature of sensing circuits. A *thermocouple* is a small thermoelectric device made from a junction of two materials that is used as a temperature sensor. It converts a small amount of energy from a temperature difference to electricity, and it can be used to measure temperature very accurately. Thermocouples are very common and often inexpensive. Thermoelectric devices are used to cool scanning electron microscopes and other types of imaging devices. Cooling is needed when imaging very small objects because heat causes atoms to vibrate, which can smear out microscopic images. Liquid nitrogen was used to cool imaging devices before thermoelectric devices became available, and it was much less convenient to use. The response of many types of sensors depend on temperature. A thermoelectric device may be part of a control circuit which keeps the sensor at a fixed temperature, so the sensitivity is accurately known.

Thermoelectric devices are used to generate power for satellites and planetary rovers because thermoelectric devices have no moving parts and do not require regular refueling. The Mars rover Curiosity is powered by NASA's Multi-Mission Radioisotope Thermoelectric Generator [119]. Figure 8.4 illustrates its major components. This power supply contains around 10 pounds of plutonium 238 in the form of plutonium dioxide. The plutonium decays naturally and produces heat. The heat interacts with a thermoelectric device and produces electricity, and the electricity is stored in a battery until use. The power supply produces around 2 kW of heat and around 120 W of electrical power, so the overall efficiency is around

8 THERMOELECTRICS

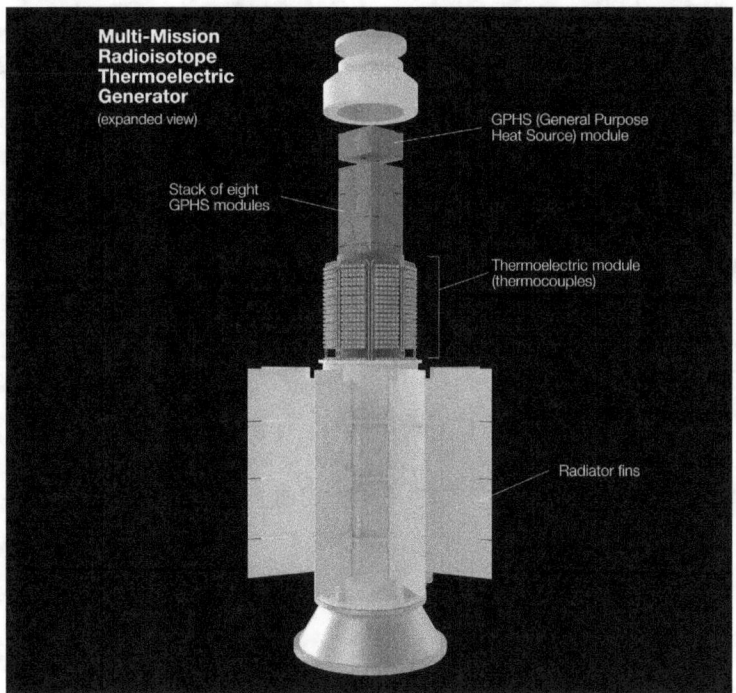

Figure 8.4: Labeled pull-apart view showing the major components of the NASA Multi-Mission Radioisotope Thermoelectric Generator. This figure is used with permission [120].

6% [119]. This technology is not new. The Apollo 12 mission in 1969 used a similar type of power supply, but that supply produced only 70 W and had a lifetime of 5-8 years. Thermoelectric devices have also been used in nuclear power plants as a secondary system to recover some electricity from heat produced [5].

While thermoelectric effects are often fundamental to the operation of sensors and power supplies, the effects are sometimes unwanted [23, p. 457]. Electrical circuits contain junctions of wires made out of different metals. Such a junction occurs, for example, when an aluminum trace on a printed circuit board meets the tin wire of a resistor or when a tin lead solder joint meets a copper wire. The Seebeck effect occurs at all of these junctions. The Seebeck coefficient at a junction of copper and tin lead solder, for example, is 2 $\frac{\mu V}{K}$ [23, p. 457]. These unwanted voltages that develop can introduce noise or distortions into sensitive circuits.

Electrical engineers often think of heat as "wasted energy". Almost every electrical circuit contains resistors which heat up when current flows through them. In some applications, this heating is the desirable outcome.

For example, some train stations have heat lamps for the use in winter, and a concert hall on a winter evening fills with people and heats up from the bodies. However, usually the heat is just considered a waste product or a nuisance.

In the long time limit, systems will reach an equilibrium temperature, but on short time scales, temperature differentials often exist. The inside of a car may be at a hotter temperature than the air outside. The air near an incandescent light bulb may be hotter than air elsewhere in a room, and so on. At one time in the past, we assumed that the earth had a nearly infinite amount of petroleum, coal, and other fossil fuels. Today, we know that these resources are finite. Recently, there has been increased interest in energy harvesting both for environmental reasons and for economic reasons, and thermoelectric devices can be used to convert this heat to usable electricity.

8.9 Problems

8.1. In a 1 mm³ volume, 10^{15} atoms of argon are at a temperature of $T = 300$ K. Calculate the pressure of the gas.

8.2. Argon gas is enclosed in a container of a fixed volume. At $T = 300$ K, the pressure of the gas is 50,000 Pa. At $T = 350$ K, calculate the pressure of the gas.

8.3. A balloon is filled with helium atoms at room temperature, 72 °F. It has a volume of $5 \cdot 10^{-5}$ m³, and the gas in the balloon has a pressure of $10^6 \; \frac{\text{N}}{\text{m}^2}$. How many helium atoms are in the balloon, and what is the mass of the gas?

8.4. A resistive heater is used to supply heat into an insulated box. The heater has current 0.04 A and resistance 1 kΩ, and it operates for one hour. Energy is either stored in the box or used to spin a shaft. If the box gains 2,500 J of energy in that one hour, how much energy was used to turn the shaft?

8.5. Qualitatively, explain the difference between each pair of related quantities.

   (a) Seebeck effect and Peltier effect
   (b) Thermal conductivity and electrical conductivity

8.6. Match the description with the quantity measured.

| A. Electrical conductivity |
| B. Peltier coefficient |
| C. Seebeck coefficient |
| D. Thermal conductivity |
| E. Thomson coefficient |

| 1. A bar is made from a junction of two metals. A current of 1 mA is placed through the bar. The temperature at each end of the bar is measured as a function of time. The rate of heat generated across the bar divided by the current is what quantity? |
| 2. A bar is made from a junction of two metals. One end of the bar is held at a temperature of 20 °C while the other is held at 45 °C. The voltage between the ends of the bar is measured with a voltmeter. This voltage divided by 15 °C is what quantity? |
| 3. One end of a metal bar of is held at 45 °C while the other end is held at 20 °C. A current of 1 mA is placed through the bar. The rate of heat generated across the bar is measured. The rate of heat generated divided by the product of 1 mA and 1 °C is which quantity? |
| 4. One end of a metal bar of cross sectional area A is heated to a temperature of 45 °C. A thermocouple is placed 3 cm down the bar away from the heater. The product of the power dissipated in the heater times 3 cm divided by the product of A and temperature difference measured is what quantity? |
| 5. A current of 1 mA is put through a metal bar of cross sectional area A. The voltage drop across the bar is measured with a voltmeter. The current times the length of the bar divided by the product of the voltage measured and A is what quantity? |

8.7. Explain how to measure each of the following quantities, and list the tools needed to make the measurement.

(a) Electrical conductivity

(b) Thermal conductivity

(c) Peltier coefficient

8 THERMOELECTRICS

8.8. A thermoelectric device has a figure of merit of $Z = 0.7 \text{ K}^{-1}$. A second device is made out of the same semiconducting materials, but it has been doped so that the electrical conductivity is 20% higher. Find the figure of merit of the second device.

8.9. A thermoelectric device is made from a material with resistivity $5 \cdot 10^{-8}$ $\Omega \cdot$m and Seebeck coefficient $8.5 \cdot 10^{-5}$ $\frac{V}{K}$. A cube, 1 cm on each side, was used to determine the thermal conductivity. One side of the cube was heated. At a steady state, the rate of energy transfer by conduction through the cube is 1.8 W. The temperature distribution through the material is linear, and a temperature difference across is measured to be 20 K across the cube. Find the thermal conductivity κ, and find the figure of merit Z for the material.

8.10. As shown in Fig. 8.3, a heater supplies heat to one side of an iron rod. The rod is cylindrical with length 30 cm and radius 2 cm. The heater supplies 2 W of power to the edge of the rod. Iron has a thermal conductivity of $\kappa = 80$ $\frac{W}{m \cdot K}$. Two thermocouples are are spaced 15 cm apart as shown in the figure. What is the difference in temperature (in degrees Celsius) measured between the two thermocouples?

8.11. A thermoelectric device is used to build a small refrigerator that can hold two pop cans. When the device is operating, the cold side of the device is at $T = 10$ °C while the hot side of the device, outside the refrigerator, is at $T = 42$ °C. What is the maximum possible efficiency of this device?

8.12. The cold side of a thermoelectric device, used to generate electricity, is at a temperature of 100 °C. What is the minimum temperature of the hot side of the device needed to achieve an efficiency of $\eta_{eff} = 15\%$?

8.13. The Carnot efficiency describes the limit of the efficiency for some devices. Does it apply to the following types of devices? (Answer yes or no.)

- Hall effect device
- Semiconductor laser
- Photovoltaic device
- Piezoelectric device
- Pyroelectric device
- Thermoelectric device used as a temperature sensor

- Thermoelectric device used as a refrigerator
- Thermoelectric device used to generate electricity for a sensor system

8.14. The figures show Seebeck coefficient $S(T)$, electrical resistivity $\rho(T) = \frac{1}{\sigma(T)}$, and thermal conductivity $\kappa(T)$ plotted versus temperature T for a family of materials known as skutterudites. These materials have the composition $Tl_xCo_{4-y}Fe_ySb_{12}$ and $Tl_xCo_4Sb_{12-y}Sn_y$ where x and y range from zero to 1. The figures used with permission from reference [121]. Recently, scientists have been studying the possibility of making thermoelectric devices from these materials. Using the data in the figures, approximate the thermoelectric figure of merit Z in units K^{-1} at a temperature of $T = 200$ K for the material with $x = 0.1$.

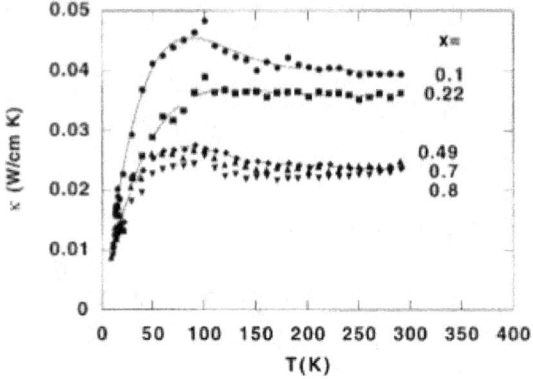

FIG. 4. Total thermal conductivity versus temperature for several Tl-doped alloys: $Tl_{0.1}Co_4Sb_{12}$, $Tl_{0.22}Co_4Sb_{12}$, $Tl_{0.49}Co_4Sn_{0.5}Sb_{11.5}$, $Tl_{0.7}Co_4Sn_{0.75}Sb_{11.25}$, $Tl_{0.8}Co_4SnSb_{11}$.

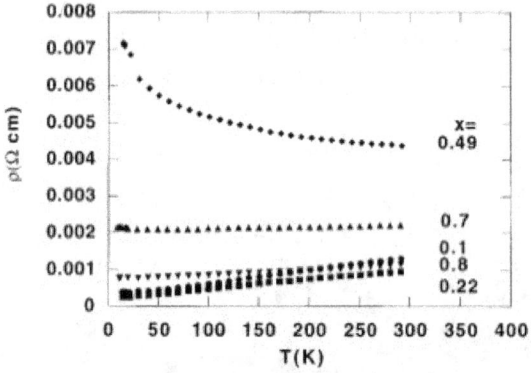

FIG. 6. Electrical resistivity versus temperature for the same Tl-filled skutterudites shown in Fig. 4.

8 THERMOELECTRICS

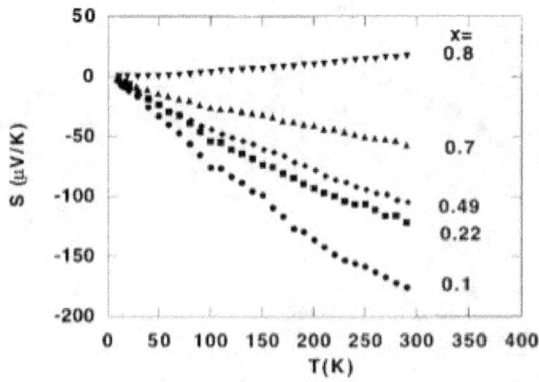

FIG. 7. Seebeck coefficient versus temperature for the same Tl-filled skutterudites shown in Figs. 6 and 4.

Figure 8.5

8.15. Consider the data in the figures of the previous problem over the range $50 < T < 300$ K for the material with $x = 0.1$. At what temperature, within this range, is the product of the figure of merit times the temperature, ZT, the largest, and what is the corresponding value of ZT? Show your work.

8.16. The figures show the Seebeck coefficient, electrical resistivity, and thermal conductivity for three different materials. Assume that we would like to use these materials to build thermoelectric devices which generate electricity where the cold side of the device is slightly below $T \approx 400$ K, and the hot side is slightly above $T \approx 400$ K. The figures are used with permission from reference [122].

(a) Approximately, calculate the Peltier coefficient and the Thomson coefficient for $CeFe_4As_{12}$ near $T \approx 400$ K.

(b) Assume you have a cube of $CeFe_4As_{12}$, 1 cm on each side. What is the resistance R of the cube?

(c) What is the thermoelectric figure of merit Z for $CeFe_4As_{12}$ near $T \approx 400$ K. Include units in your answer.

(d) All else equal, which of the three materials would produce the largest voltage for a given temperature difference. Justify your answer.

(e) Which of the three materials has the largest thermoelectric figure of merit? Justify your answer.

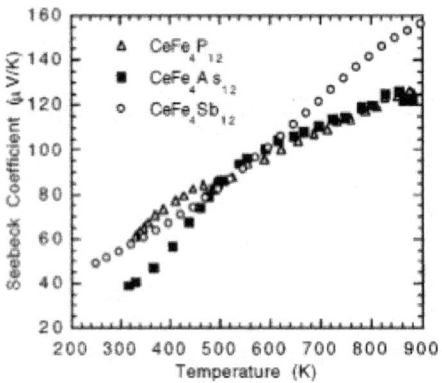

FIG. 2. Seebeck coefficient of $CeFe_4P_{12}$ (see Ref. 3) $CeFe_4As_{12}$, and $CeFe_4Sb_{12}$ (see Ref. 2) as a function of temperature.

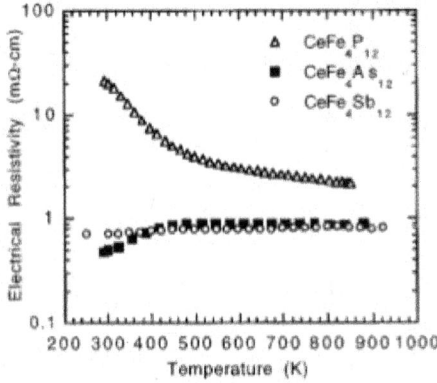

FIG. 1. Electrical resistivity of $CeFe_4P_{12}$, (see Ref. 3), $CeFe_4As_{12}$, and $CeFe_4Sb_{12}$, (see Ref. 2) as a function of temperature.

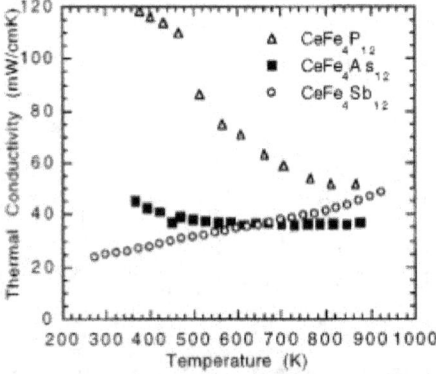

FIG. 3. Thermal conductivity of $CeFe_4P_{12}$ (see Ref. 3), $CeFe_4As_{12}$, and $CeFe_4Sb_{12}$ (see Ref. 2) as a function of temperature.

Figure 8.6

9 Batteries and Fuel Cells

9.1 Introduction

This chapter discusses two related energy conversion devices: batteries and fuel cells. A *battery* is a device which converts chemical energy to electricity, and one or both of the electrodes of the battery are consumed or deposited in the process. A *fuel cell* is a device which converts chemical energy to electricity through the oxidation of a fuel. The fuel, but not the electrodes, is consumed in the operation of a fuel cell. *Oxidation* is the process of losing an electron while *reduction* is the process of gaining an electron. Both batteries and fuel cells contain three main components: an anode, cathode, and electrolyte. The electrode which electrons flow *toward* is called the *cathode*. The electrode which electrons flow *away* from is called the anode *anode*. The *electrolyte* is a material though which ions can flow more easily than electrons.

In many ways, current technology is limited by battery technology. For example, the battery of the Apple iPhone X weighs 42 g and has a specific energy of 246 $\frac{W \cdot h}{kg}$. It accounts for 24 % of the weight of the phone [123] [124]. Similarly, the batteries of the Tesla Model S electric vehicle weigh 580 kg and have an overall specific energy of 141 $\frac{W \cdot h}{kg}$. They account for 27% of the weight of the car [125]. Relatedly, technology companies have been rocked by problems in battery manufacture. In July of 2015, more than half a million hoverboards produced by ten different companies were recalled due to battery explosions [126]. Also, Samsung recalled millions of Galaxy Note 7 smart phones in 2016, costing the company billions of dollars [127]. The batteries were manufactured by one of two different suppliers. Manufacturing issues in batteries produced by both suppliers made the phones susceptible to catching on fire [127].

Due to the importance of battery technology to the consumer product industry, electric vehicle industry, and other technology sectors, money and effort have been pouring into battery research, development, and manufacturing. Rechargeable lithium ion battery development, in particular, is an intense area of effort and investment. All of the examples in the previous paragraph involve these lithium batteries. In 2009, $13 billion worth of lithium batteries were sold, and 163 billion lithium batteries were produced [128, ch. 15]. In 2014 Tesla, one of multiple manufacturers, began construction of a new factory named the Gigafactory. Upon completion, Tesla aims for this facility to be the largest building in the world and for it to annually produce lithium batteries with a combined capacity of 35 gigawatt hours [129]. More recently, industry-wide investment has only grown larger.

9.2 Measures of the Ability of Charges to Flow

The idea of flow of charges is fundamental to both electrical engineering and chemistry. However, electrical engineers and chemists make different assumptions, and they use different notations to describe closely related phenomena. Engineers prefer to work with solids because solids are durable. Electrical engineers assume all discussions involve solids unless otherwise specified. Chemists, however, are quite interested in, and assume all discussions involve, liquids, with special focus on aqueous solutions. Batteries and fuel cells typically involve charge flow through both liquids and solids, so to understand these devices, we have to be familiar with notations and assumptions from both fields of study.

In solid conductors, valence electrons flow. Inner shell electrons are assumed to be so tightly bound to atoms that their movements can be ignored. Nuclei are so much heavier than electrons that their movements can also be ignored. In solid semiconductors, both valence electrons and holes flow. Electrical engineers measure the ability of charges to flow in materials by the electrical conductivity.

Positive and negative ions can flow more easily in liquids than solids, so chemists are concerned with the flow of both electrons and ions. Semiconductor physicists tend to use the terms electrical conductivity, resistivity, Fermi level, and energy gap. Chemists are so interested in the ability of charges to flow that they have many interrelated measures to describe it. We'll discuss the following measures:

- Mulliken electronegativity
- Ionization energy
- Electron affinity
- Electronegativity
- Chemical potential
- Chemical hardness
- Redox potential
- pH

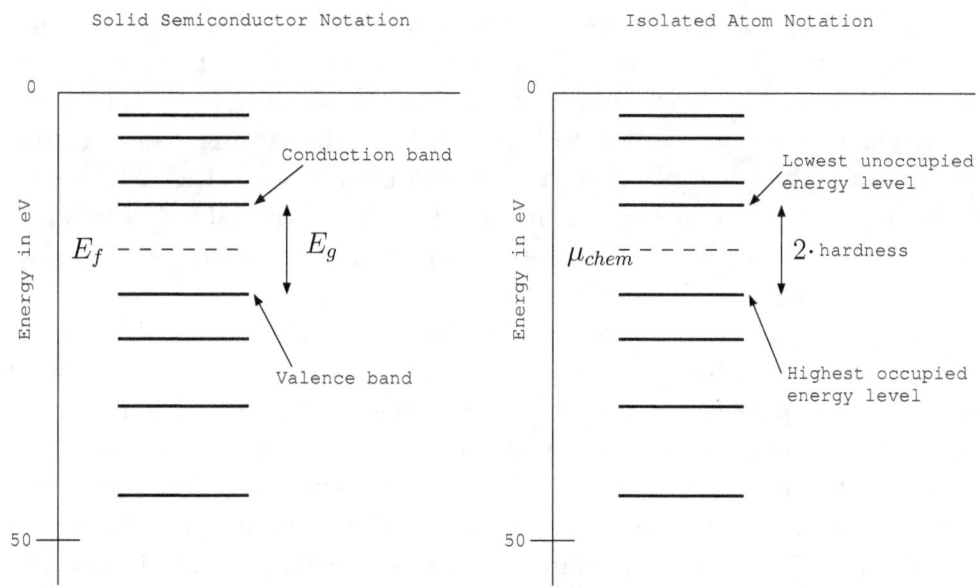

Figure 9.1: An energy level diagrams labeled in two ways.

9.2.1 Electrical Conductivity, Fermi Energy Level, and Energy Gap Revisited

Electrical conductivity σ is measured in units $\frac{1}{\Omega \cdot m}$, and it was discussed in Sec. 8.6.2. The inverse is resistivity $\rho = \frac{1}{\sigma}$ measured in $\Omega \cdot m$. *Electrical conductivity* and *resistivity* are measures of the ability of electrons to flow through a material. As described by Eq. 8.32, electrical conductivity is directly proportional to the number of charges present and the mobility of the charges. Conductors have large electrical conductivity, and insulators have small electrical conductivity. These measures can describe liquids and gases as well as solids. Also, gases, liquids, and solids can all be classified as conductors, dielectrics, or semiconductors.

Fermi energy level, energy gap, valence band, and conduction band were defined in Section 6.3. The left part of Fig. 9.1 shows an energy level diagram zoomed in so that only some levels are shown, and these terms are illustrated in the figure. At $T = 0$ K, energy levels are filled up to some level called the valence band. The energy level above it, which is unfilled or partially unfilled, is called the conduction band. The amount of energy needed to completely remove an electron from the valence band is represented by the vertical distance from that energy level to the ground state, labeled 0 eV, at the top of the figure. The *energy gap*, E_g, is the vertical distance between the valence and conduction bands. It represents the minimum amount of energy needed to excite an electron. The *Fermi energy level* represents the energy level at which the probability of finding an electron is 0.5. At $T = 0$ K, it is at the middle of the energy gap. In the figure, it is shown as a dotted line. Qualitatively, it represents the amount of energy needed to remove the next electron. No electrons have exactly that energy because there are no allowed states in the gap. For a doped semiconductor, a semiconductor with crystalline defects, or a semiconductor not at absolute zero temperature, the Fermi level is near but not quite at the middle of the gap.

The right part of the figure shows the same energy level diagram labeled using terms more commonly used by chemists to describe isolated atoms than by physicists to describe solid semiconductors. Chemists sometimes use the term highest occupied energy level instead of valence band. This term is most often used to refer to energy levels of isolated atoms or molecules because some authors reserve the term band for an energy level shared between neighboring atoms. Similarly, chemists sometimes use the term lowest unoccupied energy level in place of conduction band. As discussed below in Secs. 9.2.3 and 9.2.4, the term chemical potential μ_{chem} is used in place of Fermi energy level E_f, and the energy gap E_g may be called twice the chemical hardness.

9.2.2 Mulliken Electronegativity

One measure that chemists use to describe the ability of charges to flow is electronegativity, and this term has multiple definitions in the literature. One definition is by Mulliken in 1934 [130], and this measure will be referred to as the Mulliken electronegativity. Mulliken approximated the energy in a chemical bond by averaging the ionization energy I_{ioniz} and the electron affinity A_{aff}. *Mulliken electronegativity* is defined

$$\chi_{Mulliken} = \frac{I_{ioniz} + A_{aff}}{2}. \qquad (9.1)$$

Ionization energy is the energy needed to remove an electron from an atom or ion, and *electron affinity* is the energy change when an electron is added to an atom or ion [12]. All of these quantities, $\chi_{Mulliken}$, I_{ioniz}, and A_{aff}, are measured in the SI units of $\frac{J}{atom}$ or occasionally in other units like $\frac{eV}{atom}$ or $\frac{kJ}{mol}$.

This definition is simpler than other definitions of electronegativity, and reference [131] calls this an "operational and approximate" definition. It is useful because it involves strength of chemical bonds, and we can relate it to the measures used by semiconductor researchers. Qualitatively, ionization energy is represented by the energy needed to rip off an electron. In Fig.

9.1, it is the vertical distance from the valence band or highest occupied state to the ground state at the top of the figure. Sometimes chemists call this amount of energy the *work function* instead [60, ch. 6] [108]. In Fig. 9.1, the electron affinity is represented by the vertical distance from the conduction band or lowest unoccupied state to the ground state at the top of the figure. The magnitude of the Mulliken electronegativity is the average of these two energies, so it is the magnitude of the Fermi energy at $T = 0$ K. By convention, it has the opposite sign.

$$\chi_{Mulliken} = -E_f|_{T=0} \text{ K} \tag{9.2}$$

Fundamentally, electrical engineering is the study of flow of charges. Chemistry is the study of the strength of chemical bonds. The electrical conductivity of a material is high when the chemical bonds holding that material together are easily broken so that many free charges can flow. The electrical conductivity of a material is low when chemical bonds holding atoms together require lots of energy to break. Electronegativity is a measure of the energy required to break chemical bonds, so fundamentally, it tells us similar information to electrical conductivity.

9.2.3 Chemical Potential and Electronegativity

Another way of defining electronegativity follows the definition introduced by Pritchard in 1956 [132]. This definition is one of the more common ones, and it is used by both chemists [131] [133] and by other scientists [2, p. 124.]. The *electronegativity* of an atom is defined as

$$\chi = -\left(\frac{\partial \mathbb{U}}{\partial N}\right)\bigg|_{V,S,} \tag{9.3}$$

where $\mathbb{U}$ is the internal energy relative to a neutral atom and N represents the number of electrons around the atom. An atom is composed of a charged nucleus and charged electrons moving around the nucleus, so there is an electric field, and hence an electrical potential V in volts, around an atom. This potential significantly depends on the number of electrons around the atom. Also, when the atom is at a temperature above absolute zero, the electrons and nuclei are in motion, so the atom has some entropy S. Electronegativity involves $\frac{\partial \mathbb{U}}{\partial N}$ at constant electrical potential and entropy. It applies whether the atoms are part of a solid, liquid, or gas.

The *chemical potential* μ_{chem} is defined as the negative of this electronegativity.

$$\mu_{chem} = -\chi \tag{9.4}$$

In SI units, both chemical potential and electronegativity are measured in $\frac{J}{atom}$, but sometimes they are also expressed in $\frac{eV}{atom}$ or $\frac{kJ}{mol}$. As if the three names, chemical potential, negative of the electronegativity, and Fermi energy level, weren't enough, this quantity is also known as the *partial molar free energy* [60, p. 145].

Electronegativity is used to describe a collection of atoms, molecules, or ions all of the same ionization state [131]. Less energy is required to rip the first electron off an atom than the second or third electron. The definition of electronegativity is specific to potential V, in volts, due to the nucleus and electrons around an atom. For example, we can talk about the electronegativity, energy required to rip off the electron, of a neutral magnesium atom. We can also talk about the electronegativity, energy required to rip off an electron, from a Mg^+ ion. The electric field, and hence potential V, around a neutral Mg atom and the electric field, and hence potential V, around a magnesium ion Mg^+ are necessarily different because of the number of electrons present. The energies required to rip off the next electron from these atoms will also necessarily be different. So, electronegativity of a material always refers to a specific ionization state. Electronegativity incorporates both the energy required or gained by ripping off an electron and the energy required or gained by acquiring an electron. Qualitatively, it is the average of the ionization energy required to rip off an electron and the electron affinity released when an electron is captured. In the case of the Mg atom from the example above, the energy gained by releasing an electron is the significant term, but that is not always the case.

In most energy conversion devices, and most chemical reactions, we are interested in only the valence electrons. So, even if an atom has dozens of electrons around it and the energy to rip off each electron is different, we are just interested in the first few valence electrons. We will see that batteries and fuel cells involve energy stored in chemical bonds. Only the valence electrons are involved in the reactions of batteries and fuel cells, so in studying batteries and fuel cells, we are most interested in the electronegativity of neutral or singly ionized atoms.

Equation 9.3 defines electronegativity as the energy required to rip off the next electron from the atom. Again consider Fig. 9.1. The energy level known as the valence band to semiconductor physicists and the highest occupied state to chemists is filled with electrons. The next highest band, called either the conduction band by semiconductor physicists or lowest unoccupied state by chemists, is not filled with electrons. The electronegativity according to this definition is the energy required to rip off the next electron. On average, it is again graphically represented by the Fermi level.

9 BATTERIES AND FUEL CELLS

Both electronegativity defined by Eq. 9.3 and Mulliken electronegativity defined by Eq. 9.1 have the same units. However, multiple other definitions of electronegativity can be found in the literature. One of the oldest definitions is due to Pauling in 1932 [134], and that definition is measured instead in the units of square root of joules on a relative scale. Reference [135] expanded on Pauling's definition to show variation with ionization state and atom radius. Reference [133] also contains a different definition of electronegativity also with its own units.

9.2.4 Chemical Hardness

Chemists sometimes use the term hardness when semiconductor physicists would use the term half the energy gap. Chemical hardness has nothing to do with mechanical hardness. As with electronegativity, there are multiple related definitions of hardness. The Mulliken hardness is defined as [131]

$$\text{Mulliken hardness} = \frac{I_{ioniz} - A_{aff}}{2}. \quad (9.5)$$

A more careful definition of *chemical hardness* is [131] [136, p. 93]

$$\text{hardness} = \frac{1}{2}\left(\frac{\partial \mu_{chem}}{\partial N}\right)\bigg|_{V,S}. \quad (9.6)$$

It is half the change in chemical potential for the next electron, and qualitatively it is represented by half the energy gap. As with electronegativity, it is specified for a given potential in volts around the atom and a given entropy. Liquids may be classified as hard or soft. Hard acids and hard bases have large energy gaps, so they are electrical insulators. Soft acids and soft bases have small energy gaps, so they are electrical conductors. No additional variable will be introduced for hardness because this quantity can be represented by half the energy gap, $\frac{E_g}{2}$.

9.2.5 Redox Potential

Redox (from REDuction-OXidation) potential V_{rp} is yet another measure used by chemists to describe the ability of electrons to be ripped off their atoms and flow in the presence of an applied voltage, nearby chemical, optical field, or other energy source. As defined above, the process of ripping off electrons is called *oxidation*. The process of gaining electrons is called *reduction*. Together, they form *redox reactions*. Instead of being measured in joules like electronegativity, it is measured in volts where a volt is a joule per coulomb. Redox potential represents the energy stored in a chemical bond per unit charge. It is more often used by experimentalists than theorists, and it is often used to describe solids instead of liquids. Redox potential is a macroscopic property, describing a larger piece of material as opposed to describing just an individual atom. It is also sometimes called oxidation reduction potential or the standard electrode potential [137]. It is a relative measure of the ability of a substance to lose an electron. A list of redox potentials can be found in references [60, p. 158] and [137]. There are different ways of defining redox potential in the literature. The definitions vary in their choice of a ground reference voltage, and they vary in their sign conventions. American and European researchers tend to use different definitions.

Redox potential is measured on a relative scale. To measure redox potential [138], electrodes are put in the system being studied. A potential is applied to balance the internal voltage. By measuring this externally applied voltage, the potential of an electrode is determined with respect to a reference electrode. Often, the potential of a platinum electrode is used as a reference and said to have zero volts at standard conditions of $T = 25\ ^0\text{C}$ and $\mathbb{P} = 1$ atm. The reaction at the platinum electrode is

$$H_2 \to 2H^+ + 2e^-. \tag{9.7}$$

9.2.6 pH

pH is a unitless measure of the likelihood that a water molecule is bonded or has been ionized in a liquid solution. It is used to classify liquids as acidic or basic. When discussing pH, we assume the material under test is a liquid solution at a temperature of 25 °C and a pressure of 1 atm [12] [81]. A liquid solution is a mixture of water and another material called a *solute*. More specifically, *pH* is defined as

$$\text{pH} = \log_{10}\left(\frac{1}{[\text{H}^+]}\right). \tag{9.8}$$

The quantity $[\text{H}^+]$ is the *amount concentration of hydrogen ions* in the units of $\frac{\text{mol}}{\text{L}}$ [68, p. 39].

$$[\text{H}^+] = \frac{\text{concentration H}^+\text{ions, mol}}{\text{volume of solution in L}} \tag{9.9}$$

This quantity was formerly called molarity or molar concentration, but these terms are no longer recommended for use [68, p. 39]. pH is a measure often used by experimentalists.

The concept of pH is fundamentally related to the flow of charges, a concept which is very important to electrical engineers. Water is composed of H_2O molecules. In pure water, some of these molecules fall apart, ionizing into H^+ ions (protons) and OH^- ions. However, most of the molecules remain intact. If some solutes are mixed with the water, more of the H_2O molecules will ionize than in pure water. For example, carbon dioxide will bond with OH^- ions forming carbonic acid HCO_3^- causing an increase in H^+ ions. Since pH is the negative log of H^+ ion activity, increasing H^+ ion concentration is equivalent to a pH decrease. If ammonia, NH_3, is added to water, NH_4^+ is formed, and the number of OH^- ions increases resulting in an increase in pH. Since water is a liquid, both these positive and negative ion charge carriers can move about relatively easily. If an external voltage is applied across the liquid, ions will flow. Electrical conductivity will be higher in a liquid with more ions present than in liquids with fewer ions present.

As an example, consider what happens when neutral sodium atoms are added to water. (For obvious reasons, don't try this at home [139].) It is energetically favorable for the sodium atoms to ionize to Na^+ giving up an electron. In the process, some more water molecules ionize, and some H^+ ions become neutral H atoms.

$$Na + H_2O \rightarrow Na^+ + H + OH^- \quad (9.10)$$

By adding the solute sodium, the solution has fewer H^+ ions.

Consider what happens when neutral chlorine atoms are added to water. It is energetically favorable for a chlorine atom to acquire an electron to a form Cl^- ion.

$$Cl + H_2O \rightarrow Cl^- + H^+ + OH \quad (9.11)$$

By adding the solute chlorine, the solution has more H^+ ions. While these examples involve adding neutral atoms, the concept of pH applies to solutes which are molecules too.

Solutions with pH less than 7 are called *acidic*. If a solution has a high concentration of H^+ ions, it will have a low pH and be acidic. In strongly acidic solutions, molecules of the solute rip apart many water molecules, so lots of ions are present. Solutions with pH greater than 7 are called *alkaline* or *basic*. If a solution has a low concentration of H^+ ions, and hence a high concentration of OH^- ions, it will have a high pH and be alkaline. In strongly alkaline solutions, molecules of the solute rip apart many water molecules, so again lots of ions are present. Neutral solutions have a pH near 7, and some neutral solutions may be electrical insulators. Solutions with a pH much below or much above 7 necessarily have many ions present, and they are good electrical conductors.

As an example, let's find the pH of a solution with 10^{15} ions of H^+ in 1 L of water.

$$[H^+] = \frac{10^{15} \text{ ions}}{1 \text{ L}} \cdot \frac{1}{6.022 \cdot 10^{23} \frac{\text{ions}}{\text{mol}}} = 1.66 \cdot 10^{-9} \frac{\text{mol}}{\text{L}} \qquad (9.12)$$

$$\text{pH} = \log\left(\frac{1}{1.66 \cdot 10^{-9}}\right) \approx 9 \qquad (9.13)$$

Notice that the exponent of $[H^+]$ is -9, and the pH is 9. Qualitatively, the pH tells us the negative of the order of magnitude of the amount concentration of hydrogen ions. The solution in this example is alkaline.

As a related example, let's find the pH of a solution with 10^{20} ions of H^+ in 1 L of water.

$$[H^+] = \frac{10^{20} \text{ ions}}{1 \text{ L}} \cdot \frac{1}{6.022 \cdot 10^{23} \frac{\text{ions}}{\text{mol}}} = 1.66 \cdot 10^{-4} \frac{\text{mol}}{\text{L}} \qquad (9.14)$$

$$\text{pH} = \log\left(\frac{1}{1.66 \cdot 10^{-4}}\right) \approx 4 \qquad (9.15)$$

This example has more hydrogen ions in the solution, so it is more acidic. The pH of 4 tells us that the solution has approximately $10^{-4} \frac{\text{mol}}{\text{L}}$ of hydrogen ions.

How many hydrogen ions are found in a 1 L solution with a pH of 7?

$$7 = \log\left(\frac{1}{[H^+]}\right) \qquad (9.16)$$

$$[H^+] = 10^{-7} \frac{\text{ions}}{\text{mol}} \qquad (9.17)$$

$$10^{-7} \frac{\text{mol}}{\text{L}} \cdot 1 \text{ L} \cdot 6.022 \cdot 10^{23} \frac{\text{ions}}{\text{mol}} = 6.022 \cdot 10^{16} \text{ ions } H^+ \qquad (9.18)$$

A neutral solution, with a pH of 7, still contains H^+ ions.

9.3 Charge Flow in Batteries and Fuel Cells

9.3.1 Battery Components

The flow of both positive and negative charges must be considered to understand the operations of batteries and fuel cells. The simplest battery contains just an anode, cathode, and electrolyte. These components are illustrated in Fig. 9.2.

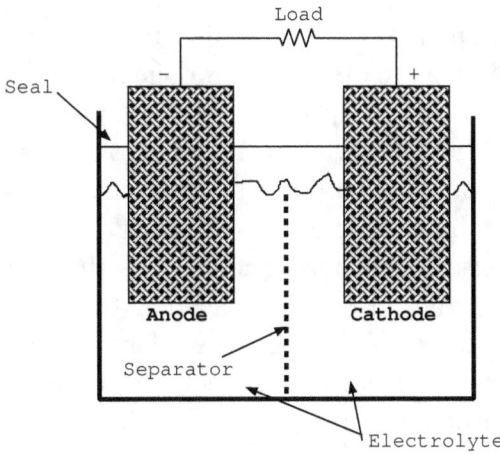

Figure 9.2: Battery components.

Both of the electrodes must be good conductors. They are often porous to increase the surface area where the reaction occurs. The cathode is a sink for electrons and positive ions, and both of these types of charges are attracted towards this terminal. The cathode is the positive electrode of a discharging battery. The anode is source for electrons and positive ions, and both of these types of charges flow away from the anode. The anode is the negative electrode of a discharging battery.

The electrolyte has high ionic conductivity but low electrical conductivity. For this reason, during discharge of a battery, ions flow from the anode to the cathode through the electrolyte. Meanwhile, electrons are forced to flow from the anode to the cathode through the load. The electrolyte is often a liquid but sometimes a thin solid. Batteries are contained in a package. If the electrolyte is liquid, a seal is included to prevent it from spilling or escaping [140].

Most batteries also contain a separator, which is typically made from a thin polymer membrane [140]. The separator allows some but not other ions to flow through, and it is a physical barrier that prevents the electrodes from contacting and shorting out the battery.

Battery components	Optional components for protection
Anode	Diode
Cathode	Fuse or circuit breaker
Electrolyte	Vent
Separator	Microcontroller
Seal	Thermocouple

Table 9.1: Battery Components.

Additional components are often added to improve device safety, and Table 9.1 lists some of these optional components. A user may mistakenly insert a battery backwards. To prevent damage due to this error, some batteries incorporate a diode [128, ch. 5.1]. The voltage across the terminals of a battery with an internal diode will necessarily be less than the voltage across an equivalent battery without the diode present. Other batteries, like typical 9V batteries, incorporate connectors that can only be attached one way. A battery may also be damaged if the terminals are shorted. Most batteries include vents so gases can safely escape when a battery is damaged due to shorting the terminals, attempting too much current draw, or overheating for other reasons [128, ch. 5.1]. Some batteries include a fuse or circuit breaker in the package to prevent damage in these cases too. Additionally, rechargeable batteries can be damaged if the recharging process is not properly controlled [128, ch. 5.1]. Some rechargeable batteries have a thermocouple and microcontroller built into the package to control the recharging process and prevent overheating during recharging [128, ch. 5.1]. Users should not try to recharge nonrechargeable batteries. While the chemical reaction can often go in either direction, the package and structure of a primary battery are not designed to withstand the charging process and will typically be damaged [128, ch. 5.1].

9.3.2 Charge Flow in a Discharging Battery

As a battery discharges, chemical energy stored in the bonds holding together the electrodes is converted to electrical energy in the form of current flowing through the load. Consider an example battery with a magnesium anode and a nickel oxide cathode. The reaction at the anode is given by

$$Mg + 2OH^- \rightarrow Mg(OH)_2 + 2e^- \qquad (9.19)$$

which has a redox potential of $V_{rp} = 2.68$ V [137] [140]. The reaction at the cathode is given by

$$NiO_2 + 2H_2O + 2e^- \rightarrow Ni(OH)_2 + 2OH^- \qquad (9.20)$$

9 BATTERIES AND FUEL CELLS

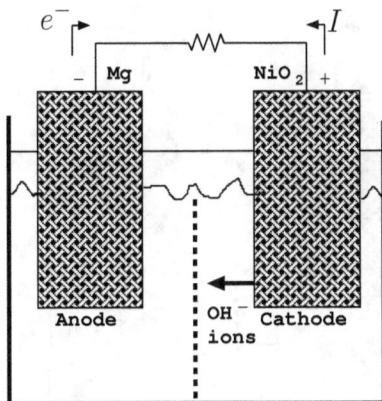

Figure 9.3: Charge flow in a discharging battery.

which has a redox potential of $V_{rp} = 0.49$ V [140]. The overall reaction is given by

$$\text{Mg} + \text{NiO}_2 + 2\text{H}_2\text{O} \rightarrow \text{Mg(OH)}_2 + \text{Ni(OH)}_2$$

This reaction occurs in alkaline solutions that contain OH^- ions available to react, so an electrolyte such as potassium hydroxide, KOH, can be used [140]. Other reactions may simultaneously occur at these electrodes [137], but for simplicity these other reactions will be ignored.

Figure 9.3 illustrates the charge flow in the battery during normal operation. A complete circuit is formed not just by the flow of electrons but by a combination of the flow of electrons and ions [128]. Electrons flow away from the negative terminal (anode) through the load. Negative OH^- ions flow away from the positive terminal (cathode) through the electrolyte. The separator should allow the OH^- to flow from the positive terminal to the negative terminal. For some electrodes, though not in this example, positive ions, instead of negative ions, complete the circuit by flowing away from the negative terminal. As shown in the figure, the direction of current flow is opposite to the direction of electron flow. The battery continues to discharge until one of the electrodes is used up [3, p. 226].

9.3.3 Charge Flow in a Charging Battery

Figure 9.4 illustrates the flow of charges when the battery is charging. During charging, energy is converted from electrical energy due to the external voltage source back to chemical energy stored in the chemical bonds holding together the electrodes. Again, the flow of both electrons and ions, not just electrons, must be considered. As above, the direction of the current

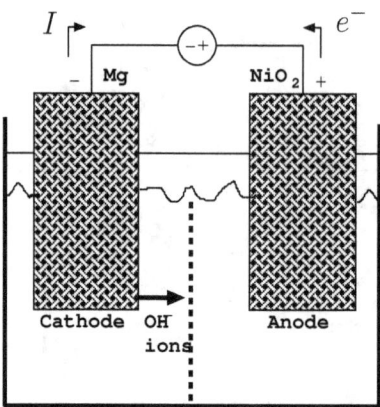

Figure 9.4: Charge flow in a charging battery.

is the opposite of the direction of the flow of electrons. Reactions occurring are the opposite of the reactions given by Eqs. 9.19 and 9.20. By definition, the cathode is the electrode which electrons flow towards, and the anode is the electrode which electrons flow away from. During charging, unlike during discharging, the cathode is the negative terminal and the anode is the positive terminal. For this example, the reaction at the cathode is

$$\text{Mg(OH)}_2 + 2e^- \rightarrow \text{Mg} + 2\text{OH}^- \qquad (9.21)$$

and the reaction at the anode is

$$\text{Ni(OH)}_2 + 2\text{OH}^- \rightarrow \text{NiO}_2 + 2\text{H}_2\text{O} + 2e^-. \qquad (9.22)$$

In this example, OH$^-$ ions flow away from the cathode during charging. However, in some reactions, both the flow of negative ions away from the cathode and positive ions away from the anode must be considered during charging.

9.3.4 Charge Flow in Fuel Cells

A fuel cell contains many of the same components as a battery [3, p. 226] [128, p. 376] [141]. Like a battery, a fuel cell contains an anode and a cathode. These electrodes must be good conductors, and they are often porous so that they have a large surface area. Electrodes are in a liquid or solid electrolyte through which ions can flow. The electrodes are often coated in a catalyst, such as platinum, to speed up chemical reactions [141]. A fuel cell contains a separator, typically called a membrane, which selectively allows ions to flow. As with the separator of a battery, it is

9 BATTERIES AND FUEL CELLS

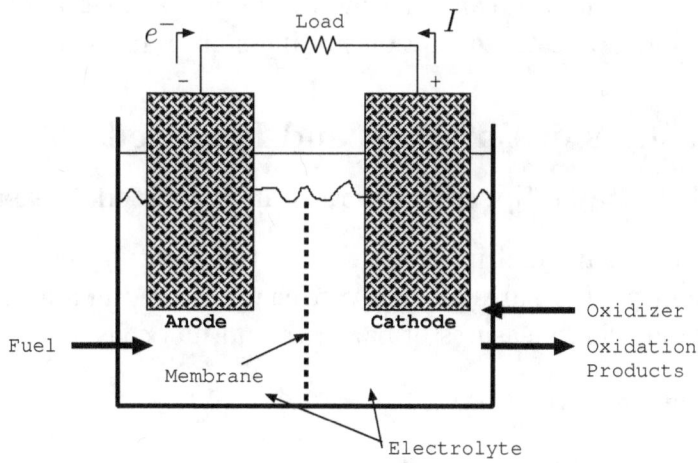

Figure 9.5: Charge flow in a fuel cell.

typically made from a thin polymer. Fuel is added at the anode, and an oxidizer is added at the cathode. Typically, both the fuel and oxidizer are liquids or gases. They get consumed during operation while the anode and cathode are not consumed as they are in a discharging battery. These components are illustrated in Fig. 9.5.

As an example, a fuel cell may use H_2 gas as the fuel and O_2 gas as the oxidizer. The anode may be carbon cloth [141], and this reaction is sped up by a platinum catalyst [108]. An alkaline solution such as KOH can be the electrolyte. For this fuel in an alkaline electrolyte, the reaction at the anode is

$$H_2 + 2OH^- \rightarrow 2H_2O + 2e^- \tag{9.23}$$

and the reaction at the cathode is

$$\frac{1}{2}O_2 + 2e^- + H_2O \rightarrow 2OH^- \tag{9.24}$$

[108].

Figure 9.5 also illustrates charge flow in an example fuel cell [3, p. 226] [128, p. 376] [141]. Oxidation, the process of ripping electrons off the fuel leaving positive ions, occurs at the anode. These electrons flow from the anode to the cathode through the load. At the cathode, the oxidizer is reduced. In other words, at the cathode, the oxidizer reacts incorporating these electrons to form negative ions. These negative ions flow from the cathode to the anode, and positive ions flow from the anode to the cathode. The membrane prevents charges from flowing in the reverse direction, and it prevents the positive ions and negative ions from combining with each

9.4 Measures of Batteries and Fuel Cells

other directly. A fuel cell can continue to operate as long as the fuel and oxidizer are added and the oxidation products are removed.

9.4 Measures of Batteries and Fuel Cells

9.4.1 Cell Voltage, Specific Energy, and Related Measures

Just as chemists have multiple measures of the ability of charges to flow, they have multiple measures of energy or charge stored in a device. In this section, the following measures of batteries and fuel cells are defined:

- Cell voltage in volts

- Specific energy in $\frac{J}{g}$ or $\frac{W \cdot h}{kg}$

- Energy density in $\frac{J}{m^3}$ or $\frac{W \cdot h}{L}$

- Capacity in mA·h or C

- Specific capacity in $\frac{mA \cdot h}{g}$ or $\frac{C}{kg}$

- Charge density in $\frac{mA \cdot h}{L}$ or $\frac{C}{L}$

Definitions throughout this section follow references [128, ch. 1] and [140].

If these measures are calculated using knowledge of chemical reactions and quantities found in the periodic table, they are called *theoretical* values. If these quantities are experimentally measured, they are called *practical* values. Practical values are necessarily less because no energy conversion device is ever completely efficient. Measures preceded by the word *specific* are given per unit mass. Measures followed by the word *density* are give per unit volume. For example, specific energy is measured in the SI units of joules per gram and energy density is measured in the SI units of joules per meter cubed. However, these rules are not closely followed, so the term energy density is sometimes used to mean energy per unit weight instead of per unit volume. It is safest to explicitly specify the units of measure to avoid this confusion.

Theoretical *cell voltage*, V_{cell} measured in volts, is the voltage between the anode and the cathode in a battery or fuel cell. It is the sum of the redox potential for the half reaction at the anode and the redox potential for the half reaction at the cathode. It represents the voltage between the terminals of a completely charged battery or fuel cell. Many authors call this measure theoretical cell potential instead of cell voltage, and symbols

E^0 or Ξ^0 are also used in the literature. As discussed in Appendix C, the word potential is overloaded with multiple meanings. The word voltage and the symbol V_{cell} are used here to emphasize that this quantity is essentially voltage. Since redox potentials for many half reactions are tabulated [128, app. B] [137], theoretical cell voltage can be quickly calculated for many reactions. While we can calculate the theoretical cell voltage, we can measure the practical cell voltage with a voltmeter. The theoretical cell voltage will always be slightly larger than the practical cell voltage because the theoretical cell voltage ignores a number of effects including internal resistance and other factors discussed in the next section. Reactions with $V_{cell} > 0$ occur spontaneously [12, ch. 18].

Three related measures are capacity, specific capacity, and charge density. *Capacity* is measured in ampere hours or coulombs. (By definition, one ampere is equal to one coulomb per second.) It is a measure of the charge stored in a battery or fuel cell. *Specific capacity* is a measure of the charge stored per unit mass. It is specified in $\frac{mA \cdot h}{g}$, $\frac{C}{kg}$, or related units. *Charge density* is a measure of the charge stored per unit volume, and it is specified in $\frac{mA \cdot h}{L}$, $\frac{C}{m^3}$, or related units. While capacity depends on the amount of material present, specific capacity and charge density do not. All of these measures may be specified as theoretical values calculated from knowledge of the chemical reactions involved or practical values measured experimentally where the theoretical values are always slightly higher. Also for all of these values, only valence electrons are considered. Batteries and fuel cells necessarily have more electrons than are included in these measures because inner shell electrons, which do not participate in the chemical reaction, are ignored. Energy is stored in the bonds holding inner shell electrons, but this energy is not converted to electricity in batteries or fuel cells. The concept of charge density, ρ_{ch} in units $\frac{C}{m^3}$, was first introduced in section 1.6.1, and it shows up in Gauss's law, one of Maxwell's equations. However, the word capacity has nothing to do with the word capacitance introduced earlier. See Appendix C for more information on this and other overloaded terms.

Theoretical *specific energy* is measured in $\frac{J}{g}$, $\frac{W \cdot h}{kg}$, or related units [128, ch. 1]. It is a measure of the energy stored in a battery or fuel cell per unit weight. It is the product of the theoretical cell voltage and the specific charge. Relatedly, theoretical *energy density*, measured in $\frac{J}{m^3}$ or $\frac{W \cdot h}{L}$, is a measure of the energy stored in a device per unit volume. Theoretical energy density is the product of theoretical cell voltage and charge density. These measures can be calculated from knowledge of the chemical reactions involved using information found in the periodic table. Practical specific energy and practical energy density are typically 25-35% below the theoretical values [128, ch. 1.5]. Specific energy and energy density are important measures of a battery. Often, high values are desired so that small and light batteries can be used to power devices for as long as possible. However, as specific energy and energy density increase, safety considerations increase.

Chemists sometimes define the charge in a mol of electrons as the Faraday constant. It has the value

$$\frac{6.022 \cdot 10^{23} \text{ atoms}}{1 \text{ mol}} \cdot \frac{1\ e^-}{\text{atom}} \cdot \frac{1.602 \cdot 10^{-19}\ \text{C}}{e^-} = 9.649 \cdot 10^4\ \frac{\text{C}}{\text{mol}} \quad (9.25)$$

[68]. This quantity will not be used below because the Avogadro constant N_a and the magnitude of the charge of an electron q are already specified and because this text already has too many variables.

We can calculate the theoretical specific capacity in $\frac{\text{A} \cdot \text{h}}{\text{g}}$ and the theoretical specific energy in $\frac{\text{J}}{\text{g}}$ for the reactions given by Eq. 9.19 and 9.20. The redox potential for the Mg half reaction is $V_{rp} = 2.68$ V, and the redox potential for the Ni half reaction is $V_{rp} = 0.49$ V [140] [137]. The overall cell voltage is

$$V_{cell} = 2.68 + 0.49 = 3.17 \text{ V}. \quad (9.26)$$

The reaction occurs spontaneously when it is set up because $V_{cell} > 0$.

By unit conversions, we can calculate the weight per unit charge for each half reaction. From the periodic table, the atomic weight of Mg is 24.31 $\frac{\text{g}}{\text{mol}}$, the atomic weight of Ni is 58.69 $\frac{\text{g}}{\text{mol}}$, and the atomic weight of O is 16.00 $\frac{\text{g}}{\text{mol}}$. First consider the Mg half reaction of Eq. 9.19 which involves two valence electrons.

$$24.31\ \frac{\text{g}}{\text{mol}} \cdot \frac{1 \text{ mol}}{6.022 \cdot 10^{23} \text{ atoms}} \cdot \frac{1 \text{ atom}}{2 \text{ valence } e^-} \cdot \frac{1\ e^-}{1.602 \cdot 10^{-19}\ \text{C}} \cdot \frac{1\ \text{C}}{1\ \text{A} \cdot \text{s}} \cdot \frac{3600\ \text{s}}{1\ \text{h}}$$
$$= 0.454\ \frac{\text{g}}{\text{A} \cdot \text{h}}$$
$$(9.27)$$

Next, consider the Ni half reaction of Eq. 9.20 which also involves two valence electrons. The weight of NiO_2 is 90.69 $\frac{\text{g}}{\text{mol}}$.

$$90.69\ \frac{\text{g}}{\text{mol}} \cdot \frac{1 \text{ mol}}{6.022 \cdot 10^{23} \text{ atoms}} \cdot \frac{1 \text{ atom}}{2 \text{ valence } e^-} \cdot \frac{1\ e^-}{1.602 \cdot 10^{-19}\ \text{C}} \cdot \frac{1\ \text{C}}{1\ \text{A} \cdot \text{s}} \cdot \frac{3600\ \text{s}}{1\ \text{h}}$$
$$= 1.69\ \frac{\text{g}}{\text{A} \cdot \text{h}}$$
$$(9.28)$$

For the overall reaction,

$$0.454 + 1.692 = 2.146\ \frac{\text{g}}{\text{A} \cdot \text{h}}. \quad (9.29)$$

9 BATTERIES AND FUEL CELLS

The overall theoretical specific capacity is the inverse of this quantity.

$$\frac{1}{2.146} = 0.466 \; \frac{\text{A·h}}{\text{g}} \tag{9.30}$$

Adding charge densities for each half reaction does not make sense, but we can sum the terms for weight per unit charge in unit $\frac{\text{g}}{\text{A·h}}$.

We can calculate the theoretical specific energy by multiplying the theoretical cell voltage and the theoretical specific capacity.

$$3.17 \text{ V} \cdot 0.466 \; \frac{\text{A·h}}{\text{g}} = 1.48 \; \frac{\text{W·h}}{\text{g}} \tag{9.31}$$

The theoretical specific energy can be converted to the units $\frac{\text{J}}{\text{g}}$.

$$1.48 \; \frac{\text{W·h}}{\text{g}} \cdot \frac{1 \text{ J}}{1 \text{ W·s}} \cdot \frac{3600 \text{ s}}{1 \text{ h}} = 5.32 \cdot 10^3 \; \frac{\text{J}}{\text{g}} \tag{9.32}$$

In the calculation above, only the electrode weights were considered. However, the package, separator, and other battery components all have some mass which contribute to the weight of the battery.

9.4.2 Practical Voltage and Efficiency

We can model both a battery and a fuel cell as an ideal voltage source. This is a useful model, but at times, it is not good enough for multiple reasons. A better model includes some internal resistance [128, p. 9.27]. However, even this model is inadequate because the voltage of any practical battery depends on temperature, the load, the current through the battery, the fraction of capacity used, the number of times it has been recharged, and other factors [128, p. 3.2]. An even better model includes these variations too, as shown in Fig. 9.6.

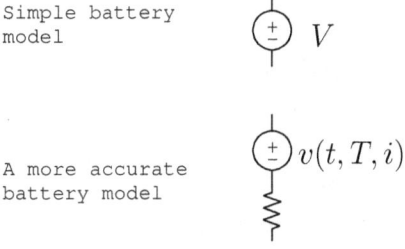

Figure 9.6: Models of a battery.

There are many measures used to describe the voltage across a battery or fuel cell. The *nominal voltage* is the typical voltage during use, and it is often the voltage printed on the label. The *end* or *cutoff voltage* is the voltage at the end of the battery's useful life. The *open circuit voltage* is the voltage under no load, and it is approximately the initial voltage of the battery. The *closed circuit voltage* is the voltage under load. It is less than the open circuit voltage due to the internal resistance of the battery [128, p. 3.2].

All batteries and fuel cells have some internal resistance. The cathode and anode are made of metals which are good, but imperfect, conductors. For example, carbon is a common electrode material, and it has an electrical conductivity between $1.6 \cdot 10^4$ and $2.0 \cdot 10^7$ $\frac{1}{\Omega \cdot m}$ [106]. Anytime current flows through a physical material with finite electrical conductivity, energy is converted to heat. Actual voltage is a function of current drawn from the battery because at high currents, this effect is larger. Also, the actual voltage is a function of temperature because ions move faster at higher temperatures, so there is less internal resistance at higher temperatures [128, p. 3.9]. However, at higher temperatures, chemical reactions may occur more quickly, so the life of the batteries may be less because reactions occur faster.

The actual voltage across a battery or fuel cell is also influenced by the accumulation of chemical reaction products. In the example given by Eqs. 9.19 and 9.20, the reactants were Mg and NiO_2 and the reaction products were $Mg(OH)_2$ and $Ni(OH)_2$. The actual voltage across the device decays with use because reactants build up in the electrolyte as the reaction occurs. These reactants inhibit further reactions from taking place [128, p. 3.2].

The effect of the accumulation of products on the voltage of a battery can be modeled by

$$V_{cell\ theor} - V_{cell\ prac} = \frac{k_B T}{N_v q} \ln\left(\frac{[\text{products}]}{[\text{reactants}]}\right) \qquad (9.33)$$

which is known as the *Nernst equation* [12, p. 750,789]. Many authors replace the Boltzmann constant in this expression using $\mathbb{R} = N_a k_B$ where N_a is the Avogadro constant and $\mathbb{R}$ is the molar gas constant. In this expression, $V_{cell\ theor}$ is the theoretical cell voltage, and $V_{cell\ prac}$ is the practical cell voltage that incorporates the effect of reaction products. The quantity N_v represents the number of valence electrons involved in the chemical reaction. For the example of Eqs. 9.19 and 9.20, two electrons are involved. So, the quantity $\frac{k_B T}{N_v}$ represents the internal energy per valence electron involved in the reaction. The quantity $\frac{[\text{products}]}{[\text{reactants}]}$ is known as the activity

9 BATTERIES AND FUEL CELLS

quotient, and its natural log is between zero and one.

$$0 \leq \ln\left(\frac{[\text{products}]}{[\text{reactants}]}\right) \leq 1 \tag{9.34}$$

When a battery is first set up, there are many reactants but few products present, and

$$\ln\left(\frac{[\text{products}]}{[\text{reactants}]}\right) \approx 0. \tag{9.35}$$

In this case, the activity quotient is very small, so the practical cell voltage between the terminals is very close to the theoretical cell voltage. After a battery has been discharging for a long time, the activity quotient is large because many products are present.

$$\ln\left(\frac{[\text{products}]}{[\text{reactants}]}\right) \approx 1 \tag{9.36}$$

As expected, this model shows that as a battery discharges, the difference between the theoretical and practical cell voltage grows. We cannot ever use the entire capacity stored in a battery. As the battery discharges, the voltage between the terminals drops. At some point, the voltage level is too low to be useful, and the end voltage is reached. At this point, the battery should be replaced even though it still has some stored charge.

The Nernst equation is useful to chemists because it can be used to solve for the amount concentration of reaction products and reactants. The theoretical cell voltage can be calculated or found in a table, and the practical cell voltage can be measured with a voltmeter. Reference [137] tabulates components of the activity quotient as a function of temperature for various reactions.

Electrical engineers may be more interested in the Nernst equation because it gives information on the efficiency of batteries and fuel cells. Efficiency is defined as the output power over the input power or the output energy over the input energy.

$$\eta_{eff} = \frac{E_{out}}{E_{in}} \tag{9.37}$$

Energy stored in an electrical component is given by Eq. 2.8 where Q is charge and V is voltage. The amount of charge involved in each reaction is given by number of electrons involved times their charge for each, $Q = qN_v$.

$$E_{in} = \frac{1}{2}qN_v V_{cell\ theor} \tag{9.38}$$

Internal energy of a reaction at temperature T is also given by

$$E_{in} = \frac{1}{2}k_B T. \tag{9.39}$$

We can model the theoretical voltage of a battery cell by combining Eqs. 9.38 and 9.39.

$$k_B T = q N_v V_{cell\ theor} \tag{9.40}$$

$$V_{cell\ theor} = \frac{k_B T}{q N_v} \tag{9.41}$$

The output energy produced by the battery is proportional to the practical cell voltage measured between the terminals.

$$E_{out} = \frac{1}{2} q N_v V_{cell\ prac} \tag{9.42}$$

The efficiency can then be rewritten.

$$\eta_{eff} = \frac{V_{cell\ prac}}{V_{cell\ theor}} \tag{9.43}$$

With some algebra, we can use the Nernst equation to write this quantity as a function of the activity quotient.

$$\eta_{eff} = \frac{V_{cell\ prac} + V_{cell\ theor} - V_{cell\ theor}}{V_{cell\ theor}} \tag{9.44}$$

$$\eta_{eff} = 1 - \left(\frac{V_{cell\ theor} - V_{cell\ prac}}{V_{cell\ te}} \right) \tag{9.45}$$

The numerator can be replaced using the Nernst equation.

$$\eta_{eff} = 1 - \frac{1}{V_{cell\ theor}} \left(\frac{k_B T}{N_v q} \ln \left(\frac{[\text{products}]}{[\text{reactants}]} \right) \right) \tag{9.46}$$

$$\eta_{eff} = 1 - \ln \left(\frac{[\text{products}]}{[\text{reactants}]} \right) \tag{9.47}$$

Equation 9.47 shows that the efficiency is a function of the activity quotient. As described above, the activity quotient is different for different reactions, and it varies with temperature. The activity quotient is a measure of the effect of the accumulation of products in the electrolyte of a battery or fuel cell.

Equation 9.47 describes the efficiency of batteries and fuel cells. It is another way of expressing the Nernst equation. It is analogous to equations we have encountered describing efficiency of other energy conversion devices. More specifically, it has a similar form to the equation for the Carnot efficiency, Eq. 8.42. Carnot efficiency describes the temperature dependence of the efficiency of all devices which convert a temperature difference to another form of energy. It was introduced in the context of thermoelectric devices, but it applies to pyroelectric devices, steam turbines, and other devices too. These equations also have a similar form to Eq. 7.23 which modeled the effect of mirror reflectivity and optical absorption on the efficiency of a laser.

9.5 Battery Types

9.5.1 Battery Variety

An ideal battery has many desirable qualities. It should:

- have high specific energy and energy density

- contain no toxic chemicals so that it is environmentally friendly and easy to dispose of safely

- be safe to use

- be inexpensive

- be rechargeable

- require no complicated procedure to recharge

- be able to output large current

- be able to withstand a wide range of temperatures

- produce a constant voltage output throughout its life (have a flat discharge curve)

- remain charged for a long time while in storage

The list above is not complete, and it is in no particular order. Tradeoffs are needed because many of these qualities inherently contradict. For example, a device with a high specific energy necessarily requires more safety precautions and controlled use than a device with low specific energy.

Batteries are used in a wide range of applications, so one type is not best in all situations. As an example, a car ignition battery must be rechargeable, have high capacity, output large current, and operate over a wide temperature range. However, car batteries do not require particularly high specific energies. As another example, tiny batteries are used to power microelectromechanical systems such as micropumps [142] [143]. These batteries must have high specific energy and be able to be produced in small packages. Some are even built into integrated circuits [144] [145].

One way to classify batteries is as primary or secondary. A *primary battery* is used once, then disposed. A *secondary battery* is a rechargeable battery. Primary batteries have the advantage of simplicity [128, ch. 8]. They do not require maintenance, so they are simple to use. Also, their construction may be simpler than secondary batteries because they do not need additional circuitry built in to monitor or control the recharging process. They often have high specific energy too [128, ch. 8]. They come in a variety of sizes and shapes, and they are made with a variety of electrode and electrolyte materials. Many alkaline and lithium ion batteries are designed to be primary batteries. Secondary batteries have the obvious advantage of not producing as much waste that ends up in a landfill. Also, the user does not need to continually purchase replacements. While secondary batteries may cost more initially, they can be cheaper in long run. They are often designed to be recharged thousands of times [128, ch. 15]. Many secondary batteries have a very flat discharge curve, so they produce a constant voltage throughout use, even upon multiple charging cycles [128, ch. 15]. Two of the most common types of secondary batteries are lead acid batteries and lithium batteries.

There are many battery types, distinguished by choice of electrolyte and electrodes. Four common battery types are discussed in this section: lead acid, alkaline, nickel metal hydride, and lithium. Not all batteries fit into one of these families. Some devices, like zinc air batteries, are even harder to categorize. Zinc air batteries are actually battery fuel cell hybrids because the zinc of the anode is consumed as in battery operation while oxygen from air is consumed as in fuel cell operation. However, by considering these four classes, we will see some of the variety available. For a more thorough and encyclopedic discussion of battery types, see reference [128].

Table 9.2 summarizes example batteries of each of these four types. The first three rows list example materials used to make the anode, cathode, and electrolyte for batteries. Materials listed in the table are just examples, so batteries of each type can be made with a variety of other materials too. The next two rows give approximate values for the specific energy in units of $\frac{W \cdot h}{kg}$. All values are approximate values for representative devices provided to give an approximate value for comparison, not necessarily values for a particular device. The fifth row lists example values for the theoretical

	Lead acid	Alkaline	Lithium	Nickel Metal Hydride
Example anode material	Pb	Zn	Li	$LaNi_5$
Example cathode material	PbO_2	MnO_2	CF or MnO_2	NiOOH
Example electrolyte	H_2SO_4	KOH or NaOH	Organic solvents and $LiBF_4$	KOH
Example applications	Car ignitions	Toys	Cellphones, Medical devices	Power tools
Theoretical specific energy, $\frac{W \cdot h}{kg}$	252	358	448	240
Practical specific energy, $\frac{W \cdot h}{kg}$	35	154	200	100
References	[128, p. 15.11] [140]	[128, p. 8.10] [140]	[128, p. 15.1, p. 31.5]	[128, p. 15.1] [146]

Table 9.2: Example material components and specific energy values for batteries based on different chemistries.

specific energy of the chemical reaction involved while the sixth row lists example specific energy values for practical devices which are necessarily lower than the theoretical values. The specific energy values in the table can be compared to specific energy of various other materials or energy conversion devices listed in Appendix D.

9.5.2 Lead Acid

Lead acid batteries are secondary batteries which typically have an anode of Pb and a cathode of PbO_2 [128, ch. 15]. The electrolyte is a liquid solution of the acid H_2SO_4 which ionizes into $2H^+$ and SO_4^{2-}. The reaction

at the anode is
$$\text{Pb} + \text{SO}_4^{2-} \rightarrow \text{PbSO}_4 + 2e^- \tag{9.48}$$
with a redox potential of $V_{rp} = 0.37$ V [140]. The reaction at the cathode is
$$\text{PbO}_2 + \text{SO}_4^{2-} + 4\text{H}^+ + 2e^- \rightarrow \text{PbSO}_4 + 2\text{H}_2\text{O} \tag{9.49}$$
with a redox potential of $V_{rp} = 1.685$ V [140]. The overall cell voltage is $V_{cell} = 2.055$ V, so in a car battery, six cells are packaged in series.

Lead acid batteries have a long history. The development of the battery dates to the work of Volta around 1795 [3, p. 2], and practical lead acid batteries were first developed around 1860 by Raymond Gaston Planté [128, p. 16.1.1]. Today, lead acid batteries are used to start the ignition system in cars and trucks, used as stationary backup power systems, and used in other applications requiring large capacity and large output current. Typically, lead acid batteries can handle relatively high current, and they operate well over a wide temperature range [128, p. 15.2]. Additionally, they have a flat discharge curve [128, p. 15.2]. Other types of batteries have a higher energy density and specific energy, so lead acid batteries are used in situations where specific energy is less of a concern than other factors.

9.5.3 Alkaline

Alkaline batteries typically have a zinc anode and a manganese dioxide MnO_2 cathode [128, p. 8.10]. Figure 9.7 shows naturally occurring manganese dioxide (the dark mineral) on feldspar (the white mineral) from Ruggles mine near Grafton, New Hampshire. The batteries are called alkaline due to the use of an alkaline electrolyte, typically a liquid potassium hydroxide KOH solution [128, p. 8.10]. Most alkaline batteries are primary batteries, but some secondary alkaline batteries are available. Alkaline batteries have many nice properties. They can handle high current outputs, they are inexpensive, and they operate well over a wide temperature range [128, p. 8.10]. One limitation, though, is that they have a sloping discharge curve [128, p. 8.10]. Alkaline batteries were originally developed for military applications during WWII [128, ch. 8]. They became commercially available in 1959, and they became popular in the 1980s with improvements in their quality [128, p.11.1]. They are commonly used today in inexpensive electronics, toys, and gadgets.

9 BATTERIES AND FUEL CELLS

Figure 9.7: Naturally occurring manganese dioxide (the dark mineral) on feldspar (the white mineral).

9.5.4 Nickel Metal Hydride

Nickel metal hydride batteries have an anode made from a nickel metal alloy saturated with hydrogen. One example alloy used is $LaNi_5$ [146]. Another rare earth atom may replace the lanthanum [146], and other alloys like $TiNi_2$ or $ZrNi_2$ saturated with hydrogen are also used as anode materials [146]. The cathode is typically made from a nickel oxide, and the electrolyte is potassium hydroxide, KOH [128, p. 15.11]. The reaction at the anode is [146]

$$\text{Alloy(H)} + OH^- \rightarrow \text{Alloy} + H_2O + e^- \qquad (9.50)$$

and the reaction at the cathode is [146]

$$NiOOH + H_2O + e^- \rightarrow Ni(OH)_2 + OH^-. \qquad (9.51)$$

This cathode reaction has a redox potential of $V_{rp} = 0.52$ V [137].

Nickel metal hydride batteries have many advantages. They have a flat discharge curve. They are secondary batteries which can be charged reliably many times [128, p. 15.1] [147]. Additionally, they are better for the environment than the related nickel cadmium batteries, so there are less constraints on how they can be safely disposed [147]. However, they do not have as high of energy density as lithium batteries [147]. Nickel metal hydride batteries were first developed in the 1960s for satellite applications, and research into them accelerated in the 1970s and 1980s. At the time, they were used in early laptops and cellphones, but lithium batteries are used in these applications today [128, p. 22.1]. They are found now in

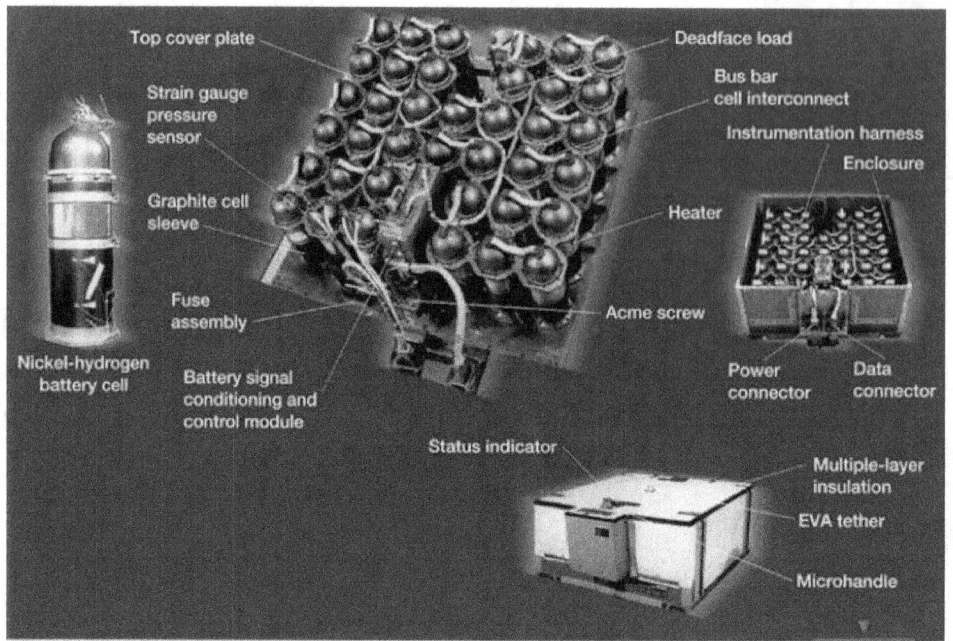

Figure 9.8: The illustration shows a nickel-hydrogen battery and orbital replacement unit which powers the International Space Station. This figure is used with permission from [148].

some portable tools, in some cameras, and in some electronics requiring repeated recharging cycles or requiring high current output. The International Space Station is powered by 48 orbital replacement units, and each orbital replacement unit contains 38 nickel-hydrogen battery cells. Figure 9.8 illustrates an orbital replacement unit [148].

9.5.5 Lithium

Lithium has a high specific energy, so it is very reactive and a good choice for battery research. For this reason, many different battery chemistries utilizing lithium have been developed. The anode may be made out of lithium or carbon [128, ch. 8,15]. Possible cathode materials include MnO_2, $LiCoO_2$, and FeS_2 [128, ch. 8,15]. Electrolytes may be liquid or solid. A possible electrolyte is the mixture of an organic solvent such as propylene carbonate and dimethoxyethane mixed with lithium salts such as $LiBF_4$ or $LiClO_4$ [128, p. 31.5]. Figure 9.9 shows lepidolite, a lithium containing ore of composition $K(Li,Al)_{2-3}(AlSi_3O_{10})(O,OH,F)_2$, from Ruggles mine near Grafton, New Hampshire.

Lithium batteries have been in development since the 1960s, and they were used in the 1970s in military applications [128, p. 14.1]. Both primary and secondary lithium batteries are available today. They are popular due to their high specific energy and energy density. They are used in many consumer goods including cellphones, laptops, portable electronics, hearing aids, and other medical devices [149]. Many lithium batteries are designed to output relatively low current to prevent damage, and secondary lithium batteries require controlled recharging to prevent damage [128, ch. 15]. Even with these limitations, over 250 million cells are produced each month [128, ch. 15].

Figure 9.9: Naturally occurring lepidolite, an ore of lithium.

9.6 Fuel Cells

9.6.1 Components of Fuel Cells and Fuel Cell Systems

A fuel cell is a device which converts chemical energy to electrical energy through the oxidation of a fuel. Like batteries, all fuel cells contain an anode from which electrons and ions flow away, a cathode from which electrons and ions flow towards, and an electrolyte. The electrodes are typically porous which makes it easier for the fuel and oxidizer to get to the reaction site, provides more surface area for the reaction to occur, allows for a higher current through the electrode, and allows for less catalyst to be used [60, ch. 5]. The electrolyte may be a liquid or a solid. Examples of liquid electrolytes include potassium hydroxide solution and phosphoric acid solution [128]. Examples of solid electrolytes include $(ZrO_2)_{0.85}(CaO)_{0.15}$ and $(ZrO_2)_{0.9}(Y_2O_3)_{0.1}$ [60]. Also like a battery, individual cells may be stacked together in a package. A single fuel cell may have a cell voltage of a few volts, but multiple cells may be packaged together in series to produce tens or hundreds of volts from the unit.

Fuel cell components	Fuel cell system components
Anode	Fuel processor
Cathode	Flow plates
Electrolyte	Heat recovery system
Membrane	Inverter
Catalyst	Other electronics
Fuel	
Oxidizer	

Table 9.3: Fuel cell components.

In addition to these components, fuel cells often contain a thin polymer membrane, and fuel cell electrodes are often coated with a catalyst which speeds up the chemical reaction. An example material used to make the membrane is a 0.076 cm layer of polystyrene [60, ch. 10]. Another example membrane is polybenzimidazole containing phosphoric acid [128, ch. 37]. Membranes allow ions, but not the fuel and oxidizer to pass through [60, ch. 10]. In addition to selectively allowing ions to pass through, membranes should be chemically stable to not break down in the presence of the often acidic or alkaline electrolyte, should be electrical insulators, and should be mechanically stable [60, ch. 10]. A useful catalyst speeds up the reaction at the electrodes. In addition, a good catalyst must not dissolve or oxidize in the presence of the electrolyte, fuel, and oxidizer [60, ch. 8]. Additionally, it should only catalyze the desired reaction, not other reactions [60, ch. 8]. Examples of catalysts used include platinum, nickel, acetylacetone, and sodium tungsten bronze Na_xWO_3 with $0.2 < x < 0.93$, [60, ch. 6].

During operation, the fuel and oxidizer are continuously supplied to the device. Fuel may be in the form of a gas such as hydrogen or carbon monoxide gas, it may be in the form of a liquid such as methanol or ammonia, or it may be in the form of a solid such as coal [60, ch. 10]. Oxygen gas or air which contains oxygen is typically used as the oxidizer [60, ch. 10].

Additional chemical, mechanical, thermal, and electrical components are often included in an entire fuel cell system. Some fuel cell systems include a fuel processor which breaks down the fuel to convert it to a usable form and which filters out impurities [141]. For example, a fuel processor may take in coal and produce smaller hydrocarbons which are used as fuel. Also, fuel cells system may contain flow plates which channel the fuel and oxidizer to the electrodes and channel away the waste products and heat [141]. Some fuel cells include heat recovery systems, built in thermoelectric devices which convert some of the heat generated back to electricity. For systems intended to be connected to the electrical grid, inverters which convert the DC power from the fuel cell to AC are included. A fuel cell system also typically include a control system that regulates the flow of the fuel and oxidizer, monitors the temperature of the device, and manages its overall operation [128, ch. 37].

9.6.2 Types and Examples

Fuel cells may be classified in different ways. One way is by operating temperature: low 25-100 °C, medium 100-500 °C, high 500-1000 °C, and very high over 1000 °C [60, ch. 1]. Chemical reactions typically occur more quickly at higher temperatures. However, one challenge of designing high temperature fuel cells is that materials must be selected that can withstand the high temperatures without melting or corroding [60, ch. 2].

As with batteries, another way to classify fuel cells is as primary or secondary [60, ch. 1]. In a *primary fuel cell*, also called *nonregenerative*, the reactants are used once then discarded. In *secondary fuel cells*, also called *regenerative*, the reactants are used repeatedly. An external source of energy is needed to refresh the fuel for reuse, and this source may supply energy electrically, thermally, photochemically, or radiochemically [60, p. 515]. Both primary and secondary fuel cells have been made with a variety of organic and inorganic fuels [60, p. 515].

Another way to classify fuel cells is as direct or indirect [60, ch. 1,7] [128, ch. 37]. In a *direct fuel cell*, the fuel is used as is. In an *indirect fuel cell*, the fuel is processed first inside the system. For example, an indirect fuel cell may take in coal and use an enzyme to break it down into smaller hydrocarbons before the reaction of the cell [60, ch. 7].

Families of fuel cells are often distinguished by the type of electrolyte used. Examples include alkaline which use a potassium hydroxide solution as the electrolyte, phosphoric acid, molten carbonate, and solid oxide which use solid ceramic electrolytes. Other times, fuel cells are categorized by the type of membrane or the type of fuel used. Two of the most common types of fuel cells are proton exchange membrane fuel cells and direct methanol fuel cells [128, ch. 37]. Proton exchange membrane fuel cells use hydrogen gas as the fuel, oxygen from air as the oxidizer, a solid electrolyte, and a platinum catalyst [128] [141]. They operate at low temperature and are used in buses, aerospace applications, and for backup power. Direct methanol fuel cells use methanol as a fuel. They also often operate at low or medium temperatures [128] and are used for similar applications.

9.6.3 Practical Considerations of Fuel Cells

The history of fuel cells goes back almost as long as the history of batteries. The concept of the fuel cell dates to around 1802 [3, p. 2,222] [60, p. v]. Working fuel cells were demonstrated in the 1830s [3, p. 222] [60, p. v], and the first practical device was built in 1959 as pure materials became commercially available [5, p. 46] [60, p. v, 26]. While both batteries and fuel cells are commercially available, batteries have found a home inside almost every every car, computer, and electronic devices while fuel cells are more specialized products. There are a number of limitations of fuel cell technology that have prevented more widespread use. One limitation is their cost. Some fuel cells use platinum as the catalyst, and platinum is not cheap. Some cells that do not use platinum catalysts have the problem that their efficiency is reduced in the presence of carbon monoxide or carbon dioxide, which are commonly found in air. Hydrogen gas or methane are used as the fuel in some cells, and the delivery and storage of these fuels pose challenges. Additionally, some of the more efficient systems are large and require fixed space, air or water cooling, and additional infrastructure, so these devices do not lend themselves to portable applications.

Fuel cells have advantages which lead to useful applications. Many fuel cells produce no harmful outputs. If hydrogen gas is used as the fuel and oxygen from the air is used as the oxidizer, the only byproduct is pure water. It is hard to find an energy conversion device which generates electricity and is easier on the environment than this type of fuel cell. The left part of Fig. 9.10 shows a photograph of a proton exchange membrane fuel cell. The right part of Fig.9.10 shows an image of the water formed during its operation. The image was obtained by the neutron radiography method, and it was taken at the National Institute of Standards and Technology Center for Neutron Research in Gaithersburg, Maryland. These figures are used with permission from [150]. In some applications, the water production is a main advantage. NASA space vehicles have used fuel cells to produce both electricity and pure water since the Gemini and Apollo projects dating to the 1960s [3, p. 250]. They have been used to produce both electricity and water on military submarines since the 1960s too [3, p. 250]. Another advantage of fuel cells is that they can be more efficient than other devices which generate electricity. High temperature and higher power units can have efficiencies up to 65% [128]. Since some of the highest efficiencies are achieved in higher temperature and higher power devices, fuel cells have found a niche in large and stationary applications generating kilowatts or megawatts of electricity.

9 BATTERIES AND FUEL CELLS

Figure 9.10: The picture on the left shows a proton exchange membrane fuel cell. The figure on the right is an image of the water formed in it during operation. The image was obtained by the neutron radiography method at the NIST Center for Neutron Research. These figures are used with permission [150].

9.7 Problems

9.1. A 50 liter solution contains $8 \cdot 10^{19}$ H^+ ions. Calculate the pH. Is this solution acidic or basic?

9.2. A bottle contains 3 liters of a chemical solution with a pH of 8.

   (a) Does the bottle contain an acid or a base?

   (b) Approximately how many H^+ ions are in the bottle?

   (c) Would a 3 liter bottle with a pH of 9 contain more or less ions of H^+ than the bottle with a pH of 8?

   (d) How many times as many/few H^+ ions are in the bottle with solution of pH 8 than in the bottle with solution of pH 9?

9.3. Consider a battery with a lithium electrode and a silver chloride (AgCl) electrode. Assume the following chemical reactions occur in the battery, and the redox potential for each reaction is shown.

$$AgCl + e^- \rightarrow Ag + Cl^- \quad V_{rp} = 0.22 \text{ V}$$
$$Li \rightarrow Li^+ + e^- \quad V_{rp} = 3.04 \text{ V}$$

   (a) Which reaction is likely to occur at the cathode, and which reaction is likely to occur at the anode? Justify your answer.

   (b) What is the overall theoretical cell voltage?

   (c) If the battery is connected to a 1 kΩ load, approximately what is the power delivered to that load?

9.4. Suppose the chemical reactions and corresponding redox potentials in a battery are given by [137]:
$$Li \rightarrow Li^+ + e^- \quad V_{rp} = 3.04 \text{ V}$$
$$S + 2e^- \rightarrow S^{2-} \quad V_{rp} = -0.57 \text{ V}$$

   (a) Find the overall theoretical specific capacity of the battery in $\frac{C}{g}$.

   (b) Find the overall theoretical specific energy of the battery in $\frac{J}{g}$.

   (c) Which material, lithium or sulfur, gets oxidized, and which material gets reduced?

9 BATTERIES AND FUEL CELLS

9.5. A battery has specific capacity 252 $\frac{C}{g}$ and mass of 50 g. Its overall density is 2.245 $\frac{g}{m^3}$.

(a) Find the specific capacity in $\frac{mA \cdot h}{g}$.

(b) Find the capacity in mA·h.

(c) Find the charge density in $\frac{mA \cdot h}{m^3}$.

9.6. A battery has a specific capacity of 55 $\frac{mA \cdot h}{g}$ and a nominal voltage of 2.4 V. The battery has a mass of 165 g. Find the energy stored in the battery in J.

9.7. A battery has a sulfur cathode where the reaction $S + 2e^- \rightarrow S^{2-}$ occurs. The anode is made from a mystery material, X, and at the anode, the reaction $X \rightarrow X^{2+} + 2e^-$ occurs. The theoretical specific capacity of the sulfur reaction is 1.76 $\frac{A \cdot h}{g}$ and the theoretical specific capacity of material X is 0.819 $\frac{A \cdot h}{g}$. The theoretical specific capacity of the materials combined is 0.559 $\frac{A \cdot h}{g}$. What is material X, and what is V_{rp}, the redox potential of the battery?
(Hint: Use a periodic table and a list of redox potentials.)

9.8. What is the difference between each of the items in the pairs below?

- A battery and a fuel cell
- A primary battery and a secondary battery
- Redox potential and chemical potential

9.9. Consider the polymer electrolyte membrane fuel cell shown below. The reactions at the electrodes are:
$H_2 \rightarrow 2H^+ + 2e^-$
$4e^- + 4H^+ + O_2 \rightarrow 2H_2O$
Match the label in the picture to the component name.

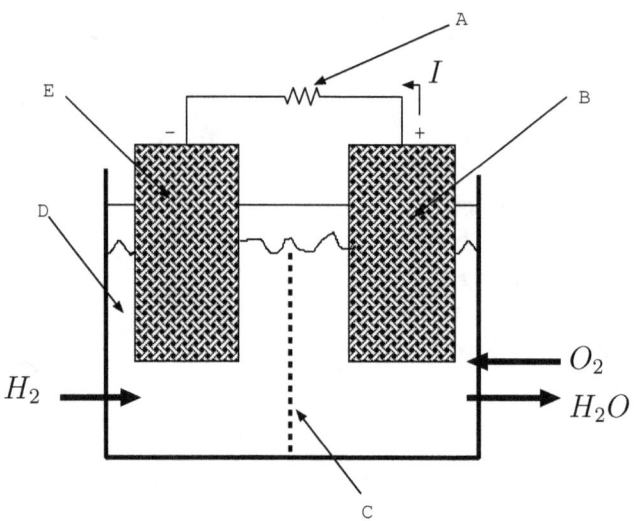

Component name	Label
1. Anode	
2. Cathode	
3. Electrolyte	
4. Load	
5. Polymer electrolyte membrane	

9.10. Match the name of the fuel cell components to a material used to make that component.

Fuel cell component name	Material
1. Anode	A. Platinum
2. Byproduct (waste produced)	B. Carbon (solid, but porous)
3. Catalyst	C. Water
4. Electrolyte	D. Sulfuric Acid
5. Fuel	E. liquid Hydrogen

10 Miscellaneous Energy Conversion Devices

10.1 Introduction

This text is limited to discussing energy conversion devices that involve relatively low powers and that involve electrical energy. Furthermore, this text excludes energy conversion devices involving magnets and coils. Even within these limitations, a wide variety of energy conversion devices have been discussed. This chapter briefly mentions a few additional devices that meet these criteria. Many more devices exist, and with continued creativity and ingenuity by scientists and engineers, more will be developed in the future.

10.2 Thermionic Devices

Thermionic devices convert thermal energy to electricity using the thermionic effect [3, p. 182]. A thermionic device consists of a vacuum tube with electrodes in it. The metal cathode is heated until electrons start evaporating off the metal. The electrons collect at the anode which is at a colder temperature. In a typical device, the cathode may be at a temperature of 1500 ^{0}C, and the distance between the anode and the cathode may be 10 μm [60]. A device based on this effect was first patented by Thomas Edison in 1883. The Carnot efficiency limits this effect because a temperature differential is converted to electricity [5]. Efficiencies up to 12% have been measured. However, for a given temperature differential, other methods of converting temperature difference to electricity are often more efficient. Cathodes have been made from tungsten, molybdenum, tantalum, and barium oxide [3]. The cathode gets used up in the process and eventually needs to be replaced. Anodes have been made from copper, cesium, nickel, barium oxide, strontium oxide, and silver [3] [60]. Some gas chromatographs use nitrogen phosphorous thermionic detectors [151].

10.3 Radiation Detectors

Radiation detectors convert energy from radioactive sources to electricity. Excessive radiation can be harmful to people, and humans cannot sense radioactivity. We can only measure it indirectly. For these reasons, radiation detectors are used as safety devices. Radiation can be classified as alpha particles, beta particles, gamma rays, or neutrons [37, p. 404]. *Alpha particles* are positively charged radiation composed of ionized nuclei of helium. *Beta particles* are high energy electrons. *Gamma rays* are high energy, short wavelength electromagnetic radiation. When these three types of radiation interact with air or another gas, they can excite or ionize the atoms of the gas. Flowing ions are a current, so this process converts the radiation to electricity. Types of radiation detectors include ionization chambers, Geiger counters, scintillation counters, and photographic film based detectors [37].

Ionization chambers and *Geiger counters* work on the same principle. In both cases, a gas is enclosed in a chamber or tube, and a voltage is applied across the gas [37]. Incoming alpha particles, beta particles, or gamma rays, ionize the gas. Due to the applied voltage, positive ions flow to one of the electrodes, and negative ions flow to the other electrode thereby forming a current. Geiger counters operate at higher voltages than ionization chambers. The voltage between the electrodes in an ionization chamber may be from a few volts to hundreds of volts while the voltage is a Geiger counter is typically from 500 V to 2000 V [37]. Many smoke detectors are ionization chambers [152]. When no smoke is present, radiation from a weak radiation source ionizes air between the electrodes, and a current is detected on the electrodes. When smoke is present, it scatters the radiation, so no current is detected [152]. In an ionization chamber, each incoming radioactive particle causes a single atom to ionize. In a Geiger counter, an incoming radioactive particle causes an atom to ionize. Then, the ions formed ionize additional atoms of the gas, and these ions ionize additional atoms forming a cascading reaction powered and maintained by the voltage gradient which accelerates and separates the ion pairs. Geiger counters are often more sensitive due to this amplification of the current produced.

Scintillation counters and photographic film based detectors involve an additional step in converting radiation to electricity. A *scintillation counter* is often made from a crystalline material such as sodium iodide [37]. Sometimes a phosphor is also used [5, p. 166]. Incoming radioactive particles excite, but do not ionize, the atoms of the material. These atoms then decay and emit a photon. Semiconductor or other types of photodetectors convert the photons to electricity [37]. In *photographic film based detectors*, incoming radioactive particles expose the film thereby changing its color [37]. Materials used in the film include Al_2O_3 and lithium fluoride [153]. Again, photodetectors are used to convert the information recorded on the film to a measurable signal. Scintillation counters can be higher sensitivity than other types of radiation detectors, and they can be used to determine the energy of incoming radiation by spectroscopy [154]. The film based detectors can be worn as a ring or badge. These type of detectors are used, for example, by radiology technicians and by nuclear power plant employees. These detectors must be sent in to a lab to be analyzed, and both the amount and the type of radiation can be determined [153].

10.4 Biological Energy Conversion

The human body can be considered an energy conversion device. Humans take in chemical energy in the form of food and convert it to kinetic energy, heat, and other forms of energy. Some components of the human body are also energy conversion devices. Muscles convert chemical energy to kinetic energy. Photoreceptors in the retina of the eye convert optical energy of photons to electrical energy of neurons. The ear converts sound waves to energy stored in the pressure of the fluid of the inner ear, kinetic energy of moving hairs that line the cochlea of the inner ear, and electrical energy of neurons. The human body also stores energy. Muscles can store energy as they stretch and contract. Human fat cells store energy in chemical form. When you walk, your center of mass moves up and down storing energy in pendulum-like motion [155]. Additionally, bone, skin, and collagen exhibit piezoelectricity [156].

Neurons are nerve cells that convert chemical energy to electrical energy. The human brain has around 10^{11} neurons [157, p. 135]. They are composed of a cell body, an axon, dendrites, and synapses [158]. The *axon* is the fibrous part that transmits information to other neurons. The *dendrites* are the fibrous part that receives information from other neurons. A *synapse* is a gap between neurons. Ions, such as Na^+, K^+, or Cl^-, build up on the membrane or in the gap between two neurons, and the charge separation of the ions causes an electrical potential [157]. Ions sometimes cross the gap between neurons. Neurons may be classified as sensory afferents, interneurons, or motoneurons [157]. Sensory afferents transmit a signal from sensory receptors to the nervous system. Interneurons transmit the signal throughout the nervous system, and motoneurons transmit the signal from the nervous system to muscles [157]. Electrical signals transmitted along the nervous system involve pulses with a duration of a few milliseconds [157]. The information is encoded in the frequency rate of the pulses [157].

10.5 Resistive Sensors

Sensors may be made from capacitive, inductive, or resistive materials. These sensors may involve direct energy conversion or may involve multiple energy conversion processes. In Chapters 2 and 3 capacitive energy conversion devices were discussed. The capacitance C of a parallel plate capacitor is given by

$$C = \frac{\epsilon A}{d_{thick}}. \tag{10.1}$$

If the permittivity ϵ, cross sectional area A, or separation of the plates d_{thick} change with respect to any effect, we can make a capacitive sensor. Capacitive sensors are calibrated devices which involve energy conversion between electricity and material polarization. While most inductive energy conversion devices are outside the scope of this book, a few such devices were discussed in Chapters 4 and 5. The inductance L of a single turn inductor is given by

$$L = \frac{\mu d_{thick}}{w}. \tag{10.2}$$

If the permeability μ, thickness d_{thick}, or width w change with respect to any effect, we can make an inductive sensor which utilizes energy conversion between electricity and magnetic energy. Similarly, the resistance R of a uniform resistive device is given by

$$R = \frac{\rho l}{A}. \tag{10.3}$$

If the resistivity ρ, length l, or cross sectional area A change with respect to any effect, we can make a resistive sensor. When a current is applied through a resistive sensor, energy is converted from electricity to heat, and a resistive sensor is calibrated so that a given voltage drop corresponds to a known change in some parameter.

Many resistive senors are available. A *potentiometer* is a variable resistor. As current flows through it, energy is converted from electricity to heat. When the knob of a potentiometer is turned, the length of the material through which the current flows is changed, so the rate of energy conversion through the device changes. A *resistance temperature detector* converts a temperature difference to electricity [37, p. 88]. Resistance temperature detectors work based on the idea of the Thomson effect discussed in Section 8.6.1. In these devices, the resistivity varies with temperature. When a strain is applied to a resistive *strain gauge*, both the length and cross sectional area of the device change. *Pirani hot wire gauges* are used to measure pressure in low pressure environments [37, p. 97]. In a Pirani gauge, current is applied through a metallic filament, and the filament heats up. As air molecules hit the filament, heat is transferred away from it. The resistance of the filament depends on temperature, and the filament cools more quickly in an environment with more air molecules than in an environment at a lower pressure. By monitoring the resistance of the filament, the pressure can be determined.

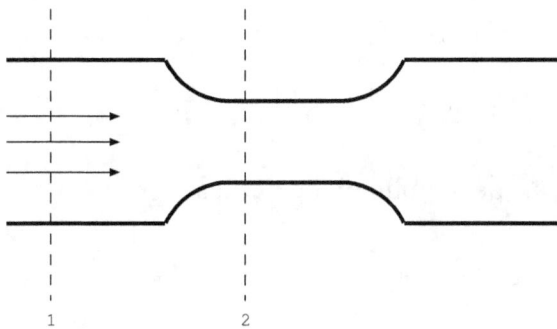

Figure 10.1: A constricted pipe used to illustrate Bernoulli's equation.

10.6 Electrofluidics

Electrohydrodynamic devices (EHDs) convert between electrical energy and fluid flow. These devices are also known as *electrokinetic devices*. *Microfluidic devices* are EHD devices that are patterned on a single silicon wafer or other substrate, and length scales are often less than a millimeter [159]. Engineers have built EHD pumps, valves, mixers, separators, and other EHD devices [159] [160]. Electrohydrodynamic or microfluidic devices have been used in products including ink jet printers, chemical detectors, machines for DNA sequencing or protein analysis, and insulin pumps [61] [160] [161].

Some EHD devices operate based on the idea of Bernoulli's equation, and this relationship is a direct consequence of energy conservation. To illustrate the fundamental physics of this idea, begin by considering a simpler device, a constricted pipe. This pipe converts energy from a pressure differential to kinetic energy [103, ch. 3] [162, p. 346]. Consider a fluid with zero viscosity and zero thermal conductivity flowing through a horizontal pipe (so gravity can be ignored). Figure 10.1 illustrates this geometry. The velocity $\vec{v}$ and pressure $\mathbb{P}$ are different at locations with different pipe diameter, for example locations 1 and 2 in the figure. Consider a small amount of water with mass $m = \rho_{dens}\Delta \mathrm{V}$ where ρ_{dens} is density and $\Delta \mathrm{V}$ is the small volume. Assume there are two, and only two, components of energy: kinetic energy and energy due to the compressed fluid. In going from location 1 to location 2, the pressure of this little mass of fluid changes. Change in energy due to compressing this drop of water is equal to $(\mathbb{P}_1 - \mathbb{P}_2)\Delta \mathrm{V}$. The kinetic energy also changes, and change in kinetic energy is given by

$$\frac{1}{2}m|\vec{v_1}|^2 - \frac{1}{2}m|\vec{v_2}|^2 = \frac{1}{2}\left(\rho_{dens}\Delta \mathrm{V}\right)\left(|\vec{v_1}|^2 - |\vec{v_2}|^2\right). \qquad (10.4)$$

However, energy is conserved, so

$$(\mathbb{P}_1 - \mathbb{P}_2)\Delta V + \frac{1}{2}(\rho_{dens}\Delta V)\left(|\vec{v_1}|^2 - |\vec{v_2}|^2\right) = 0. \tag{10.5}$$

This expression can be simplified algebraically.

$$\mathbb{P}_1 - \mathbb{P}_2 + \frac{1}{2}\rho_{dens}|\vec{v_1}|^2 - \frac{1}{2}\rho_{dens}|\vec{v_2}|^2 = 0 \tag{10.6}$$

Both pressure $\mathbb{P}$ and velocity $\vec{v}$ are functions of location. The only way this expression can be true for all locations is if it is true for each location and a constant.

$$\mathbb{P}_1 + \frac{1}{2}\rho_{dens}|\vec{v_1}|^2 = \mathbb{P}_2 + \frac{1}{2}\rho_{dens}|\vec{v_2}|^2 = \text{constant} \tag{10.7}$$

Bernoulli's equation with the rather severe assumptions above becomes

$$\mathbb{P} + \frac{1}{2}\rho_{dens}|\vec{v}|^2 = \text{constant}. \tag{10.8}$$

Bernoulli's equation is also used to describe the lift of an air foil or the path of a curve ball in baseball [162, p. 350]. In some EHDs electricity induces changes in the pressure or volume of a microfluidic channel. The fluid in these devices may be conductive or insulating. As seen by Eq. 10.8, this change in pressure induces a change in fluid velocity.

In other EHDs, applied voltages exert forces on conductive fluids. A charged object in an external electric field $\vec{E}$ feels a force in the direction of the electric field. A current in an external magnetic field $\vec{B}$ feels a force. The direction of this force is perpendicular to both the direction of the current and the direction of the external magnetic field. These effects are summarized by the Lorentz force equation

$$\vec{F} = Q\left(\vec{E} + \vec{v} \times \vec{B}\right) \tag{10.9}$$

which was discussed in Chapter 5. In that chapter, Hall effect devices and magnetohydrodynamic devices were discussed, both of which can be understood by the Lorentz force equation with an external magnetic field but no electrical field. This type of EHD can be understood by the Lorentz force equation with an external electric field but no magnetic field. The liquid in these devices must be conductive. When a voltage is applied across this type of EHD, an electric field is induced which causes the liquid to flow, and this effect is said to be due to a *streaming potential* [159]. A related effect called *electrophoresis* occurs in liquids which contain charged particles [161]. If an electric field is applied, these particles will move. This effect has been demonstrated with charged DNA molecules and charged protein molecules in solutions [161].

Other EHD devices operate by changing material polarization of an insulating liquid, and this effect is called *dielectrophoresis*. The concept of material polarization was discussed in Section 2.2.1. If we apply an electric field across a conductor, whether that conductor is a solid or a liquid, charges will flow. If we apply an electric field across a dielectric, the material may polarize. In other words, there will be some net charge displacement even if all electrons remain bound to atoms. The external electric field causes both the atoms of the liquid to polarize and these polarized atoms to flow.

There are a number of other interrelated EHD effects. *Electroosmosis* can occur in fluids with a surface charge. In some liquids, ions build up on the surface due to unpaired chemical bonds, due to ions adsorbed onto the surface, or for other reasons. If an electric field is applied across this layer of charges, the fluid will flow, and this effect is called electroosmosis [161] [159]. Also, an external electric field applied across a fluid may heat up part of the fluid and cause a temperature gradient. Fluid may flow due to the temperature gradient, and this effect is called *electrothermal flow* [161]. Another effect, known as *electrowetting*, occurs in conductive liquids. At the interface of a solid conductor and a conductive liquid, charges build up [61]. Again, if an electric field is applied, the liquid will flow.

Part II
Theoretical Ideas

11 Calculus of Variations

11.1 Introduction

The previous chapters surveyed various energy conversion devices. The purpose of Chapters 11 and 12 is to establish a general framework to describe any energy conversion process. By placing energy conversion processes in a larger framework, we may be able to see relationships between processes or identify additional energy conversion processes to study. Establishing this framework requires some abstraction and hence some mathematics. In the next section, we define the Principle of Least Action and the idea of calculus of variations. In the following sections, we apply these ideas to two example energy conversion systems: a mass spring system and a capacitor inductor system.

An advantage of using calculus of variations over other techniques is that the analysis is based on energy, which is a scalar, instead of the potential, which may be a scalar or vector. Working with a scalar quantity like energy instead of a vector can make the mathematics quite a bit more manageable.

11.2 Lagrangian and Hamiltonian

Consider a process which converts energy from one form to another. We are interested in how some quantity evolves during the energy conversion process, and we call this quantity the *generalized path, $y(t)$*. For simplicity, we consider only the case where this path has one independent variable t and one dependent variable y. In this chapter, t represents time, but it can also represent position or another independent variable. These ideas generalize directly to situations with multiple independent and dependent variables [163] [164], but the multiple variable problem requires more involved mathematics. The units of generalized path depend on the energy conversion process under consideration. In the mass spring example of Sec. 11.5, it represents position of a mass. In the capacitor inductor example of Sec. 11.6, it represents the charge built up on the plates of the capacitor. Aside from the energy conversion process under consideration, assume that no other energy conversion processes occur, even though this situation is unlikely. The system goes from having all energy in the first form to having all energy in the second form following the path $y(t)$.

Define the *Lagrangian* $\mathcal{L}$ as the difference between the first and second forms of energy under consideration. The Lagrangian is a function of t, y, and $\frac{dy}{dt}$, and it has the units of joules.

$$\mathcal{L}\left(t, y, \frac{dy}{dt}\right) = \text{(First form of energy)} - \text{(Second form of energy)} \quad (11.1)$$

At any time, the total energy of the system is the sum. Define the *Hamiltonian* H, also in joules, as the total energy.

$$H\left(t, y, \frac{dy}{dt}\right) = \text{(First form of energy)} + \text{(Second form of energy)} \quad (11.2)$$

Some forms of energy cannot be described by a Lagrangian of the form $\mathcal{L}\left(t, y, \frac{dy}{dt}\right)$ and instead require a Lagrangian of the form

$$\mathcal{L}\left(t, y, \frac{dy}{dt}, \frac{d^2y}{dt^2}, \frac{d^3y}{dt^3}, ...\right) \quad (11.3)$$

[163, p. 56]. Such forms of energy will not be considered here. Energy is conserved in any energy conversion process. Conservation of energy can be expressed as

$$\frac{dH}{dt} = \frac{\partial H}{\partial t} = 0. \quad (11.4)$$

Derivatives of the Lagrangian will be useful in the discussion below. Define the *generalized potential* as the partial derivative of the Lagrangian with respect to the path, $\frac{\partial \mathcal{L}}{\partial y}$. The units of the generalized potential depend on the units of the path. More specifically, the units of the generalized potential are joules divided by the units of the path. Note that generalized potential and potential energy are different ideas. Potential energy has units of joules while the units of generalized potential vary. Some authors use the term potential as a synonym for voltage, but this definition of generalized potential is more broad. For more information on the distinction between potential, generalized potential, and potential energy see Appendix C.

Define the *generalized momentum* $\mathbb{M}$ as the partial derivative of the Lagrangian with respect to the time derivative of the path.

$$\mathbb{M} = \frac{\partial \mathcal{L}}{\partial \left(\frac{dy}{dt}\right)}. \quad (11.5)$$

11 CALCULUS OF VARIATIONS

Many authors use the variable p for generalized momentum. However, $\mathbb{M}$ will be used here because the variable p is already too overloaded. Define the *generalized capacity* as the ratio of the generalized path to the generalized potential.

$$\text{Generalized capacity} = \frac{\text{Generalized path}}{\text{Generalized potential}} \tag{11.6}$$

Capacity is also discussed in Appendix C.

11.3 Principle of Least Action

Define the *action* $\mathbb{S}$ as the magnitude of the integral of the Lagrangian along the path.

$$\mathbb{S} = \left| \int_{t_0}^{t_1} \mathcal{L}\left(t, y, \frac{dy}{dt}\right) dt \right| \tag{11.7}$$

Assuming the independent variable t represents time in seconds, the action will have the units joule seconds. For energy conversion processes, the path found in nature experimentally is the path that minimizes the action. This idea is known as the *Principle of Least Action* or sometimes as *Hamilton's principle* [163, p. 11]. The idea of conservation of energy is contained in this principle.

To find a minimum or maximum of a function, find where the derivative of the function is zero. Here, $\mathcal{L}$ and H are not quite functions. Instead, they are *functionals*. A *function* takes a scalar quantity as an input and returns a scalar quantity. A *functional* takes a function as an input and returns a scalar quantity. Both $\mathcal{L}$ and H take the function $y(t)$ as input and return a scalar quantity in joules. The idea of taking a derivative and setting it to zero to find a minimum is still useful, but we have to take the derivative with respect to the function $y(t)$. The process of finding the maximum or minimum of a functional described by an integral relationship is known as *calculus of variations*.

It is often easier to work with differential relationships than integral relationships. We can express the Principle of Least Action as differential equation, and it is called the *Euler-Lagrange equation*.

$$\frac{\partial \mathcal{L}}{\partial y} - \frac{d}{dt} \frac{\partial \mathcal{L}}{\partial \left(\frac{dy}{dt}\right)} = 0 \tag{11.8}$$

If the Lagrangian $\mathcal{L}$ is known, we can simplify the Euler-Lagrange equation to an equation involving only the unknown path. The resulting equation in terms of path $y(t)$ is called the *equation of motion*.

The Lagrangian provides a ton of information about an energy conversion process. If we can describe the difference between two forms of energy by a Lagrangian $\mathcal{L}\left(t, y, \frac{dy}{dt}\right)$, we can set up the Euler-Lagrange equation. From the Euler-Lagrange equation, we may be able to find the equation of motion and solve it. The resulting path minimizes the action and describes how the energy conversion process evolves with time. We can find the generalized potential of the system as a function of time too. The Euler-Lagrange equation is a conservation law for the generalized potential. The symmetries of the equation of motion may lead to further conservation laws and invariants. These last two ideas, and the math behind them, are often known as *Noether's theorem*. Noether's theorem says that there is a very close relationship between symmetries of either the path or the equation of motion and conservation laws [165] [166]. These ideas are discussed further in Sec. 14.5.

Notice the mix of partial and total derivative symbols in Eq. 11.8. Since $y(t)$ depends on only one independent variable, there is no need to use partial derivatives in expressing $\frac{dy}{dt}$. The derivative $\frac{dy}{dt}$ is written in shorthand notation as $\dot{y}$, and $\ddot{y}$ may be used in place of $\frac{d^2y}{dt^2}$. The Lagrangian $\mathcal{L}$ depends on three independent-like variables: t, y, and $\frac{dy}{dt}$. Thus, the partial derivative symbols are used to indicate which partial derivative of $\mathcal{L}$ is being considered.

The first term of the Euler-Lagrange equation, $\frac{\partial \mathcal{L}}{\partial y}$, is the generalized potential defined above. The units of the generalized potential are joules over units of path, $\frac{\text{J}}{\text{units of path}}$. Each term of the Euler-Lagrange equation has these units. For example, if $y(t)$ is in the units of meters, the generalized potential is in $\frac{\text{J}}{\text{m}}$ or newtons. Each term of the Euler-Lagrange equation represents a force, and the Euler-Lagrange equation is a conservation relationship about forces. As another example, if the path $y(t)$ represents charge in coulombs, then the generalized potential has the units $\frac{\text{J}}{\text{C}}$ which is volts. The Euler-Lagrange equation in this case is a conservation relationship about voltages.

11.4 Derivation of the Euler-Lagrange Equation

In this section, we use the Principle of Least Action to derive a differential relationship for the path, and the result is the Euler-Lagrange equation. This derivation closely follows [163, p. 23-33], so see that reference for a more rigorous derivation. Assume that we know the Lagrangian which describes the difference between two forms of energy, and we know the action. We want to find a differential relationship for the path $y(t)$ which minimizes the action. This path has the smallest integral over t of the difference between the two forms of energy.

Suppose that the path $y(t)$ minimizes the action and is the path found in nature. Consider a path $\tilde{y}(t)$ which is very close to the path $y(t)$. Path $\tilde{y}(t)$ is equal to path $y(t)$ plus a small difference.

$$\tilde{y} = y + \varepsilon \eta \tag{11.9}$$

In Eq. 11.9, ε is a small parameter, and $\eta = \eta(t)$ is a function of t. We can evaluate the Lagrangian at this nearby path.

$$\mathcal{L}\left(t, \tilde{y}, \frac{d\tilde{y}}{dt}\right) = \mathcal{L}\left(t,\ y + \varepsilon\eta,\ \dot{y} + \varepsilon\frac{d\eta}{dt}\right) \tag{11.10}$$

The Lagrangian of the nearby path $\tilde{y}(t)$ can be related to the Lagrangian of the path $y(t)$.

$$\mathcal{L}\left(t, \tilde{y}, \frac{d\tilde{y}}{dt}\right) = \mathcal{L}\left(t, y, \dot{y}\right) + \varepsilon \left(\eta \frac{\partial \mathcal{L}}{\partial y} + \frac{d\eta}{dt}\frac{\partial \mathcal{L}}{\partial \dot{y}}\right) + O(\varepsilon^2) \tag{11.11}$$

Equation 11.11 is written as an expansion in the small parameter ε. The lowest order terms are shown, and $O(\varepsilon^2)$ indicates that all additional terms are multiplied by ε^2 or higher powers of this small parameter.

We can also express the difference in the action for paths $\tilde{y}$ and y as an expansion in the small parameter ε.

$$\mathbb{S}(\hat{y}) - \mathbb{S}(y) = \varepsilon \left[\int_{t_0}^{t_1} \eta \frac{\partial \mathcal{L}}{\partial y} + \frac{d\eta}{dt}\frac{\partial \mathcal{L}}{\partial \dot{y}} dt\right] + O(\varepsilon^2) \tag{11.12}$$

The term in brackets is called the *first variation* of the action, and it is denoted by the symbol δ.

$$\delta \mathbb{S}(\eta, y) = \int_{t_0}^{t_1} \eta \frac{\partial \mathcal{L}}{\partial y} + \frac{d\eta}{dt}\frac{\partial \mathcal{L}}{\partial \dot{y}} dt \tag{11.13}$$

Path y has the least action, and all nearby paths $\tilde{y}$ have larger action. Therefore, the small difference $\mathbb{S}(\tilde{y}) - \mathbb{S}(y)$ is positive for *all* possible choices of $\eta(t)$. The only way this can occur is if the first variation is zero.

$$\delta \mathbb{S}(\eta, y) = 0 \tag{11.14}$$

$$\int_{t_0}^{t_1} \eta \frac{\partial \mathcal{L}}{\partial y} + \frac{d\eta}{dt} \frac{\partial \mathcal{L}}{\partial \dot{y}} dt = 0 \tag{11.15}$$

If the action is a minimum for path y, then Eq. 11.15 is true. However, if path y satisfies Eq. 11.15, the action may or may not be at a minimum.

Use integration by parts on the second term to put Eq. 11.15 in a more familiar form.

$$u = \frac{\partial \mathcal{L}}{\partial \dot{y}}$$

$$du = \frac{d}{dt} \frac{\partial \mathcal{L}}{\partial \dot{y}} dt$$

$$v = \eta$$

$$dv = \frac{d\eta}{dt} dt$$

$$\int_{t_0}^{t_1} \frac{d\eta}{dt} \frac{\partial \mathcal{L}}{\partial \dot{y}} dt = \left[\eta \frac{\partial \mathcal{L}}{\partial \dot{y}} \right]_{t_0}^{t_1} - \int_{t_0}^{t_1} \eta \frac{d}{dt} \left(\frac{\partial \mathcal{L}}{\partial \dot{y}} \right) dt \tag{11.16}$$

Assume the endpoints of path y and $\tilde{y}$ align

$$\eta(t_0) = \eta(t_1) = 0. \tag{11.17}$$

$$\int_{t_0}^{t_1} \frac{d\eta}{dt} \frac{\partial \mathcal{L}}{\partial \dot{y}} dt = - \int_{t_0}^{t_1} \eta \frac{d}{dt} \left(\frac{\partial \mathcal{L}}{\partial \dot{y}} \right) dt \tag{11.18}$$

Combine Eq. 11.18 with Eq. 11.15.

$$0 = \int_{t_0}^{t_1} \eta \frac{\partial \mathcal{L}}{\partial y} - \eta \frac{d}{dt} \left(\frac{\partial \mathcal{L}}{\partial \dot{y}} \right) dt \tag{11.19}$$

$$0 = \int_{t_0}^{t_1} \eta \left[\frac{\partial \mathcal{L}}{\partial y} - \frac{d}{dt} \left(\frac{\partial \mathcal{L}}{\partial \dot{y}} \right) \right] dt \tag{11.20}$$

For Eq. 11.20 to be true for all functions η, the term in brackets must be zero, and the result is the Euler-Lagrange equation.

$$\frac{\partial \mathcal{L}}{\partial y} - \frac{d}{dt} \left(\frac{\partial \mathcal{L}}{\partial \dot{y}} \right) = 0 \tag{11.21}$$

We have completed the derivation. Using the Principle of Least Action, we have derived the Euler-Lagrange equation. If we know the Lagrangian for an energy conversion process, we can use the Euler-Lagrange equation to find the path describing how the system evolves as it goes from having energy in the first form to the energy in the second form.

The Euler-Lagrange equation is a second order differential equation. The relationship can be written instead as a pair of first order differential equations,

$$\frac{d\mathbb{M}}{dt} = \frac{\partial \mathcal{L}}{\partial y} \quad (11.22)$$

and

$$\mathbb{M} = \frac{\partial \mathcal{L}}{\partial \dot{y}}. \quad (11.23)$$

The Hamiltonian can be expressed as a function of the generalized momentum, [167, ch. 3].

$$H(t, y, \mathbb{M}) = |\mathbb{M}\dot{y} - \mathcal{L}| \quad (11.24)$$

Using the Hamiltonian, the Euler-Lagrange equation can be written as [167]

$$\frac{d\mathbb{M}}{dt} = -\frac{\partial H}{\partial y} \quad (11.25)$$

and

$$\frac{dy}{dt} = \frac{\partial H}{\partial \mathbb{M}}. \quad (11.26)$$

This pair of first order differential equations is called *Hamilton's equations*, and they contain the same information as the second order Euler-Lagrange equation. They can be used to solve the same types of problems as the Euler-Lagrange equation, for example finding the path from the Lagrangian.

11.5 Mass Spring Example

Examples in this section and the next section will illustrate how we can use the Euler-Lagrange equation to find the equation of motion describing an energy conversion process. Consider a system comprised of a mass and a spring where energy is transfered between spring potential energy stored in the compressed spring and kinetic energy of the mass. The mass is specified by m in kg. It is attached to a spring with spring constant K in $\frac{J}{m^2}$. The position of the mass is specified by $x(t)$ where x is the dependent variable in meters and t is the independent variable time in seconds. Assume this mass and spring are either fixed on a level plane or in some other way not influenced by gravity. This mass spring system is illustrated on the left side of Fig. 11.1. When the spring is compressed, the system gains spring potential energy. When the spring is released, energy is converted from spring potential energy to kinetic energy. Assume no other energy conversion processes, such as heating due to friction, occur.

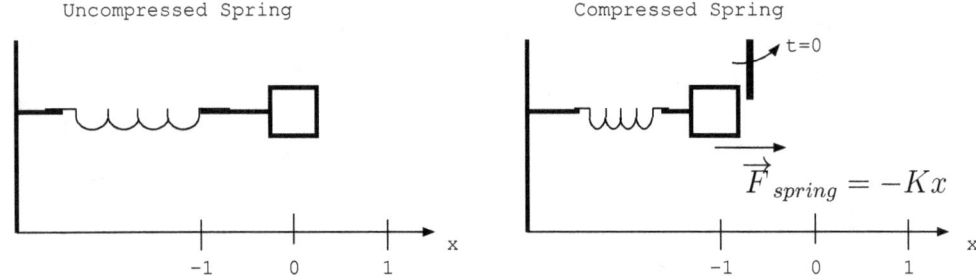

Figure 11.1: A mass spring system.

The right side of Fig. 11.1 shows the compressed spring held in place by a restraint. For $t < 0$, the system has no kinetic energy because the mass is not moving, and the system has potential energy in the compressed spring. At this time, the mass is at position x where $x < 0$. The spring exerts a force on the mass,

$$\vec{F}_{spring} = -Kx\hat{a}_x \qquad (11.27)$$

which is in the $\hat{a}_x$ direction.

At $t = 0$, the restraint is removed, and the spring potential energy is converted to kinetic energy. The first form is spring potential energy.

$$E_{potential\ energy} = \frac{1}{2}Kx^2 \qquad (11.28)$$

The second form is kinetic energy of the mass.

$$E_{kinetic} = \frac{1}{2}m\left(\frac{dx}{dt}\right)^2 \qquad (11.29)$$

At any instant of time, when the mass is at location $x(t)$, the total energy is represented by the Hamiltonian.

$$H = E_{total} = E_{potential\ energy} + E_{kinetic} \qquad (11.30)$$

$$H = \frac{1}{2}Kx^2 + \frac{1}{2}m\left(\frac{dx}{dt}\right)^2 \qquad (11.31)$$

11 CALCULUS OF VARIATIONS

The Lagrangian represents the difference between the forms of energy.

$$\mathcal{L} = E_{potential\ energy} - E_{kinetic} \tag{11.32}$$

$$\mathcal{L}\left(t, x, \frac{dx}{dt}\right) = \frac{1}{2}Kx^2 - \frac{1}{2}m\left(\frac{dx}{dt}\right)^2 \tag{11.33}$$

Both the Hamiltonian and Lagrangian have units of joules. The generalized potential is

$$\frac{\partial \mathcal{L}}{\partial x} = Kx \tag{11.34}$$

in units of newtons. Note that $Kx = -\vec{F}_{spring}$. The generalized momentum is

$$\mathbb{M} = \frac{\partial \mathcal{L}}{\partial \left(\frac{dx}{dt}\right)} = -m\frac{dx}{dt} \tag{11.35}$$

in units of $\frac{\text{kg·m}}{\text{s}}$ which is the units of momentum.

At $t = 0$, the restraint is removed. The mass follows the path $x(t)$. If we know the Lagrangian, we can find the path by trial and error. To find the path in this way, guess a path $x(t)$ that the mass follows and calculate the action.

$$\mathbb{S} = \left|\int_{t_1}^{t_2} \frac{1}{2}m\left(\frac{dx}{dt}\right)^2 - \frac{1}{2}Kx^2\, dt\right| \tag{11.36}$$

Repeatedly guess another path, and calculate the action. The path with the least action of all possible paths is the path that the mass follows. This path has the smallest difference between the potential energy and the kinetic energy integrated over time.

We can think of many possible, but not physical, paths $x(t)$ that the mass can follow. Figure 11.2 illustrates two nonphysical paths as well as the physical path derived below. Paths are considered over the time interval $0 < t < 1$. All three paths assume that initially, at $t = 0$, the spring is compressed so that the mass is at location $x(0) = -1$. Also, they assume that at the end of the interval, at $t = 1$, the spring has expanded so that the mass is at location $x(1) = 1$. The possible paths illustrated in the figure are

$$x_1(t) = 2t - 1 \quad \text{(not physical)}$$

$$x_2(t) = 2t^2 - 1 \quad \text{(not physical)}$$

and

$$x_3(t) = -\cos(\pi t) \quad \text{(physical)}.$$

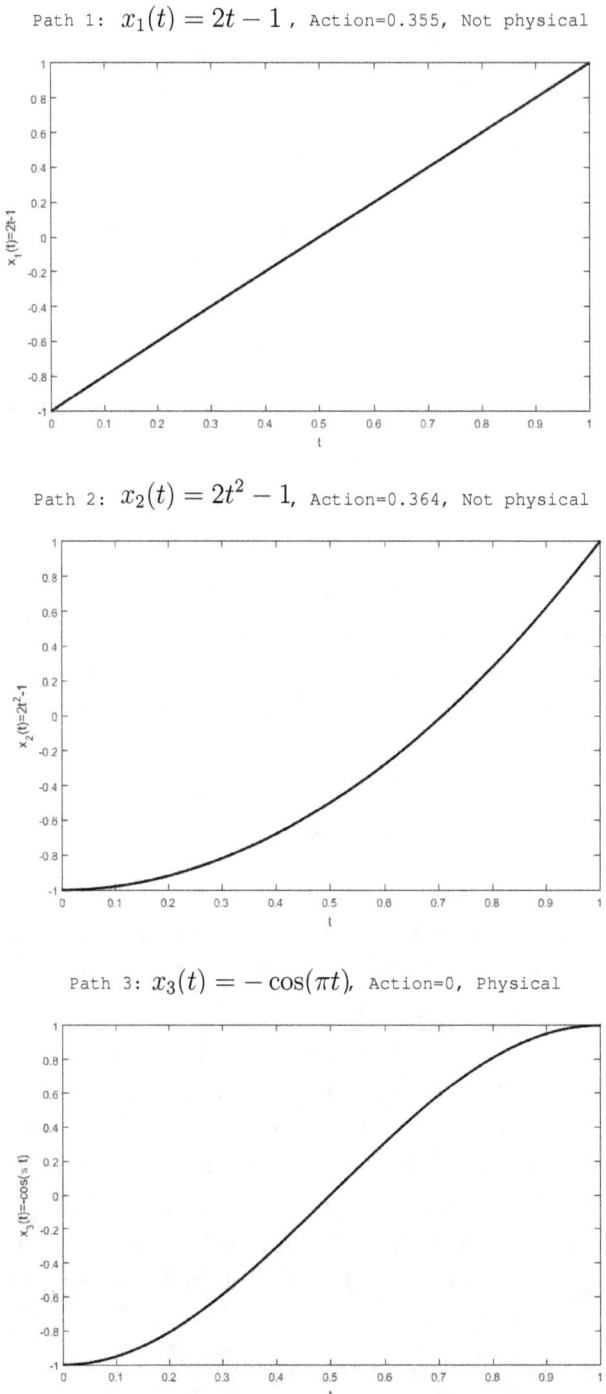

Figure 11.2: Possible paths taken by the mass and their corresponding action.

The path $x_1(t)$ describes a case where the mass travels at a constant speed. The path $x_2(t)$ describes a case where the mass accelerates when the restraint is removed, and the path $x_3(t)$ describes a case where the mass first accelerates then slows. The action of each path can be calculated using Eq. 11.36. For example purposes, the values of $m = 1$ kg and $K = \pi^2 \frac{J}{m^2}$ are used. The path $x_1(t)$ has $\mathbb{S} = 0.355$, the path $x_2(t)$ has $\mathbb{S} = 0.364$, and the physical path $x_3(t)$ has zero action $\mathbb{S} = 0$.

We can derive the path that minimizes the action and that is found in nature using the Euler-Lagrange equation.

$$\frac{\partial \mathcal{L}}{\partial x} - \frac{d}{dt}\frac{\partial \mathcal{L}}{\partial \left(\frac{dx}{dt}\right)} = 0 \qquad (11.37)$$

The first term is the generalized potential. The second term is the time derivative of the generalized momentum. The equation of motion is found by putting these pieces together.

$$Kx + m\frac{d^2x}{dt^2} = 0 \qquad (11.38)$$

The first term of the equation of motion is $-\left|\vec{F}_{spring}\right|$. The second term represents the acceleration of the mass. We have just found the equation of motion, and it is a statement of Newton's second law, force is mass times acceleration. It is also a statement of conservation of force on the mass.

Equation 11.38 is a second order linear differential equation with constant coefficients. It is the famous *wave equation*, and its solution is well known

$$x(t) = c_0 \cos\left(\sqrt{\frac{K}{m}}t\right) + c_1 \sin\left(\sqrt{\frac{K}{m}}t\right) \qquad (11.39)$$

where c_0 and c_1 are constants determined by the initial conditions. If we securely attach the mass to the spring, as opposed to letting the mass get kicked away, it will oscillate as described by the path $x(t)$.

Energy is conserved in this system. To verify conservation of energy, we can show that the total energy does not vary with time. The total energy is given by the Hamiltonian of Eq. 11.31. In this example, both the Hamiltonian and the Lagrangian do not explicitly depend on time, $\frac{\partial H}{\partial t} = 0$ and $\frac{\partial \mathcal{L}}{\partial t} = 0$. Instead, they only depend on *changes* in time. For this reason, we say both the total energy and the Lagrangian have *time translation symmetry*, or we say they are *time invariant*. The spring and mass behave the same today, a week from today, and a year from today.

We can also verify conservation of energy algebraically by showing that $\frac{dH}{dt} = 0$.

$$\frac{dH}{dt} = \frac{\partial H}{\partial t} + \frac{\partial H}{\partial x}\frac{dx}{dt} + \frac{\partial H}{\partial \left(\frac{dx}{dt}\right)}\frac{d^2x}{dt^2} \tag{11.40}$$

$$\frac{dH}{dt} = 0 + Kx\frac{dx}{dt} + m\frac{dx}{dt}\frac{d^2x}{dt^2} \tag{11.41}$$

$$\frac{dH}{dt} = \frac{dx}{dt}\left(Kx + m\frac{d^2x}{dt^2}\right) = 0 \tag{11.42}$$

Notice that the quantity in parentheses in the line above must be zero from the equation of motion.

The Euler-Lagrange equation can be split into a pair of first order differential equations called Hamilton's equations.

$$\frac{d\mathbb{M}}{dt} = -\frac{\partial H}{\partial x} \quad \text{and} \quad \frac{dx}{dt} = \frac{\partial H}{\partial \mathbb{M}} \tag{11.43}$$

This example is summarized in Table 11.1. In analogy to language used to describe circuits and electromagnetics, the relationship between the generalized path and the generalized potential is referred to as the *constitutive relationship*. Following Eq. 11.6, the ratio of the generalized path to generalized potential is the generalized capacity, and in this example, it is the inverse of the spring constant. While displacement x is assumed to be scalar, the vector $\vec{x}$ is used in the table for generality.

11 CALCULUS OF VARIATIONS

Energy storage device	Linear Spring
Generalized Path	Displacement $\vec{x}$ in m
Generalized Potential	$\vec{F}$ Force in $\frac{J}{m} = N$
Generalized Capacity	$\frac{1}{K}$ in $\frac{m^2}{J}$
Constitutive relationship	$\vec{x} = \frac{1}{K}\vec{F}$
Energy	$\frac{1}{2}K|\vec{x}|^2 = \frac{1}{2}\frac{1}{K}|\vec{F}|^2$
Law for potential	Newton's Second Law $\vec{F} = m\vec{a}$

Table 11.1: Summary of the mass spring system in the language of calculus of variations.

11.6 Capacitor Inductor Example

The ideas of calculus of variations apply to energy conversion processes in electrical systems too. Consider a circuit with a capacitor and an inductor as shown in Figure 11.3. The current i_L, the current i_c, and the voltage v are defined in the figure. Assume that wires and components have no resistance. While this is not completely physical, it will allow us to simplify the problem. Assume that the capacitor is charged for $t < 0$, and the switch is open. At $t = 0$, the switch is closed, and the capacitor begins discharging. In this example, the generalized path will be the charge built up on the plates of the capacitor. We can derive the equation of motion that describes this path.

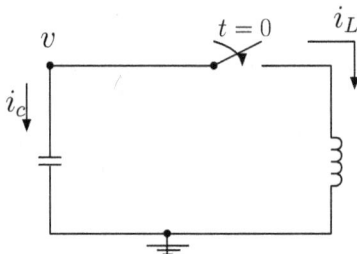

Figure 11.3: A capacitor inductor system.

Energy is converted between two forms. The first form of energy in this system is electrical energy stored in the capacitor. The voltage v in volts across a capacitor is proportional to the charge Q in coulombs across the plates of the capacitor. Capacitance C, measured in farads, is the constant of proportionality between the two measures.

$$Q = Cv \tag{11.44}$$

The current-voltage relationship across the capacitor can be found by taking the derivative with respect to time.

$$\frac{dQ}{dt} = C\frac{dv}{dt} \tag{11.45}$$

The change in charge build up with respect to time is the current. More specifically,

$$\frac{dQ}{dt} = i_c = -i_L. \tag{11.46}$$

Equations 11.45 and 11.46 can be combined.

$$-i_L = C\frac{dv}{dt}. \tag{11.47}$$

11 CALCULUS OF VARIATIONS

The energy stored in a capacitor is

$$E_{cap} = \frac{1}{2}Cv^2. \tag{11.48}$$

The second form of energy in this system is the energy stored in the magnetic field of the inductor. The current i_L through the inductor, measured in amperes, is proportional to the magnetic flux Ψ, measured in webers, around the inductor. Inductance L, measured in henries, is the constant of proportionality between the current and magnetic flux.

$$\Psi = Li_L \tag{11.49}$$

The current voltage relationship across this inductor can be found by taking the derivative with respect to time.

$$\frac{d\Psi}{dt} = v = L\frac{di_L}{dt} \tag{11.50}$$

The energy stored in the inductor is given by

$$E_{ind} = \frac{1}{2}Li_L^2. \tag{11.51}$$

We describe the energy conversion process by keeping track of a the generalized path $Q(t)$, the charge stored on the capacitor. The variable t represents the independent variable time in seconds, and Q is the dependent variable charge in coulombs. The Hamiltonian and Lagrangian, H and $\mathcal{L}$, will be considered functions of three independent-like variables: t, Q, and $\frac{dQ}{dt}$.

The Hamiltonian is the sum of the energy in the capacitor and the energy in the inductor. The Lagrangian is the difference between these energies.

$$H = E_{total} = E_{cap} + E_{ind} \tag{11.52}$$

$$\mathcal{L} = E_{cap} - E_{ind} \tag{11.53}$$

Electrical engineers typically describe physical circuits using the most easily measured quantities: current and voltage. However, here to illustrate the use of the calculus of variations formalism, we write expressions for both the total energy and the Lagrangian in terms of the specified variables: t, Q, and $\frac{dQ}{dt}$.

$$H\left(t, Q, \frac{dQ}{dt}\right) = \frac{1}{2C}Q^2 + \frac{1}{2}L\left(\frac{dQ}{dt}\right)^2 \tag{11.54}$$

$$\mathcal{L}\left(t, Q, \frac{dQ}{dt}\right) = \frac{1}{2C}Q^2 - \frac{1}{2}L\left(\frac{dQ}{dt}\right)^2 \tag{11.55}$$

We can find the path, charge on the capacitor as a function of time, by solving for the least action

$$\delta \left| \int_{t_1}^{t_2} \mathcal{L}\left(t, x, \frac{dx}{dt}\right) dt \right| = 0 \tag{11.56}$$

or by solving the Euler-Lagrange equation,

$$\frac{\partial \mathcal{L}}{\partial Q} - \frac{d}{dt}\frac{\partial \mathcal{L}}{\partial \left(\frac{dQ}{dt}\right)} = 0. \tag{11.57}$$

In Eq. 11.56, δ indicates the first variation as defined by Eq. 11.13. Solutions depend on initial conditions such as the charge stored in the capacitor and the current in the inductor at the initial time. We can use the Euler-Lagrange equation to find the equation of motion. The first term of Eq. 11.57 is the generalized potential,

$$\frac{\partial \mathcal{L}}{\partial Q} = \frac{Q}{C} \tag{11.58}$$

which is the voltage v in volts. The next term is the derivative of the generalized momentum.

$$\mathbb{M} = \frac{\partial \mathcal{L}}{\partial \left(\frac{dQ}{dt}\right)} = -L\frac{dQ}{dt} \tag{11.59}$$

We can put the pieces together to find an expression of conservation of the generalized potential.

$$\frac{Q}{C} + L\frac{d^2Q}{dt^2} = 0 \tag{11.60}$$

This is a statement of Kirchhoff's voltage law. It looks more familiar if it is written in terms of voltage $v = \frac{Q}{C}$ and current $i_L = -\frac{dQ}{dt}$.

$$v - L\frac{di_L}{dt} = 0 \tag{11.61}$$

We can solve the equation of motion, Eq. 11.60, using appropriate initial conditions, to find the path. As in the mass spring example, Eq. 11.60 is the wave equation, and its solutions are sinusoids. As expected, a circuit made of only a capacitor and inductor is an oscillator.

11 CALCULUS OF VARIATIONS

Energy storage device	Capacitor	Linear Spring
Generalized Path	Charge Q in C	Displacement $\vec{x}$ in m
Generalized Potential	Voltage v in $\frac{J}{C} = V$	$\vec{F}$ Force in $\frac{J}{m} = N$
Generalized Capacity	Capacitance C in $F = \frac{C^2}{J}$	$\frac{1}{K}$ in $\frac{m^2}{J}$
Constitutive relationship	$Q = Cv$	$\vec{x} = \frac{1}{K}\vec{F}$
Energy	$\frac{1}{2}Cv^2$	$\frac{1}{2}K|\vec{x}|^2 = \frac{1}{2}\frac{1}{K}|\vec{F}|^2$
Law for potential	KVL	Newton's Second Law $\vec{F} = m\vec{a}$

Table 11.2: Summary of the capacitor inductor system in the language of calculus of variations.

Furthermore, we can show that energy is conserved in this energy conversion process because the partial derivative of both the total energy and the Lagrangian with respect to time are zero.

$$\frac{\partial \mathcal{L}}{\partial t} = \frac{\partial H}{\partial t} = 0 \tag{11.62}$$

$$\frac{d\mathcal{L}}{dt} = \frac{dH}{dt} = 0 \tag{11.63}$$

Table 11.2 summarizes this example. It also illustrates the relationship between parameters of this example and parameters of the mass spring example.

11.7 Schrödinger's Equation

Quantum mechanics is the study of microscopic systems such as electrons or atoms. Calculus of variations and the idea of a Hamiltonian are fundamental ideas of quantum mechanics [136]. In Chapter 13, we apply the ideas of calculus of variations to an individual atom in a semiclassical way.

We can never say with certainty where an electron or other microscopic particle is located or its energy. However, we can discuss the probability of finding it with a specific energy. The probability of finding an electron, for example, in a particular energy state is specified by $|\psi|^2$ where ψ is called the *wave function* [136]. As with any probability $0 \leq |\psi|^2 \leq 1$.

For example, suppose that as an electron moves, kinetic energy is converted to potential energy. The quantum mechanical Hamiltonian H_{QM} is then the sum of the kinetic energy $E_{kinetic}$ and potential energy $E_{potential\ energy}$.

$$H_{QM} = E_{kinetic} + E_{potential\ energy} \quad (11.64)$$

Kinetic energy is expressed as

$$E_{kinetic} = \frac{1}{2m}(M_{QM})^2 \quad (11.65)$$

where m is the mass of an electron. In the expression above, M_{QM} is the quantum mechanical momentum operator, and

$$(M_{QM})^2 = M_{QM} \cdot M_{QM}. \quad (11.66)$$

The quantum mechanical momentum operator is defined by

$$M_{QM} = j\hbar \overrightarrow{\nabla} \quad (11.67)$$

where the quantity $\hbar$ is the Planck constant divided by 2π. The *del operator*, $\overrightarrow{\nabla}$, was introduced in Sec. 1.6.1, and it represents the spatial derivative of a function. The quantities H_{QM}, M_{QM}, and $\overrightarrow{\nabla}$ are all operators, not just values. An *operator*, such as the derivative operator $\frac{d}{dt}$, acts on a function. It itself is not a function or value.

Using the of momentum definition of Eq. 11.67 and the vector identity of Eq. 1.10, we can rewrite the Hamiltonian.

$$H_{QM} = \frac{-\hbar^2}{2m}\nabla^2 + E_{potential\ energy} \quad (11.68)$$

In quantum mechanics, the Hamiltonian is related to the total energy.

$$H_{QM}\psi = E_{total}\psi \quad (11.69)$$

The above two equations can be combined algebraically.

$$\left(\frac{-\hbar^2}{2m}\nabla^2 + E_{potential\ energy}\right)\psi = E_{total}\psi \qquad (11.70)$$

With some more algebra, Eq. 11.70 can be rewritten.

$$\nabla^2\psi + \frac{2m}{\hbar^2}\left(E_{total} - E_{potential\ energy}\right)\psi = 0 \qquad (11.71)$$

Equation 11.71 is the *time independent Schrödinger equation*, and it is one of the most fundamental equations in quantum mechanics. Energy level diagrams were introduced in Section 6.3. Allowed energies illustrated by energy level diagrams satisfy the Schrödinger equation. At least for simple atoms and ground state energies, energy level diagrams can be derived by solving Schrödinger equation.

11.8 Problems

11.1. In the examples below, identify whether f is a function or a functional.

- A parabola is described by $f(x) = x^2$.
- Given two forms of energy and a path $y(t)$, f is the Lagrangian of the system $\mathcal{L}\left(t, y, \frac{dy}{dt}\right)$.
- Given the magnitude of the velocity $|\vec{v}(t)|$ of an object, f represents the distance that the object travels from time 0 to time 3600 seconds.
- Given the position (x, y, z) in space, $f(x, y, z)$ represents the distance from that point to the origin.

11.2. A system has the Lagrangian $\mathcal{L}\left(t, y, \frac{dy}{dt}\right) = \left(\frac{dy}{dt}\right)^3 + e^{3y}$. Find an equation for the path $y(t)$ that minimizes the action $\int_{t_1}^{t_2} \mathcal{L}\left(t, y, \frac{dy}{dt}\right) dt$. (The result is nonlinear, so don't try to solve it.)

11.3. A system has Lagrangian $\mathcal{L}\left(t, y, \frac{dy}{dt}\right) = \frac{1}{2}\left(\frac{dy}{dt}\right)^2 + \frac{1}{2} \cdot y^{-2}$. Find the corresponding equation of motion. (The result is nonlinear, so don't try to solve it.)

11.4. Figure 11.2 illustrates three possible paths for the mass spring system and their corresponding actions. The paths considered are:

$$x_1(t) = 2t - 1$$

$$x_2(t) = 2t^2 - 1$$

$$x_3(t) = -\cos(\pi t)$$

For each path, calculate the action using Eq. 11.36 to verify the values shown in the figure. Assume a mass of $m = 1$ kg and a spring constant of $K = \pi^2 \; \frac{\text{J}}{\text{m}^2}$.

11.5. The figure shows a torsion spring. It can store potential energy $\frac{1}{2}\mathbb{K}\theta^2$, and it can convert potential energy to kinetic energy $\frac{1}{2}\mathbb{I}\left(\frac{d\theta}{dt}\right)^2$. In these expressions, $\theta(t)$ is the magnitude of the angle the spring turns in radians, and $\omega = \frac{d\theta}{dt}$ is the magnitude of the angular velocity in radians per second. $\mathbb{K}$ is the torsion spring constant, and $\mathbb{I}$ is the (constant) moment of inertia.

(a) Find the Lagrangian.

(b) Use the Euler-Lagrange equation to find a differential equation describing $\theta(t)$.

(c) Show that energy is conserved in this system by showing that $\frac{dH}{dt} = 0$.

(d) Set up Hamilton's equations.

11.6. The purpose of this problem is to derive the shortest path $y(x)$ between the points (x_0, y_0) and (x_1, y_1). Consider an arbitrary path between these points as shown in the figure. We can break the path into differential elements $\vec{dl} = dx\hat{a}_x + dy\hat{a}_y$. The magnitude of each differential element is

$$|\vec{dl}| = \sqrt{(dx)^2 + (dy)^2} = dx\sqrt{1 + \left(\frac{dy}{dx}\right)^2}.$$

The distance between the points can be described by the action

$$S = \int_{x_0}^{x_1} \sqrt{1 + \left(\frac{dy}{dx}\right)^2}\, dx.$$

To find the path $y(x)$ that minimizes the action, we can solve the Euler-Lagrange equation, with $\mathcal{L} = \sqrt{1 + \left(\frac{dy}{dx}\right)^2}$ as the Lagrangian, for this shortest path $y(x)$. This approach can be used because we want to minimize the integral of some functional $\mathcal{L}$ even though this functional does not represent an energy difference [163, p. 33].

Set up the Euler-Lagrange equation, and solve it for the shortest path, $y(x)$.

Hint 1: The answer to this problem is that the shortest path between two points is a straight line. Here, you will derive this result.
Hint 2: In the examples of this chapter, the Lagrangian had the form $\mathcal{L}\left(t, y, \frac{dy}{dt}\right)$ with independent variable t and path $y(t)$. Here, the Lagrangian has the form $\mathcal{L}\left(x, y, \frac{dy}{dx}\right)$ where the independent variable is position x, and the path is $y(x)$.
Hint 3: If $\frac{d}{dx}(something) = 0$, then you know that $(something)$ is constant.

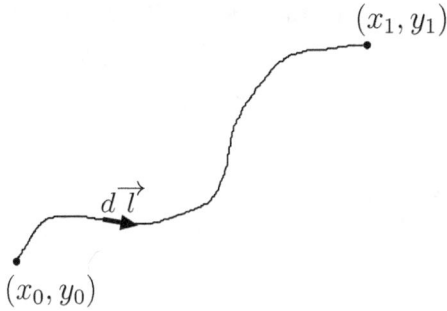

11.7. Light travels along the quickest path between two points. This idea is known as *Fermat's principle*. In a material with relative permittivity ϵ_r and permeability μ_0, light travels at the constant speed $\frac{c}{\sqrt{\epsilon_r}}$ where c is the speed of light in free space. In Prob. 11.6, we showed that the shortest path between two points is a straight line, so in a uniform material, light will travel along a straight line between two points. However, what if light travels across a junction between two materials? In this problem, we will answer this question and derive a famous result known as *Snell's law*.

Consider the figure below. Assume that a ray of light travels from (x_0, y_0) to (x_1, y_1) along the path which takes the shortest time. Material 1 has relative permittivity ϵ_{r1}, so the light travels in that material at a constant speed $\frac{c}{\sqrt{\epsilon_{r1}}}$. Material 2 has relative permittivity ϵ_{r2}, so the light travels in that material at a constant speed $\frac{c}{\sqrt{\epsilon_{r2}}}$. As we derived in the Prob. 11.6, the light travels along a straight line in material 1, and it travels along a straight line in material 2. However, the lines have different slopes as shown in the figure. Assume that the junction of the two materials occurs at $x = 0$.

(a) Find an equation for the total time it takes the light to travel as a function of h, the vertical distance at which the path crosses the y axis. Note that you are finding a function here, $F(h)$, not a functional. You can use the fact that you know the light follows a straight line inside each material to find this function.

(b) The path followed by the light takes the minimum time, so the derivative $\frac{dF}{dh} = 0$. Use this idea to find an equation for the unknown vertical height h. Your answer can be written as a function of the known constants ϵ_{r1}, ϵ_{r2}, x_0, y_0, x_1, y_1, and c. You do not need to solve for h here, but instead just evaluate the derivative and set it to zero.

(c) Use your result in part b above to derive *Snell's law*:

$$\sqrt{\epsilon_{r1}} \sin \theta_1 = \sqrt{\epsilon_{r2}} \sin \theta_2$$

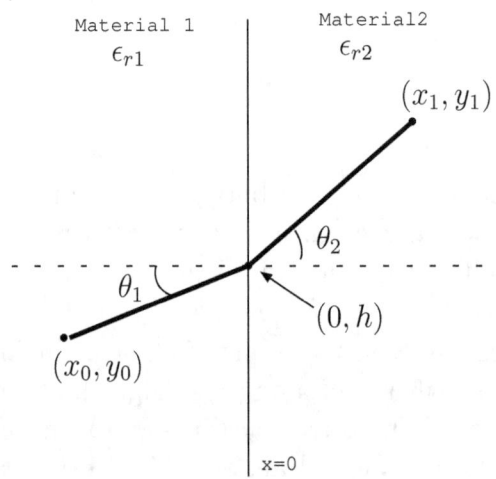

11.8. A pendulum converts kinetic energy to and from gravitational potential energy. As shown in the figure, a ball of mass m is hung by a string 1 m long. The pendulum is mounted on a base that is 3 m high. As shown in the figure, $\theta(t)$ is the angle of the pendulum. The kinetic energy of the ball is given by $E_{kinetic} = \frac{1}{2}m\left(\frac{d\theta}{dt}\right)^2$, and the gravitational potential energy is given by $E_{p.e.} = mg(3 - \cos\theta)$. The quantity g is the gravitational constant, $g = 9.8 \frac{m}{s^2}$.

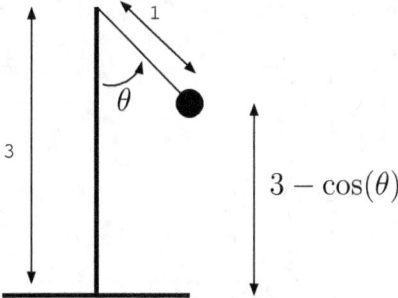

(a) Find $\mathcal{L}$, the Lagrangian of the system.

(b) Find H, the Hamiltonian of the system.

(c) Set up the Euler-Lagrange equation, and use it to find the equation of motion for $\theta(t)$, the angle of the pendulum as a function of time.

(d) Show that energy is conserved in this system by showing that $\frac{dH}{dt} = 0$.

The equation of motion found in part c is nonlinear, so don't try to solve it. Interestingly, it does have a closed form solution [164, Ch. 6]. (This problem is a modified version of an example in reference [163].)

11.9. As shown in the figure, an object of charge Q_1 and mass m moves near a stationary object with charge Q_2. Assume the mass and the charges are constants, and assume the objects are surrounded by free space. The kinetic energy of the moving object is converted to or from energy stored in the electric field between the objects. The kinetic energy of the moving object is given by $\frac{1}{2}m\left(\frac{dx}{dt}\right)^2$. The energy of the electric field is given by $\frac{Q_1 Q_2}{4\pi\epsilon_0 x}$ where ϵ_o is the permittivity of free space. The distance between the objects is given by $x(t)$.

(a) Find the Lagrangian of the system.

(b) Find the generalized momentum.

(c) Find the generalized potential.

(d) Find the equation of motion for the path $x(t)$ of the system. (Don't try to solve this nonlinear equation.)

(e) Find the total energy of the system.

(f) Show that energy is conserved in this system.

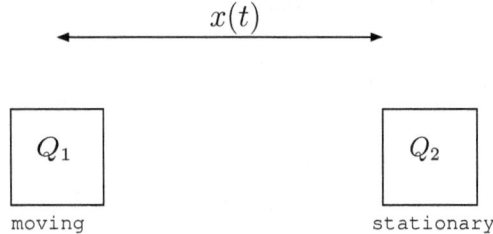

12 Relating Energy Conversion Processes

12.1 Introduction

In the previous chapter, the concept of calculus of variations was introduced. The purpose of this chapter is to draw relationships between a wide range of energy conversion processes. Processes in electrical engineering, mechanics, thermodynamics, and chemistry are described using the language of calculus of variations. Similarities between the processes are highlighted and summarized into tables.

This chapter illustrates how to apply calculus of variations ideas to disparate branches of science and engineering. Electrical engineers typically use current and voltage to describe circuits. Chemists use temperature, pressure, entropy, and volume when describing chemical reactions. Engineers and scientists in each discipline have their own favorite quantities. However, energy conversion is a common topic of study. Calculus of variations provides a unifying language. Scientists and engineers typically specialize, becoming experts in a particular area. However, open questions are more often found at the boundary between disciplines, where there is less expertise. Comparing ideas between different disciplines is useful because ideas from one discipline may answer questions in another, and challenges in one discipline may pose interesting research questions in another.

By studying the mass spring system of Sec. 11.5, the resulting equation of motion was Newton's second law. By studying the capacitor inductor system of Sec. 11.6, the resulting equation of motion was Kirchoff's voltage law. In this chapter we identify the equation of motion for multiple other systems. Through this procedure, we encounter some of the most fundamental laws of physics including including Gauss's laws, conservation of momentum, conservation of angular momentum, and the second law of thermodynamics.

The discussion in this chapter is necessarily limited. Entire texts have been written about each energy conversion processes discussed. Additionally, the idea of applying calculus of variations to these energy conversion processes is not novel. Other authors have compared electrical, mechanical, and other types of energy conversion processes too [168] [169].

Some rather drastic assumptions are made in this chapter. We assume energy is converted between one form and another with no other energy conversion process occurring. For example in a mass spring system, energy is converted between kinetic energy and spring potential energy while ignoring heating due to friction, energy conversion due to gravitational potential energy, and so on that might occur in a real system.

12.2 Electrical Energy Conversion

Electrical can be described either in circuits language or electromagnetics language. Using circuits language, electrical systems are described by four fundamental parameters: charge in coulombs Q, voltage in volts v, magnetic flux in webers Ψ, and current in amperes i. For circuits described in this language, resistors, capacitors, and other electrical energy storage and conversion devices are treated as point-like with no length or extent, and forces and fields outside the path of the circuit are ignored. An alternative approach is to use electromagnetics language the electrical properties of materials are studied as a function of position and forces and fields outside of the path of a circuit are studied.

We can use circuits language to describe a number of energy conversion devices. Resistors convert electrical energy to thermal energy, and thermoelectric devices convert thermal energy to or from electrical energy. A charging capacitor converts electrical energy to energy stored in a material polarization, and a discharging capacitor converts the energy of the material polarization back to electrical energy. In an inductor, electrical energy is converted to and from energy of a magnetic field.

In Sec. 11.6, energy storage in a capacitor was studied in detail and described in the language of calculus of variations. Table 11.2 summarized the use of calculus of variations language to describe the energy conversion process, and it is repeated in the second column of the Table 12.1. In that example, charge built up on the capacitor plates, Q, was the generalized path. The generalized potential was v, the voltage across the capacitor. From these choices, other parameters were selected.

Instead of choosing charge Q as the generalized path, we could have chosen the generalized path to be one of the other fundamental variables of circuit analysis, voltage v, magnetic flux Ψ, or current i. Table 12.1 summarizes parameters that result when we describe energy conversion processes occurring in a capacitor or inductor in the language of calculus of variations with these choices of generalized path. More specifically, the third column shows parameters when voltage is chosen as the generalized path. The fourth column shows parameters when magnetic flux is chosen as the generalized path, and the fifth column shows parameters when current is chosen as the generalized path. By reading down a column of the table, we see how to describe a process with this choice of generalized path. By reading across the rows of the table, we can draw analogies between parameters of energy conversion processes.

To describe the energy conversion processes occurring in a capacitor, we can choose either the charge or voltage to be the generalized path then use the language of calculus of variations. Notice that if charge is chosen as the generalized path as seen in column two of Table 12.1, voltage becomes the generalized potential. However, when voltage is chosen as the generalized path as seen in column three, charge becomes the generalized potential. The path found in nature minimizes the action, and we saw in Sec. 11.6 that we could use the Euler-Lagrange equation to set up an equation of motion for the system. Each term of the equation of motion has the same units as the generalized potential. The equation of motion found when using Q as generalized path is Kirchoff's Voltage Law (KVL), which says the sum of all voltage drops around a closed loop in a circuit is zero. The equation of motion found when using v as the generalized path is the law of conservation of charge. Both of these concepts are fundamental ideas in circuit theory, and they are shown in the second to last row of the table.

Similarly, to describe the energy conversion processes occurring in an inductor, we may choose either magnetic flux or current as the generalized path. If we choose magnetic flux as the generalized path, the generalized potential is current. If we choose current as the generalized path, the generalized potential is magnetic flux. From the first choice, the equation of motion found is Kirchoff's Current Law (KCL). From the second choice, the equation of motion found is conservation of magnetic flux.

The relationship between the generalized path and the generalized potential is known as the constitutive relationship [168, p. 30]. For a capacitor, it is given as

$$Q = Cv. \qquad (12.1)$$

The constant C that shows up in this equation is the capacitance in farads. Analogously for an inductor, the constitutive relationship is

$$\Psi = Li \qquad (12.2)$$

where L is the inductance in henries. We will see that we can identify constitutive relationships for other energy conversion processes, and we similarly can come up with a parameter describing the ability to store energy in the device. In analogy to the capacitor, we will call this parameter the *generalized capacity*. Capacitance C represents the ability to store energy in the device, so generalized capacity represents the ability to store energy in other devices. Overloading of the term capacity is discussed in Appendix C.

Energy storage device	Capacitor	Capacitor	Inductor	Inductor
Generalized Path	Charge Q in C	Voltage v in V	Mag. Flux Ψ in Wb	Current i in A
Generalized Potential	Voltage v in $\frac{J}{C} = V$	Charge Q in C	Current i in $A = \frac{J}{Wb}$	Mag. Flux Ψ in Wb
Generalized Capacity	Capacitance C in $F = \frac{C^2}{J}$	$\frac{1}{C}$	Inductance L in $H = \frac{Wb^2}{J}$	$\frac{1}{L}$
Constitutive relationship	$Q = Cv$	$v = \frac{Q}{C}$	$\Psi = Li$	$i = \frac{\Psi}{L}$
Energy	$\frac{1}{2}Cv^2$	$\frac{1}{2}\frac{Q^2}{C}$	$\frac{1}{2}Li^2$	$\frac{1}{2}\frac{\Psi^2}{L}$
Law for potential	KVL	Conservation of Charge	KCL	Conservation of Mag. Flux
This column assumes	AC current and voltage	AC current and voltage	AC current and voltage	AC current and voltage

Table 12.1: Describing electrical circuits in the language of calculus of variations.

Circuit Quantity		Electromagnetic Field	
Q	Charge in C	$\vec{D}$	Displacement flux density in $\frac{C}{m^2}$
v	Voltage in V	$\vec{E}$	Electric field intensity in $\frac{V}{m}$
Ψ	Magnetic flux in Wb	$\vec{B}$	magnetic flux density in $\frac{Wb}{m^2}$
i	Current in A	$\vec{H}$	Magnetic field intensity in $\frac{A}{m}$

Table 12.2: Quantities used to describe circuits and electromagnetic fields.

Using electromagnetics language, four vector fields describe systems: $\vec{D}$ displacement flux density in $\frac{C}{m^2}$, $\vec{E}$ electric field intensity in $\frac{V}{m}$, $\vec{B}$ magnetic flux density in $\frac{Wb}{m^2}$, and $\vec{H}$ magnetic field intensity in $\frac{A}{m}$. These electromagnetic fields are generalizations of the circuit parameters charge Q, voltage v, magnetic flux Ψ, and current i respectively as shown in Table 12.2. However, the electromagnetic fields are functions of position x, y, and z in addition to time, and they are vector instead of scalar quantities. More specifically, displacement flux density is the charge built up on a surface per unit area, and magnetic flux density is the magnetic flux through a surface. Similarly, electric field intensity is the negative gradient of the voltage, and magnetic field intensity is the gradient of the current. We encountered these electromagnetic fields when discussing antennas in Chapter 4.

A capacitor can store energy in the charge built up between the capacitor plates. Analogously, an insulating material with permittivity greater than the permittivity of free space, $\epsilon > \epsilon_0$, can store energy in the distributed charge separation throughout the material. We can describe the energy conversion processes occurring in a capacitor using the language of calculus of variations by choosing either charge Q or voltage v as the generalized path. Parameters resulting from these choices are shown in the second and third column of Table 12.1. Analogously, we can describe the energy conversion processes occurring in an insulating material with $\epsilon > \epsilon_0$ using the language of calculus of variations by choosing either $\vec{D}$ or $\vec{E}$ as the generalized path. Parameters resulting from these choices are shown in the second and third column of Table 12.3. The equation of motion that results in either case is Gauss's law for the electric field,

$$\vec{\nabla} \cdot \vec{D} = \rho_{ch} \qquad (12.3)$$

where ρ_{ch} is charge density. The derivation is beyond the scope of this text, however, because it involves applying calculus of variations to quantities with multiple independent and dependent variables. Gauss's law is one of

Maxwell's equations, and it was introduced in Section 1.6.1. In Chapter 2, piezoelectric energy conversion devices were discussed, and in Chapter 3, pyroelectric and electro-optic energy conversion devices were discussed. All of these devices involved converting electrical energy to and from energy stored in a material polarization of an insulating material with $\epsilon > \epsilon_0$. Calculus of variations can be used to describe energy conversion in all of these devices with either displacement flux density or electric field intensity as the generalized path. For a device made from a material of permittivity ϵ with an external electric field intensity across it given by $\vec{E}$, the energy density stored is $\frac{1}{2}\epsilon|\vec{E}|^2$ in $\frac{J}{m^3}$. The energy stored in a volume $\mathbb{V}$ is found by integrating this energy density with respect to volume, and this energy stored in a volume is listed in the second to last row of Table 12.3. Notice the similarity of the equation for the energy stored in a capacitor (second column, second to last box of Table 12.1) and this equation for the energy density stored in a material with $\epsilon > \epsilon_0$ (second column second to last box of the Table 12.3).

Energy can also be stored in materials with permeability greater than the permeability of free space, $\mu > \mu_0$. Hall effect devices and magneto-hydrodynamic devices were discussed in Chapter 5. These devices are all inductor-like, and the parameters used to describe inductive energy conversion processes in the language of calculus of variations are summarized in the last two columns of the Table 12.3. Calculus of variations can be used to describe energy conversion processes in these devices with either magnetic flux density or magnetic field intensity as the generalized path and the other choice as the generalized potential. The equation of motion resulting from using calculus of variations to describe inductive systems corresponds to Gauss's law for the magnetic field,

$$\vec{\nabla} \cdot \vec{B} = 0. \tag{12.4}$$

The physics of antennas is described by electric and magnetic fields, and any of the columns of Table 12.3 can be used to describe energy conversion between electricity and electromagnetic waves in antennas using the language of calculus of variations.

12 RELATING ENERGY CONVERSION PROCESSES

Energy storage device	Dielectric Material, $\epsilon > \epsilon_0$	Dielectric Material, $\epsilon > \epsilon_0$	Magnetic Material, $\mu > \mu_0$	Magnetic Material, $\mu > \mu_0$
Generalized Path	Displacement Flux Density $\vec{D}$ in $\frac{C}{m^2}$	Electric Field Intensity $\vec{E}$ in $\frac{V}{m} = \frac{J}{C \cdot m}$	Magnetic Flux Density $\vec{B}$ in $\frac{Wb}{m^2}$	Magnetic Field Intensity $\vec{H}$ in $\frac{A}{m} = \frac{J}{Wb \cdot m}$
Generalized Potential	Electric field Intensity $\vec{E}$ in $\frac{V}{m} = \frac{J}{C \cdot m}$	Displacement Flux Density $\vec{D}$ in $\frac{C}{m^2}$	Magnetic Field Intensity $\vec{H}$ in $\frac{A}{m} = \frac{J}{Wb \cdot m}$	Magnetic Flux Density $\vec{B}$ in $\frac{Wb}{m^2}$
Generalized Capacity	Permittivity ϵ in $\frac{F}{m} = \frac{C^2}{J \cdot m}$	$\frac{1}{\epsilon}$	Permeability μ in $\frac{H}{m} = \frac{Wb^2}{J \cdot m}$	$\frac{1}{\mu}$
Constitutive relationship	$\vec{D} = \epsilon \vec{E}$	$\vec{E} = \frac{1}{\epsilon} \vec{D}$	$\vec{B} = \mu \vec{H}$	$\vec{H} = \frac{1}{\mu} \vec{B}$
Energy	$\int_V \frac{1}{2}\epsilon |\vec{E}|^2 dV$	$\int_V \frac{1}{2}\frac{1}{\epsilon} |\vec{D}|^2 dV$	$\int_V \frac{1}{2}\mu |\vec{H}|^2 dV$	$\int_V \frac{1}{2}\frac{1}{\mu} |\vec{B}|^2 dV$
Law for potential	Gauss's Law for Elec. $\vec{\nabla} \cdot \vec{D} = \rho_{ch}$	Gauss's Law for Elec. $\vec{\nabla} \cdot \vec{E} = \epsilon \rho_{ch}$	Gauss's Law for Mag. $\vec{\nabla} \cdot \vec{B} = 0$	Gauss's Law for Mag. $\vec{\nabla} \cdot \vec{H} = 0$

Table 12.3: Describing electromagnetic systems in the language of calculus of variations.

12.3 Mechanical Energy Conversion

The previous section summarized how the language of calculus of variations can be applied to electrical and electromagnetic energy conversion devices. Similarly, this language can be used to describe energy conversion processes occurring in linear springs, torsion springs, moving masses, and flywheels.

We can convert energy to and from spring potential energy by compressing and releasing a spring. Similarly, we can store or release energy from a moving mass by changing its velocity. A *flywheel* is a device that stores energy in a spinning mass. Flywheels are used, in addition to batteries, in some electric and hybrid vehicles because storing rotational kinetic energy in a flywheel requires fewer energy conversion processes than storing energy in a battery. All of these energy conversion devices can be described in the language of calculus of variations with some parameter chosen as the generalized path.

Tables 12.4 and 12.5 summarize the parameters resulting from describing mechanical energy conversion processes in the language of calculus of variations. While electromagnetic systems are described by four vector fields, mechanical systems are described by eight possible vector fields, and they are listed along with their units in Table 12.6. Each column of Tables 12.4 and 12.5 describes the case of choosing a different vector field from Table 12.6 as the generalized path. By comparing across the rows of these tables as well as the electrical tables, comparisons can be made between the different energy conversion processes.

In Sec. 11.5, energy conversion in a linear spring was discussed in the language of calculus of variations. That example considered the displacement of a point mass m in kg where the generalized path was chosen to be displacement x in m. The resulting Euler-Lagrange equation was Newton's second law. Section 11.5 concluded with Table 11.1 summarizing the resulting parameters. The third column of the Table 12.4 repeats that information.

Circuit devices are often assumed to be point-like while electromagnetic properties of materials, like permittivity and permeability, are specified as functions of position. Similarly, mechanical devices can be treated as point-like or as functions of position. For example, mass is used to describe a point-like device while density is used to describe a device that varies with position. Researchers studying aerodynamics and fluid dynamics typically prefer the latter description. However, in Tables 12.4 and 12.5, point-like devices of mass m are assumed. Ideas in these tables can be generalized to situations where energy conversion devices are not treated as point-like and instead mass and other material properties vary with position.

12 RELATING ENERGY CONVERSION PROCESSES

Energy storage device	Linear Spring	Linear Spring	Flywheel	Flywheel
Generalized Path	$\vec{F}$ Force in $\frac{J}{m} = N$	Displacement $\vec{x}$ in m	$\vec{\omega}_{ang}$ Angular Velocity in $\frac{rad}{s}$	$\vec{L}_{am}$ Angular Momentum in J·s
Generalized Potential	Displacement $\vec{x}$ in m	$\vec{F}$ Force in $\frac{J}{m} = N$	$\vec{L}_{am}$ Angular Momentum in J·s	$\vec{\omega}_{ang}$ Angular Velocity in $\frac{rad}{s}$
Generalized Capacity	K in $\frac{J}{m^2}$	$\frac{1}{K}$ in $\frac{m^2}{J}$	$\frac{1}{\mathbb{I}}$ in $\frac{1}{kg \cdot m^2}$	$\mathbb{I}$ in $kg \cdot m^2$
Constitutive relationship	$\vec{F} = K\vec{x}$	$\vec{x} = \frac{1}{K}\vec{F}$	$\vec{\omega}_{ang} = \frac{1}{\mathbb{I}}\vec{L}_{am}$	$\vec{L}_{am} = \mathbb{I}\vec{\omega}_{ang}$
Energy		$\frac{1}{2}K|\vec{x}|^2 = \frac{1}{2}\frac{1}{K}|\vec{F}|^2$	$\frac{1}{2}\mathbb{I}|\vec{\omega}_{ang}|^2 = \frac{1}{2}\frac{1}{\mathbb{I}}|\vec{L}_{am}|^2$	
Law for potential	Newton's Second Law $\vec{F} = m\vec{a}$	Newton's Second Law $\vec{F} = m\vec{a}$	Conservation of Angular Momentum	Conservation of Angular Momentum

Table 12.4: Describing mechanical systems in the language of calculus of variations.

Energy storage device	Moving Mass	Moving Mass	Torsion Spring	Torsion Spring
Generalized Path	$\vec{M}$ Momentum in $\frac{\text{kg}\cdot\text{m}}{\text{s}} = \frac{\text{J}\cdot\text{s}}{\text{m}}$	$\vec{v}$ Velocity in $\frac{\text{m}}{\text{s}}$	$\vec{\tau}$ torque in $\frac{\text{N}\cdot\text{m}}{\text{rad}} = \frac{\text{J}}{\text{rad}}$	Angular Displacement $\vec{\theta}$ in radians
Generalized Potential	$\vec{v}$ Velocity in $\frac{\text{m}}{\text{s}}$	$\vec{M}$ Momentum in $\frac{\text{kg}\cdot\text{m}}{\text{s}} = \frac{\text{J}\cdot\text{s}}{\text{m}}$	Angular Displacement $\vec{\theta}$ in radians	$\vec{\tau}$ torque in $\frac{\text{N}\cdot\text{m}}{\text{rad}} = \frac{\text{J}}{\text{rad}}$
Generalized Capacity	m in kg	$\frac{1}{m}$ in $\frac{1}{\text{kg}}$	$\mathbb{K}$ in $\frac{\text{J}}{\text{rad}^2}$	$\frac{1}{\mathbb{K}}$ in $\frac{\text{rad}^2}{\text{J}}$
Constitutive relationship	$\vec{M} = m\vec{v}$	$\vec{v} = \frac{1}{m}\vec{M}$	$\vec{\tau} = \mathbb{K}\vec{\theta}$	$\vec{\theta} = \frac{1}{\mathbb{K}}\vec{\tau}$
Energy	$\frac{1}{2}m|\vec{v}|^2 = \frac{1}{2}\frac{|\vec{M}|^2}{m}$	$\frac{1}{2}m|\vec{v}|^2 = \frac{1}{2}\frac{|\vec{M}|^2}{m}$	$\frac{1}{2}\mathbb{K}|\vec{\theta}|^2 = \frac{1}{2}\mathbb{K}|\vec{\tau}|^2$	$\frac{1}{2}\mathbb{K}|\vec{\theta}|^2 = \frac{1}{2}\mathbb{K}|\vec{\tau}|^2$
Law for potential	Conservation of Momentum	Conservation of Momentum	Conservation of Torque	Conservation of Torque

Table 12.5: Describing more mechanical systems in the language of calculus of variations.

12 RELATING ENERGY CONVERSION PROCESSES

Symbol	Quantity	Units
$\vec{F}$	Force	N
$\vec{M}$	Momentum	$\frac{\text{kg·m}}{\text{s}}$
$\vec{v}$	Velocity	$\frac{\text{m}}{\text{s}}$
$\vec{x}$	Positional displacement	m
$\vec{L_{am}}$	Angular momentum	J·s
$\vec{\theta}$	Angular displacement vector	rad
$\vec{\tau}$	Torque	N·m
$\vec{\omega_{ang}}$	Angular velocity	$\frac{\text{rad}}{\text{s}}$

Table 12.6: Vector fields for describing mechanical displacement and fluid flow.

The vector fields listed in Table 12.6 are related by constitutive relationships:

$$\vec{M} = m\vec{v} \quad (12.5)$$

$$\vec{F} = K\vec{x} \quad (12.6)$$

$$\vec{\tau} = \mathbb{K}\vec{\theta} \quad (12.7)$$

$$\vec{L_{am}} = \mathbb{I}\vec{\omega_{ang}} \quad (12.8)$$

Equation 12.6 is more familiarly known as Hooke's law. By analogy to the capacitance of Eq. 12.1, the coefficients in these equations are referred to in Tables 12.4 and 12.5 as generalized capacity, and they represent the ability to store energy in the device. The constant m in Eq. 12.5 is mass in kg. The constant K in Eq. 12.6 is spring constant in $\frac{\text{J}}{\text{m}^2}$. The constant $\mathbb{K}$ in Eq. 12.7 is torsion spring constant in $\frac{\text{J}}{\text{radians}^2}$. The constant $\mathbb{I}$ in Eq. 12.8 is moment of inertia in units kg·m². A point mass rotating around the origin has a moment of inertia $\mathbb{I} = m|\vec{r}|^2$ where $|\vec{r}|$ is the distance from the mass to the origin. A solid shape has moment of inertia

$$\mathbb{I} = \int d\mathbb{I} = \int_0^m |\vec{r}|^2 dm. \quad (12.9)$$

Interestingly, there is a close relationship between the quantities in Tables 12.3 and 12.5. Maxwell's equations, first introduced in Section 1.6.1, relate the four electromagnetic field parameters. Assuming no sources, $\vec{J} = 0$ and $\rho_{ch} = 0$, Maxwell's equations can be written:

$$\vec{\nabla} \times \vec{E} = -\frac{\partial \vec{B}}{\partial t} \tag{12.10}$$

$$\vec{\nabla} \times \vec{H} = \frac{\partial \vec{D}}{\partial t} \tag{12.11}$$

$$\vec{\nabla} \cdot \vec{D} = 0 \tag{12.12}$$

$$\vec{\nabla} \cdot \vec{B} = 0 \tag{12.13}$$

The last two relationships, Gauss's laws, result directly from using calculus of variations to set up the Euler-Lagrange equation and solving for the corresponding equation of motion. We can replace electromagnetic vector fields in the source-free version of Maxwell's equations by mechanical fields according to the transformation:

$$\vec{D} \to \vec{M} \tag{12.14}$$

$$\vec{E} \to \vec{v} \tag{12.15}$$

$$\vec{B} \to \vec{\tau} \tag{12.16}$$

$$\vec{H} \to \vec{\theta} \tag{12.17}$$

The transformation of Eqs. 12.14 - 12.17 leads to set of equations accurately describing relationships between these mechanical fields.

$$\vec{\nabla} \times \vec{v} = -\frac{\partial \vec{\theta}}{\partial t} \tag{12.18}$$

$$\vec{\nabla} \times \vec{\tau} = \frac{\partial \vec{M}}{\partial t} \tag{12.19}$$

$$\vec{\nabla} \cdot \vec{M} = 0 \tag{12.20}$$

$$\vec{\nabla} \cdot \vec{\theta} = 0 \tag{12.21}$$

12 RELATING ENERGY CONVERSION PROCESSES

The last rows of Tables 12.4 and 12.5 list the relationship that results when an energy conversion device is described in the language of calculus of variations, the Euler-Lagrange equation is set up, and the Euler-Lagrange equation is solved for the equation of motion. The laws that result, Newton's second law, conservation of momentum, conservation of angular momentum, and conservation of torque, are fundamental ideas of mechanics.

12.4 Thermodynamic Energy Conversion

Four fundamental thermodynamic properties were introduced in Section 8.2: volume $\mathbb{V}$, pressure $\mathbb{P}$, temperature T, and entropy S. Many devices convert between some form of energy and either energy stored in a confined volume, energy stored in a material under pressure, energy in a temperature difference, or energy of a disordered system. We can describe energy conversion processes in these devices using the language of calculus of variations with one of these parameters, $\mathbb{V}$, $\mathbb{P}$, T, or S, as the generalized path and another as the generalized potential. Table 12.7 summarizes the results.

Many sensors convert energy between electrical energy and energy stored in a volume, pressure, or temperature difference. A capacitive gauge can measure the volume of liquid fuel versus vapor in the tank of an aircraft. Strain gauges and Piranhi hot wire gauges (Sec. 10.5), for example, are sensors that can measure pressure on solids or in gases. Pyroelectric detectors (Sec. 3.2), thermoelectric detectors (Sec. 8.8), thermionic devices (Sec. 10.2), and resistance temperature devices (Sec. 10.5) can be used to sense temperature changes.

Many other energy conversion devices convert between energy stored in a confined volume, energy stored in a material under pressure, or energy in a temperature difference and another form of energy without involving electricity. For example, if you tie a balloon to a toy car then release the air in the balloon, the toy car will move forward. Energy stored in the confined volume of the balloon, as well as in the stretched rubber of the balloon, is converted to kinetic energy of the toy car. An aerator or squirt bottle converts energy of a pressure difference to kinetic energy of a liquid. An eye dropper converts energy of a pressure difference to gravitational potential energy. An airfoil converts a pressure difference to kinetic energy in the form of lift. A piston converts energy of a gas under pressure to kinetic energy. As discussed in Sec. 10.6, a constricted pipe, or a weir, converts energy of a pressure difference in a flowing liquid to kinetic energy of the liquid. A baseball thrown as a curve ball converts the rotational energy of the rotating ball into a pressure differential to deflect the ball's path [162, p. 350]. A Sterling engine converts a temperature difference to kinetic energy.

Calculus of variations can be used to gain insights into thermodynamic energy conversion processes in these devices. The first step in applying the ideas of calculus of variations is to identify an initial and final form of energy. The Lagrangian is the difference between these forms of energy as a function of time. Some authors choose the Lagrangian as an entropy

12 RELATING ENERGY CONVERSION PROCESSES

Energy storage device	A balloon filled with air confined to a finite volume	A compressed piston	A cup of hot liquid (hot compared to the temp of the room)	A container with two pure gases separated by a barrier
Generalized Path	Volume $\mathbb{V}$ in m^3	Pressure $\mathbb{P}$ in Pa	Temperature T in K	Entropy S in $\frac{J}{K}$
Generalized Potential	Pressure $\mathbb{P}$ in Pa = $\frac{J}{m^3}$	Volume $\mathbb{V}$ in m^3 = $\frac{J}{Pa}$	Entropy S in $\frac{J}{K}$	Temperature T in K
Generalized Capacity	$\frac{\mathbb{V}}{\mathbb{B}} = -\frac{\partial \mathbb{V}}{\partial \mathbb{P}}$ in $\frac{m^6}{J}$	$\frac{\mathbb{B}}{\mathbb{V}} = -\frac{\partial \mathbb{P}}{\partial \mathbb{V}}$ in $\frac{J}{m^6}$	$\frac{T}{C_v} = \frac{\partial S}{\partial T}$ in $\frac{g \cdot K^2}{J}$	$\frac{C_v}{T} = \frac{\partial S}{\partial T}$ in $\frac{J}{g \cdot K^2}$
Constitutive relation-ship	$\Delta \mathbb{V} = -\frac{\mathbb{V}}{\mathbb{B}} \Delta \mathbb{P}$	$\Delta \mathbb{P} = -\frac{\mathbb{B}}{\mathbb{V}} \Delta \mathbb{V}$	$\Delta T = \frac{T}{c_v} \Delta S$	$\Delta S = \frac{C_v}{T} \Delta T$
Energy (int expression)	$\int \mathbb{V} d\mathbb{P}$	$\int \mathbb{P} d\mathbb{V}$	$\int T dS$	$\int S dT$
Energy (const. potential)	$\mathbb{V} \Delta \mathbb{P}$	$\mathbb{P} \Delta \mathbb{V}$	$T \Delta S$	$S \Delta T$
Law for potential	Bernoulli's Equation		Second Law of Thermo-dynamics	
This column assumes	constant S, T	constant S, T	constant $\mathbb{P}, \mathbb{V}$	constant $\mathbb{P}, \mathbb{V}$

Table 12.7: Describing thermodynamic systems in the language of calculus of variations.

instead of an energy [170] [171], but throughout this text Lagrangian is assumed to represent an energy as described in Ch. 11.

Assume that only one energy conversion process occurs in a device. Also assume that if we know three (not two) of the four thermodynamic parameters, we can calculate the fourth. Additionally, assume small amounts of energy are involved, and the energy conversion process occurs in the presence of a large external thermodynamic reservoir of energy.

As with the discussion of the previous tables, each column of Table 12.7 details the parameters of calculus of variations for a different choice of generalized path. In order, the columns can be used to describe energy storage in a gas confined to a finite volume, a material under pressure, a temperature differential, or an ordered system. The rows are labeled in the same way as in the previous tables of this chapter so that analogies between the systems can be drawn.

Energy can be stored and released from a gas confined to a finite volume and a gas under pressure. These related energy conversion processes are detailed in the second and third columns of Table 12.7 respectively. The second column specifies parameters of calculus of variations with volume chosen as the generalized path and pressure as the generalized potential. The third column specifies parameters with pressure chosen as the generalized path and volume as the generalized potential. In reality, energy conversion processes involving changes in the pressure and volume of a gas are unlikely to occur without a change in temperature or entropy of the system simultaneously occurring. Resistive heating, friction, gravity, and all other energy conversion processes that could simultaneously occur are ignored. Temperature and entropy are explicitly assumed to remain fixed, and these assumptions are listed in the last row of the table for emphasis. These columns can apply to energy conversion in liquids and solids in addition to gases. Using the choice of variables in the second column, the capacity to store energy is given by $\frac{V}{\mathbb{B}}$ where $\mathbb{B}$ is the bulk modulus in units pascals, and it is a measure of the ability of a compressed material to store energy [103]. Bulk modulus was introduced in Section 8.2. Using volume as the generalized path, the Euler-Lagrange equation can be set up and solved for the equation of motion. All terms of the resulting equation of motion have the units of pressure, and the equation of motion is a statement of Bernoulli's equation, an idea discussed in Section 10.6.

The fourth and fifth columns of Table 12.7 specify parameters of calculus of variations with temperature and entropy chosen as the generalized path respectively. A cup of hot liquid stores energy. Similarly, a container with two pure gases separated by a barrier stores energy. The system is in a more ordered state before the barrier is removed than after, and it

12 RELATING ENERGY CONVERSION PROCESSES

would take energy to restore the system to the ordered state. Both of these systems can be described by the language of calculus of variations. As detailed in the fourth column, temperature can be chosen as the generalized path and entropy can be chosen as the generalized potential. Alternatively as detailed in the fifth column, entropy can be chosen as the generalized path and temperature can be chosen as the generalized potential. Both of these columns assume that the pressure and volume remain constant. The quantity C_v, which shows up in these columns, is the specific heat at constant volume in units $\frac{J}{g \cdot K}$, and it was introduced in Sec. 8.3.

The equation of motion that results when temperature is chosen as the path and entropy is chosen as the generalized potential is a statement of conservation of entropy, and each term of this equation has the units of entropy. This relationship is more commonly known as the second law of thermodynamics, and it shows up in the second to last row of Table 12.7. More commonly, the law is written for a closed system as [109, p. 236],

$$\Delta S = \int \frac{\delta Q}{T} + S_{produced}. \tag{12.22}$$

In words, it says the change in entropy within a control mass is equal to the sum of the entropy out of the control mass due to heat transfer plus the entropy produced by the system.

(change in entropy) = (entropy out due to heat) + (entropy produced)

A system can become more organized or more disordered, so ΔS may be positive or negative. If energy is supplied in or out, entropy can be transfered in or out of a system, so the quantity $\int \frac{\delta Q}{T}$ may be positive or negative.

Energy is listed in the third to last row of Table 12.7 in two different forms. The first expression is an integral expression. For example, you can integrate the volume with respect to pressure to find the energy of a system.

$$E = \int \mathbb{V} d\mathbb{P} \tag{12.23}$$

Alternatively, the second expression

$$\Delta E = \mathbb{V} \Delta \mathbb{P} \tag{12.24}$$

can be used to find change in energy in the case when volume is not a strong function of pressure over a small element.

12.5 Chemical Energy Conversion

Batteries and fuel cells store energy in the chemical bonds of atoms. These devices were studied in Chapter 9. Table 12.8 details how to describe the physics of these chemical energy storage devices using the language of calculus of variations.

Sometimes chemists discuss macroscopic systems and describe charge distribution in a material by charge density ρ_{ch} in units $\frac{C}{m^3}$. In other cases, chemists study microscopic systems, where they are more interested in the number of electrons N and the distribution of these electrons around an atom. The second and third columns of Table 12.8 specify how to describe the macroscopic systems in the language of calculus of variations while the last two columns specify how to describe the microscopic systems.

In the second column of Table 12.8, the generalized path is ρ_{ch} and the generalized potential is the redox potential V_{rp} in volts. There is a close relationship between the choice of variables specified in the second column of Table 12.8 and the choices specified in the second columns of Table 12.1 and 12.3. More specifically, the generalized path described in the second column of Table 12.1 is charge Q in coulombs, where charge is the integral of the charge density with respect to volume.

$$Q = \int \rho_{ch} d\mathbb{V} \tag{12.25}$$

The generalized path described in the second column of Table 12.3 is displacement flux density $\vec{D}$ in units $\frac{C}{m^2}$. In the third column of Table 12.8, the opposite choice is made with V_{rp} for the generalized path and ρ_{ch} for the generalized potential. In Chapter 13, we consider a calculus of variations problem with this choice of variables in more detail to solve for the electron density around an atom.

Another way to apply the language of calculus of variations to chemical energy storage systems is to choose the number of electrons N as the generalized path and the chemical potential μ_{chem} as the generalized potential [172]. This situation is described in the fourth column of the Table 12.8. We could instead choose μ_{chem} as the generalized path and N as the generalized potential, and this situation is detailed in the last column of Table 12.8. Reference [172] details using calculus of variations with this choice of variables. Chemical potential is also known as the Fermi energy at $T = 0$ K, and it was discussed in Sections 6.3 and 9.2.3. It represents the average between the highest occupied and lowest unoccupied energy levels. The quantity E_g, which shows up in the fourth row of the table, is the energy gap in joules.

Energy storage device	Battery, fuel cell	Battery, fuel cell	Battery, fuel cell, chemical bonds of an atom	Battery, fuel cell, chemical bonds of an atom
Generalized Path	Charge density ρ_{ch} in $\frac{C}{m^3}$	Redox potential (voltage) V_{rp} in volts	Number of electrons N	Chemical potential μ_{chem} in $\frac{J}{atom}$
Generalized Potential	Redox potential (voltage) V_{rp} in volts	Charge Density ρ_{ch} in $\frac{C}{m^3}$	Chemical potential μ_{chem} in $\frac{J}{atom}$	Number of electrons N
Generalized Capacity	Capacitance C in farads	$\frac{1}{C}$	Inverse of energy gap $\frac{1}{E_g} = \frac{\partial N}{\partial \mu_{chem}}$	Energy gap $E_g = \frac{\partial \mu_{chem}}{\partial N}$ in J
Constitutive relationship	$\int \rho_{ch} d\mathbb{V} = CV_{rp}$	$V_{rp} = \frac{1}{C} \int \rho_{ch} d\mathbb{V}$	$\Delta N = \frac{1}{E_g}\Delta\mu_{chem}$	$\Delta\mu_{chem} = E_g \Delta N$
Energy	$\int_{\mathbb{V}} \rho_{ch} V_{rp} d\mathbb{V}$	$\int_{\mathbb{V}} \rho_{ch} V_{rp} d\mathbb{V}$	$N\mu_{chem}$	$N\mu_{chem}$
Law for potential	Nernst eq. (KVL)	Conservation of Charge	Nernst eq. (KVL)	Conservation of Charge
This column assumes	no variation in θ or ϕ	no variation in θ or ϕ	no variation in θ or ϕ	no variation in θ or ϕ

Table 12.8: Describing chemical systems in the language of calculus of variations.

12.6 Problems

12.1. Match each device, or device component, with the material or materials it is often made from.

Device or device component	Material
1. Photovoltaic device	**A.** Lead zirconium titanate
2. Piezoelectric device	**B.** Bismuth telluride
3. Battery anode	**C.** Cadmium telluride
4. Thermoelectric device	**D.** Mica, Quartz
5. Dielectric between capacitor plates	**E.** Zinc, Lithium

12.2. For each device or device component listed in the problem above, indicate whether it is typically made from a conductor, dielectric, or semiconductor.

12.3. For each of the devices below, list a material that the device is commonly made from.

- Photovoltaic Device
- Hall Effect Device
- Piezoelectric Device
- Capacitor

12.4. Appendix B lists multiple units along with whether or not they are SI base units. The joule, volt, and pascal are all SI derived units. Express each of these units in terms of SI base units.

12.5. Match the effect with the definition.

1. When an optical field is applied to a dielectric material, a material polarization develops in the material.	A. Hall effect
2. When an optical field is applied to a semiconductor junction, a voltage develops across the junction.	B. Electro- optic effect
3. When a current passes through a uniform material which has a temperature gradient, heating or cooling will occur	C. Photovoltaic effect
4. When a mechanical stress is applied to a dielectric material, a material polarization develops in the material.	D. Seebeck effect
5. When the different sides of a junction are held at different temperatures, a voltage develops across the junction.	E. Piezoelectric effect
6. When an external magnetic field is applied across a conductor or semiconductor with current flowing through it, a charge builds up perpendicular to the current and external magnetic field.	F. Thomson effect

12.6. Match the device with its definition.

1. A device which converts electromagnetic (often optical) energy directly to electricity	**A.** Fuel Cell
2. A device made from diodes of two dissimilar materials which converts a temperature differential to electricity	**B.** Photovoltaic Device
3. A device which converts chemical energy to electrical energy through the oxidation of a fuel	**C.** Piezoelectric Device
4. A device which converts mechanical stress directly to electricity	**D.** Pyroelectric Device
5. A device made from a crystal without a center of symmetry which converts a temperature differential to electricity	**E.** Thermoelectric Device

13 Thomas Fermi Analysis

13.1 Introduction

Where are the electrons found around an atom? This question is difficult for a few reasons. First, at temperatures above absolute zero, electrons are in continual motion. Second, the Heisenberg uncertainty principle tells us that we can never know the position and momentum of electrons simultaneously with complete accuracy. However, this question isn't hopeless. We can find the charge density ρ_{ch} which tells us, statistically on average, where the electrons are most likely to be found. Understanding the distribution of electrons in a material is vital to understanding the chemical properties, such as the strength of chemical bonds, as well as the electrical properties, such as how much energy is required to remove electrons.

To answer this question, we will use calculus of variations. The generalized path will be voltage V, and the generalized potential will be charge density ρ_{ch}. A Lagrangian describes an energy difference, and the Lagrangian will have the form

$$\mathcal{L} = \mathcal{L}\left(r, V, \frac{dV}{dr}\right). \tag{13.1}$$

The path found in nature is the one that minimizes the action.

$$\delta \int_{r_1}^{r_2} \mathcal{L} dr = 0 \tag{13.2}$$

In this problem, the independent variable is *position*, not time. We will set up the Euler-Lagrange equation then solve it to find the equation of motion.

Most of this chapter consists of a derivation of the resulting equation of motion called the Thomas Fermi equation. With a bit of algebra, we can find both the voltage and the charge density around the atom from the solution to the Thomas Fermi equation. The procedure is as follows.

- Describe the first form of energy, $E_{Coulomb\ e\ nucl} + E_{e\ e\ interact}$, in terms of path V. The resulting energy density is

$$\frac{E_{Coulomb\ e\ nucl}}{\mathbb{V}} + \frac{E_{e\ e\ interact}}{\mathbb{V}} = \frac{1}{2}\epsilon \left|\vec{\nabla}V\right|^2 \tag{13.3}$$

where ϵ represents permittivity and $\mathbb{V}$ represents volume.

- Describe the second form of energy $E_{kinetic\ e}$ in terms of path V. The resulting energy density is

$$\frac{E_{kinetic\ e}}{\mathbb{V}} = c_0 V^{5/2} \qquad (13.4)$$

where c_0 is a constant. This step will require the idea of reciprocal space.

- Write down the Hamiltonian $H\left(r, V, \frac{dV}{dr}\right)$ and Lagrangian $\mathcal{L}\left(r, V, \frac{dV}{dr}\right)$.

- Set up the Euler-Lagrange equation.

$$\frac{\partial \mathcal{L}}{\partial V} - \vec{\nabla} \cdot \left(\frac{\partial \mathcal{L}}{\partial \left(\frac{dV}{dr}\right)}\right) \hat{a}_r = 0 \qquad (13.5)$$

- Solve the Euler-Lagrange equation for the equation of motion. The result is

$$\frac{5}{2} c_0 V^{3/2} - \epsilon \nabla^2 V = 0. \qquad (13.6)$$

- Change variables to clean up the equation of motion. The resulting equation is called the *Thomas Fermi equation*.

$$\frac{d^2 y}{dt^2} = t^{-1/2} y^{3/2} \qquad (13.7)$$

- Voltage and charge density are algebraically related to the quantity y in the equation above.

To attempt to find charge density and voltage as a function of position r from the center of the atom, we will have to make some rather drastic assumptions. This analysis follows works of Thomas [173] and Fermi [174] which were originally completed around 1927. This derivation is discussed by numerous other authors as well [6] [46] [136] [175]. Because of the severe assumptions made below, the results will not be very accurate. However, more accurate numerical calculations are based on improved versions of the techniques established by Thomas and Fermi. We are discussing the most simplified version of the derivation, but this is the basis of more accurate approaches.

13.2 Preliminary Ideas

13.2.1 Derivatives and Integrals of Vectors in Spherical Coordinates

The derivation of the Thomas Fermi equation involves derivatives of vectors in spherical coordinates. For more details on derivatives and vectors see [11, ch. 1]. Consider a scalar function described in spherical coordinates,

$$V = V(\vec{r}) = V(r, \theta, \phi). \tag{13.8}$$

The *gradient* of $V(r, \theta, \phi)$ is defined

$$\vec{\nabla} V = \frac{\partial V}{\partial r} \hat{a}_r + \frac{1}{r} \frac{\partial V}{\partial \theta} \hat{a}_\theta + \frac{1}{r \sin \theta} \frac{\partial V}{\partial \phi} \hat{a}_\phi. \tag{13.9}$$

Gradient was introduced in Section 1.6.1. It returns a vector which points in the direction of largest change in the function. The Laplacian is defined in spherical coordinates as

$$\nabla^2 V = \frac{1}{r^2} \frac{\partial}{\partial r} \left(r^2 \frac{\partial V}{\partial r} \right) + \frac{1}{r^2 \sin \theta} \frac{\partial}{\partial \theta} \left(\sin \theta \frac{\partial V}{\partial \theta} \right) + \frac{1}{r^2 \sin^2 \theta} \frac{\partial^2 V}{\partial \phi^2}. \tag{13.10}$$

Qualitatively, the Laplacian of a scalar is the second derivative with respect to spatial position. In the derivations of this chapter, we encounter only functions which are uniform with respect to θ and ϕ. For functions of the form $V = V(r)$, the formulas for gradient and Laplacian simplify significantly.

$$\vec{\nabla} V = \frac{\partial V}{\partial r} \hat{a}_r \tag{13.11}$$

$$\nabla^2 V = \frac{1}{r^2} \frac{\partial}{\partial r} \left(r^2 \frac{\partial V}{\partial r} \right) \tag{13.12}$$

We will also need the vector identity of Eq. 1.10,

$$\nabla^2 V = \vec{\nabla} \cdot \vec{\nabla} V. \tag{13.13}$$

A differential volume element in spherical coordinates is given by

$$d\mathbb{V} = r^2 \sin \theta \, dr \, d\theta \, d\phi. \tag{13.14}$$

A volume integral of the function $V(r, \theta, \phi)$ over a sphere of radius 1 centered at the origin is denoted

$$\int_{r=0}^{1} \int_{\theta=0}^{\pi} \int_{\phi=0}^{2\pi} V(r, \theta, \phi) r^2 \sin \theta \, dr \, d\theta \, d\phi. \tag{13.15}$$

Assuming V doesn't depend on θ or ϕ, the integral is separable.

$$\left(\int_{\theta=0}^{\pi}\int_{\phi=0}^{2\pi}\sin\theta\, d\theta\, d\phi\right)\int_{r=0}^{1}V(r)r^2 dr = 4\pi\int_{r=0}^{1}V(r)r^2 dr \quad (13.16)$$

A sphere of radius r has volume $\frac{4}{3}\pi r^3$.

13.2.2 Notation

Writing this text without overloading variables has been a challenge. For example, V is the logical choice for denoting voltage, volume, and velocity. Up until now, the context offered clues to the meaning of symbols. However in this chapter, we will encounter equations involving both energy and electric field, equations involving both voltage and volume, and equations involving both mass and momentum. To help avoid confusion from the notation, Table 13.1 shows an excerpt of the variable list from Appendix A. This table does not list all quantities we will encounter. However, it highlights some of the more confusing ones.

In this chapter, we will encounter many quantities which vary with position. We will not encounter any quantities which vary with time. Therefore, voltage is denoted by a capital letter, not a lowercase letter. Voltage is a function of r, which denotes position in spherical coordinates. Assume that the origin of the coordinate system is at the center of the atom under consideration. Voltage is always specified with respect to some reference level called ground, so assume this zero volt reference level occurs at $r = \infty$. Also assume there is no θ or ϕ dependence of the voltage. Therefore, $V(\vec{r}) = V(r)$ represents voltage.

13 THOMAS FERMI ANALYSIS

Symbol	Quantity	SI Units	S/V/C	Comments
E	Energy	$J = Nm$	S	
$\vec{E}$	Electric field intensity	$\frac{V}{m}$	V	
E_f	Fermi energy level	J	S	Also called Fermi level
$\vec{k}$	Wave vector	m^{-1}	V	
k_f	Fermi wave vector	m^{-1}	S	
m	Mass	kg	S	
$\mathbb{M}$	Generalized momentum	*	S	Many authors use p
$\vec{M}$	Momentum	$\frac{kg \cdot m}{s}$	V	Many authors use $\vec{p}$
N	(Total) number of electrons per atom	$\frac{electrons}{atom}$	S	
v	Voltage (AC or time varying)	V	S	
$\vec{v}$	Velocity	$\frac{m}{s}$	V	
V	Voltage (DC)	V	S	
$\mathbb{V}$	Volume	m^3	S	
μ_{chem}	Chemical potential	$\frac{J}{atom}$	S	
ρ_{ch}	Charge density	$\frac{C}{m^3}$	S	

Table 13.1: Variable list.

13.2.3 Reciprocal Space Concepts

The idea of reciprocal space was introduced in Section 6.4 in the context of crystalline materials. We can describe the location of atoms in a crystal, for example, as a function of position where position $\vec{r}$ is measured in meters. In this chapter, we are interested in individual atoms instead of crystals composed of many atoms. We can plot quantities like energy $E(\vec{r})$ or voltage $V(\vec{r})$ as a function of position. Figure 6.11, for example, plots energy versus position inside a diode. In Section 6.4, the idea of wave vector $\vec{k}$ in units of m^{-1} was introduced. Wave vector represents the spatial frequency. We saw that we could plot energy or other quantities as a function of wave vector, and Fig. 6.8 is an example of such a plot. We will need the idea of wave vector in this chapter because we describe a situation where we do not know how the energy varies with position, but we do know something about how the energy varies with wave vector.

13.3 Derivation of the Lagrangian

The purpose of this chapter is to find the voltage $V(r)$ and the charge density $\rho_{ch}(r)$ around an atom, and we will use calculus of variations to accomplish this task. We need to make some rather severe assumptions to make this problem manageable. Consider an isolated neutral atom with many electrons around it. Assume $T \approx 0$ K, so all electrons occupy the lowest possible energy levels. Assume the atom is spherically symmetric. All of the quantities we encounter, such as voltage, charge density, and Lagrangian, vary with r but do not vary with θ or ϕ. We will use spherical coordinates with the origin at the nucleus of the atom. While quantities vary with position, assume no quantities vary with time. The charge density $\rho_{ch}(r)$ tells us where the electrons are most likely on average to be found. It is related to the quantum mechanical wave function, ψ, by

$$\rho_{ch} = -q \cdot |\psi|^2 \tag{13.17}$$

where q is the magnitude of the charge of an electron. Assume that all of the electrons surrounding the atom are distributed uniformly and can be treated as if they were a uniform electron cloud of some charge density.

Pick one of the electrons of the atom, and consider what happens when the electron is moved radially in and out. Figure 13.1 illustrates this situation. As the electron moves, energy conversion occurs. The goal of this section is to write down the Hamiltonian and Lagrangian for this energy conversion process. We write these quantities in the units of energy per unit volume per valence electron under consideration.

13 THOMAS FERMI ANALYSIS

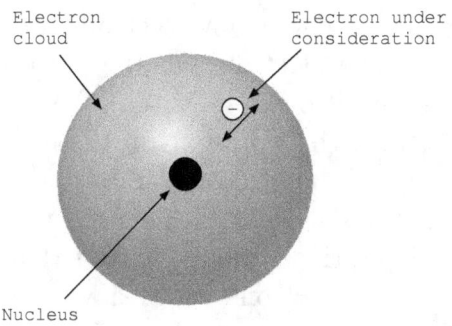

Figure 13.1: Illustration of an atom.

To understand what happens when the electron is moved, consider the energy of the atom in more detail. Coulomb's law, introduced in Eq. 1.4, tells us that charged objects exert forces on other charged objects. More specifically, the electric field intensity $\vec{E}$ due to a point charge of Q coulombs a distance r away surrounded by a material with permittivity ϵ is given by

$$\vec{E} = \frac{Q\hat{a}_r}{4\pi\epsilon r^2}. \tag{13.18}$$

The atom is composed of N positively charged protons. The electron under consideration feels an attractive Coulomb force due to these protons. Additionally, the atom has N electrons, and $N-1$ of these exert a repulsive Coulomb force on the electron under consideration. Since a charge separation and electric field exist, energy is stored. Call the component of the energy of the atom due to the Coulomb interaction between the protons of the nucleus and the electron under consideration $E_{Coulomb\ e\ nucl}$. Call the Coulomb interaction between the electron under consideration and all other electrons $E_{e\ e\ interact}$. The atom also has kinetic energy. Call the kinetic energy of the nucleus $E_{kinetic\ nucl}$ and the kinetic energy of all of the electrons $E_{kinetic\ e}$. The energy of the atom is the sum of all of these terms.

$$E_{atom} = E_{Coulomb\ e\ nucl.} + E_{kinetic\ nucl} + E_{e\ e\ interact} + E_{kinetic\ e} \tag{13.19}$$

Energy due to spin of the electrons and protons is ignored as is energy due to interaction with any other nearby charged objects. At $T \approx 0$ K, the kinetic energy of the nucleus will be close to zero, so we can ignore the term, $E_{kinetic\ nucl} \approx 0$. The quantity $E_{kinetic\ e}$ cannot be exactly zero. In Chapter 6 we plotted energy level diagrams for electrons around an atom. Even at $T = 0$ K, electrons have some internal energy, and this energy is denoted by the energy level occupied.

If we have a large atom with many electrons around it, the Coulomb interaction between any one electron and the nucleus is shielded by the Coulomb interaction from all other electrons. More specifically, suppose we have an isolated atom with N protons in the nucleus and N electrons around it. If we pick one of the electrons, $E_{Coulomb\ e\ nucl}$ for that electron describes the energy stored in the electric field due to the charge separation between the nucleus of positive charge Nq and that electron. However, there are also $N-1$ other electrons which have a negative charge. The term $E_{e\ e\ interact}$ describes the energy stored in the electric field due to the charge separation between the $N-1$ other electrons and the electron under consideration. These terms somewhat cancel each other out because the electron under consideration interacts with N protons each of positive charge q and $N-1$ electrons each of negative charge $-q$. However, the terms do not go away completely. Calculating

$$E_{Coulomb\ e\ nucl} + E_{e\ e\ interact} \tag{13.20}$$

is complicated because the electrons are in motion, and we do not really know where they are or even where they are most likely to be found. In fact, we are trying to solve for where they are likely to be found.

As we move the electron under consideration in and out radially, energy is transferred between $(E_{Coulomb\ e\ nucl} + E_{e\ e\ interact})$ and $E_{kinetic\ e}$. The Hamiltonian is the sum of these two forms of energy per unit volume, and the Lagrangian is the difference of these two forms of energy per unit volume. Both quantities have the units $\frac{J}{m^3}$. Choose voltage $V(r)$ as the generalized path and charge density $\rho_{ch}(r)$ as the generalized potential. The independent variable of these quantities is radial position r, not time. We can now write the Hamiltonian and Lagrangian.

$$H\left(r, V, \frac{dV}{dr}\right) = \left(\frac{E_{Coulomb\ e\ nucl}}{\mathbb{V}} + \frac{E_{e\ e\ interact}}{\mathbb{V}}\right) + \frac{E_{kinetic\ e}}{\mathbb{V}} \tag{13.21}$$

$$\mathcal{L}\left(r, V, \frac{dV}{dr}\right) = \left(\frac{E_{Coulomb\ e\ nucl}}{\mathbb{V}} + \frac{E_{e\ e\ interact}}{\mathbb{V}}\right) - \frac{E_{kinetic\ e}}{\mathbb{V}} \tag{13.22}$$

The next step is to write

$$\frac{E_{Coulomb\ e\ nucl}}{\mathbb{V}} + \frac{E_{e\ e\ interact}}{\mathbb{V}} \tag{13.23}$$

in terms of the path V. As detailed in Table 12.3, the energy density due to an electric field $\vec{E}$ is given by

$$\frac{E}{\mathbb{V}} = \frac{1}{2}\epsilon|\vec{E}|^2. \tag{13.24}$$

Remember that E represents energy while $\vec{E}$ represents electric field. Electric field is the negative gradient of the voltage $V(r)$.

$$\vec{E} = -\vec{\nabla}V. \tag{13.25}$$

We can combine these expressions and Eq. 13.13 to write the first term of the Hamiltonian and the Lagrangian in terms of the generalized path.

$$\frac{E_{Coulomb\ e\ nucl}}{\mathbb{V}} + \frac{E_{e\ e\ interact}}{\mathbb{V}} = \frac{1}{2}\epsilon\left|\vec{\nabla}V\right|^2 \tag{13.26}$$

$$H\left(r, V, \frac{dV}{dr}\right) = \left(\frac{1}{2}\epsilon\left|\vec{\nabla}V\right|^2\right) + \frac{E_{kinetic\ e}}{\mathbb{V}} \tag{13.27}$$

$$\mathcal{L}\left(r, V, \frac{dV}{dr}\right) = \left(\frac{1}{2}\epsilon\left|\vec{\nabla}V\right|^2\right) - \frac{E_{kinetic\ e}}{\mathbb{V}} \tag{13.28}$$

The next task is to describe the remaining term $\frac{E_{kinetic\ e}}{\mathbb{V}}$ as a function of the generalized path too. This task is a bit more challenging. We continue to take the approach of making severe approximations until it is manageable. We need to express $\rho_{ch}(r)$ as a function of $V(r)$. Then with some algebra, $\frac{E_{kinetic\ e}}{\mathbb{V}}$ can be written purely as a function of $V(r)$.

We want to generalize about the kinetic energy of the electrons. However, each electron has its own velocity $\vec{v}$ and momentum $\vec{M}$. These quantities depend on position

$$\vec{r} = r\hat{a}_r + \theta\hat{a}_\theta + \phi\hat{a}_\phi \tag{13.29}$$

in some unknown way. Furthermore, the calculation of $\frac{E_{kinetiic\ e}}{\mathbb{V}}$ depends on charge density $\rho_{ch}(r)$, which is the unknown quantity we are trying to find. We have more luck by describing these quantities in reciprocal space, introduced in Sec. 6.4. Position is denoted in reciprocal space by a wave vector

$$\vec{k} = \tilde{r}\hat{a}_r + \tilde{\theta}\hat{a}_\theta + \tilde{\phi}\hat{a}_\phi. \tag{13.30}$$

We can describe the properties of a material by describing how they vary with position in real space. For example, $\rho_{ch}(r)$ represents the charge density of electrons as a function of distance r from the center of the atom. We may be interested in how other quantities, such as the energy required to rip off an electron or the kinetic energy internal to an electron, vary with position in real space too. Instead of describing how quantities vary with position in real space, we can describe how quantities vary with spatial frequency of electrons. This is the idea behind representing quantities in reciprocal space. We may be interested in how the charge density of electrons varies as a function of the spatial frequency of charges in a crystal or other material, and this is the idea represented by functions of wave vector such as $\rho_{ch}\left(\vec{k}\right)$. We are trying to solve for charge density $\rho_{ch}(r)$. We expect that electrons are more likely to be found at certain distances r from the center of the atom than at other distances. However, there is no pattern to the charge density as a function of wave vector, $\rho_{ch}\left(\vec{k}\right)$. Assume that ρ_{ch} is roughly constant with respect to $|\vec{k}|$ up to some level. With some more work, this assumption will allow us to solve for charge density $\rho_{ch}(r)$.

The kinetic energy of a single electron is given by

$$\frac{E_{kinetic\ e}}{e^-} = \frac{1}{2}m|\vec{v}|^2 \qquad (13.31)$$

where m is the mass of the electron. We can write this energy in terms of momentum, $\vec{M} = m\vec{v}$. (Note that momentum $\vec{M}$ and generalized momentum $\mathbb{M}$ are different and have different units.)

$$\frac{E_{kinetic\ e}}{e^-} = \frac{|\vec{M}|^2}{2m} \qquad (13.32)$$

We do not know how the energy varies as a function of position r. Instead, we can write the energy as a function of the crystal momentum $\vec{M}_{crystal}$ or the wave vector $\vec{k}$, and we know something about the variation of these quantities. *Crystal momentum* is equal to the wave vector scaled by the Planck constant.

$$\vec{M}_{crystal} = \hbar\vec{k} \qquad (13.33)$$

It has the units of momentum $\frac{\text{kg·m}}{\text{s}}$, and it was introduced in Sec. 6.4.2. The kinetic energy of one electron as a function of the crystal momentum is given by

$$\frac{E_{kinetic\ e}}{e^-} = \frac{\left(\vec{M}_{crystal}\right)^2}{2m} = \frac{\left(\hbar|\vec{k}|\right)^2}{2m}. \qquad (13.34)$$

A vector in reciprocal space is represented Eq. 13.30, and Eq. 13.34 can be simplified because we are assuming spherical symmetry $\tilde{\theta} = \tilde{\phi} = 0$. The magnitude of the wave vector becomes $|\vec{k}| = \tilde{r}$, and we can write the energy as

$$\frac{E_{kinetic\ e}}{e^-} = \frac{\hbar^2\tilde{r}^2}{2m}. \qquad (13.35)$$

Just as each electron has its own momentum $m|\vec{v}|$, each electron has its own crystal momentum $\hbar|\vec{k}|$. However, we know some information about the wave vector $|\vec{k}|$ of the electrons in the atom. At $T = 0$ K, electrons occupy the lowest allowed energy states. Energy states are occupied up to some highest occupied state called the *Fermi energy* E_f. While electrical engineers use the term Fermi energy, chemists sometimes use the term *chemical potential* μ_{chem}. The lowest energy states, are occupied while the higher ones are empty. Similarly, wave vectors are occupied up to some highest occupied wave vector called the *Fermi wave vector* k_f.

$$|\vec{k}| = \begin{cases} \text{filled state} & \tilde{r} < k_f \\ \text{empty state} & \tilde{r} > k_f \end{cases} \quad (13.36)$$

The Fermi energy and the Fermi wave vector are related by

$$E_f = \frac{\hbar^2 k_f^2}{2m}. \quad (13.37)$$

We use the idea of reciprocal space to write an expression for the kinetic energy of the electrons per unit volume [136, p. 49]. The kinetic energy due to any one electron as a function of position in reciprocal space is given by Eq. 13.35. Note that at each value of $|\vec{k}| = \tilde{r}$, the electron has a different kinetic energy. To find the kinetic energy per unit volume due to all electrons, we integrate over all $|\vec{k}| = \tilde{r}$ in spherical coordinates that are occupied by electrons, and then we divide by the volume occupied in $\vec{k}$ space.

$$\frac{E_{kinetic\ e}}{\mathbb{V}} = \frac{1}{\text{vol. occupied in k space}} \cdot \int_{\text{filled k levels}} \left(\frac{E_{kinetic\ e}}{e^-}\right) \left(\frac{e^-}{\text{volume}}\right) d(\text{vol. all k space}) \quad (13.38)$$

The number of electrons per unit volume is given by

$$\left(\frac{e^-}{\text{volume}}\right) = \frac{-\rho_{ch}}{q}. \quad (13.39)$$

The volume occupied in reciprocal space is $\frac{4}{3}\pi k_f^3$, the volume of a sphere of radius k_f.

$$\frac{E_{kinetic\ e}}{\mathbb{V}} = \frac{1}{\frac{4}{3}\pi k_f^3} \cdot \int_{\text{filled k levels}} \left(\frac{\hbar^2 \tilde{r}^2}{2m}\right) \left(\frac{-\rho_{ch}}{q}\right) d(\text{vol. all k space}) \quad (13.40)$$

A differential element of the volume is expressed as

$$d^3\left|\vec{k}\right| = \tilde{r}^2 \sin\tilde{\theta}\ d\tilde{r}\ d\tilde{\theta}\ d\tilde{\phi}. \quad (13.41)$$

$$\frac{E_{kinetic\ e}}{\mathbb{V}} = \frac{1}{\frac{4}{3}\pi k_f^3} \cdot \int_{\text{filled k levels}} \left(\frac{\hbar^2 |\vec{k}|^2}{2m}\right) \left(\frac{-\rho_{ch}}{q}\right) \left(\tilde{r}^2 \sin\tilde{\theta}\ d\tilde{r}\ d\tilde{\theta}\ d\tilde{\phi}\right) \tag{13.42}$$

As described above, electrons occupy states in reciprocal space only with $0 \leq \tilde{r} \leq k_f$.

$$\frac{E_{kinetic\ e}}{\mathbb{V}} = \frac{1}{\frac{4}{3}\pi k_f^3} \cdot \int_{\tilde{r}=0}^{k_f} \int_{\tilde{\theta}=0}^{\pi} \int_{\tilde{\phi}=0}^{2\pi} \left(\frac{\hbar^2 \tilde{r}^2}{2m}\right) \left(\frac{-\rho_{ch}}{q}\right) \left(\tilde{r}^2 \sin\tilde{\theta}\ d\tilde{r}\ d\tilde{\theta}\ d\tilde{\phi}\right) \tag{13.43}$$

The integral above can be evaluated directly. The first step to evaluate it is to pull constants outside. As described above, ρ_{ch} varies with r but not $\tilde{r}$, so it can be pulled outside the integral too.

$$\frac{E_{kinetic\ e}}{\mathbb{V}} = \frac{-1}{\frac{4}{3}\pi k_f^3} \cdot \frac{\hbar^2 \rho_{ch}}{2mq} \int_{\tilde{r}=0}^{k_f} \int_{\tilde{\theta}=0}^{\pi} \int_{\tilde{\phi}=0}^{2\pi} \tilde{r}^4 \sin\tilde{\theta}\ d\tilde{r}\ d\tilde{\theta}\ d\tilde{\phi} \tag{13.44}$$

The integral separates and can be evaluated.

$$\frac{E_{kinetic\ e}}{\mathbb{V}} = \frac{-1}{\frac{4}{3}\pi k_f^3} \cdot \frac{\hbar^2 \rho_{ch}}{2mq} \left(\int_{\tilde{\theta}=0}^{\pi} \int_{\tilde{\phi}=0}^{2\pi} \sin\tilde{\theta}\ d\tilde{\theta}\ d\tilde{\phi}\right) \left(\int_{\tilde{r}=0}^{k_f} \tilde{r}^4 d\tilde{r}\right) \tag{13.45}$$

$$\frac{E_{kinetic\ e}}{\mathbb{V}} = \frac{-1}{\frac{4}{3}\pi k_f^3} \cdot \frac{\hbar^2 \rho_{ch}}{2mq} 4\pi \left(\frac{k_f^5}{5}\right) \tag{13.46}$$

$$\frac{E_{kinetic\ e}}{\mathbb{V}} = \frac{-3\rho_{ch} k_f^2 \hbar^2}{10mq} \tag{13.47}$$

Charge density is a function of position in real space r, and we are in the process of solving for this function, $\rho_{ch}(r)$. However, it also depends on the Fermi energy E_f, and hence Fermi wave vector k_f, for the atom. Next, we find the relationship between ρ_{ch} and k_f. Two electrons are allowed per energy level (spin up and spin down), hence per filled k state. The number of filled states per atom in reciprocal space is related to the charge density.

$$\rho_{ch} = -2q \left(\frac{\text{no. filled } k \text{ states}}{\text{unit vol. in } k \text{ space}}\right) \tag{13.48}$$

In Sec. 6.4.1, we saw that a primitive cell in reciprocal space was $(2\pi)^3$ times the primitive cell in real space, so

$$(\text{unit vol. } k \text{ space}) = (2\pi)^3 \cdot (\text{unit vol. real space}) = (2\pi)^3. \tag{13.49}$$

We know something about the wave vectors of filled states in reciprocal space. At $T = 0$ K, the lowest states are filled, and all others are empty, and they are filled up to a radius of k_f. The volume of a sphere of radius k_f is given by $\frac{4}{3}\pi k_f^3$, and this represents the number of filled k states per volume of reciprocal space. We can therefore simplify the expression above.

$$\rho_{ch} = -2q \cdot \frac{4}{3}\pi k_f^3 \cdot \frac{1}{(2\pi)^3} \tag{13.50}$$

$$\rho_{ch} = \frac{-q}{3\pi^2} k_f^3 \tag{13.51}$$

$$k_f = \left(\frac{-3\pi^2}{q}\rho_{ch}\right)^{1/3} \tag{13.52}$$

We want to write $\frac{E_{kinetic\ e}}{V}$ as a function of generalized path V. We can now achieve this task by combining Eqs. 13.47 and 13.52.

$$\frac{E_{kinetic\ e}}{V} = \frac{-3\hbar^2}{10mq}\rho_{ch}\left(\frac{-3\pi^2}{q}\rho_{ch}\right)^{2/3} \tag{13.53}$$

$$\frac{E_{kinetic\ e}}{V} = \frac{-3\hbar^2}{10mq}\left(\frac{-3\pi^2}{q}\right)^{2/3}\rho_{ch}^{5/3} \tag{13.54}$$

Electrical energy is the product of charge and voltage. More specifically, from Eq. 2.8, it is given by

$$E = \frac{1}{2}QV. \tag{13.55}$$

Electrical energy density is then given by

$$\frac{E}{V} = \frac{1}{2}\rho_{ch}V. \tag{13.56}$$

Use Eq. 13.56 to relate ρ_{ch} and V.

$$\frac{E_{kinetic\ e}}{V} = \frac{1}{2}\rho_{ch}V = \frac{-3\hbar^2}{10mq}\left(\frac{-3\pi^2}{q}\right)^{2/3}\rho_{ch}^{5/3} \tag{13.57}$$

We have now related the generalized path and the generalized potential.

$$V = \frac{-3\hbar^2}{5mq}\left(\frac{-3\pi^2}{q}\right)^{2/3}\rho_{ch}^{2/3} \tag{13.58}$$

$$\rho_{ch} = \left(\frac{-5mq}{3\hbar^2} \cdot \left(\frac{-3\pi^2}{q}\right)^{-2/3}\right)^{3/2} V^{3/2} \tag{13.59}$$

$$\rho_{ch} = \left[\left(\frac{-5mq}{3\hbar^2}\right)^{3/2}\left(\frac{-q}{3\pi^2}\right)\right] \cdot V^{3/2} \qquad (13.60)$$

Finally, we can write $\frac{E_{kinetic\ e}}{V}$ as a function of V.

$$\frac{E_{kinetic\ e}}{V} = \left[\left(\frac{-5mq}{3\hbar^2}\right)^{3/2}\left(\frac{-q}{3\pi^2}\right)\right] V^{5/2} \qquad (13.61)$$

Notice that the quantity in brackets above is constant. The coefficient c_0 is defined from the term in brackets.

$$c_0 = \left(\frac{-5mq}{3\hbar^2}\right)^{3/2}\left(\frac{-q}{3\pi^2}\right) \qquad (13.62)$$

$$\frac{E_{kinetic\ e}}{V} = c_0 V^{5/2} \qquad (13.63)$$

We now can describe all of the terms of the Lagrangian in terms of our generalized path.

$$\frac{E_{Coulomb\ e\ nucl}}{V} + \frac{E_{e\ e\ interact}}{V} = \frac{1}{2}\epsilon\left|\vec{\nabla}V\right|^2 \qquad (13.64)$$

$$\frac{E_{kinetic\ e}}{V} = c_0 V^{5/2} \qquad (13.65)$$

The Hamiltonian represents the total energy density, and the Lagrangian represents the energy density difference of these forms of energy. The Hamiltonian and Lagrangian have the form $H = H\left(r, V, \frac{dV}{dr}\right)$ and $\mathcal{L} = \mathcal{L}\left(r, V, \frac{dV}{dr}\right)$ where r is position in spherical coordinates. There is no θ or ϕ dependence of H or $\mathcal{L}$. Everything is spherically symmetric.

$$H = \frac{1}{2}\epsilon|\vec{\nabla}V|^2 + c_0 V^{5/2} \qquad (13.66)$$

$$\mathcal{L} = \frac{1}{2}\epsilon|\vec{\nabla}V|^2 - c_0 V^{5/2} \qquad (13.67)$$

As an aside, let us consider the Fermi energy $E_f = \mu_{chem}$ once again. With some algebra, we can write it as a function of voltage. Use Eqs. 13.37, 13.52, and 13.62.

$$E_f = \frac{\hbar^2 k_f^2}{2m} = \frac{\hbar^2}{2m}\left(\frac{-3\pi^2 \rho_{ch}}{q}\right)^{2/3} \qquad (13.68)$$

$$E_f = \frac{\hbar^2}{2m}\left(\frac{-3\pi^2}{q}\right)^{2/3}\left[\left(\frac{-5mq}{3\hbar^2}\cdot\left(\frac{-3\pi^2}{q}\right)^{-2/3}\right)^{3/2} V^{3/2}\right]^{2/3} \qquad (13.69)$$

$$E_f = \frac{-5q}{6}V \qquad (13.70)$$

Notice that the Fermi energy is just a scaled version of the voltage V with respect to a ground level at $r = \infty$. Electrical engineers often use the word voltage synonymously with potential. When chemists use the term chemical potential, they are referring to the same quantity just scaled by a constant. Just as voltage is a fundamental quantity of electrical engineering that represents how difficult it is to move electrons around, chemical potential is fundamental quantity of chemistry that represents how difficult it is to move electrons around.

13.4 Deriving the Thomas Fermi Equation

As the electron around an atom moves, energy is converted between energy of the Coulomb interaction and kinetic energy of the electron. The action is

$$\mathbb{S} = \left| \int_{r_1}^{r_2} \mathcal{L} dr \right|. \qquad (13.71)$$

The path found in nature minimizes the action.

$$\delta \left| \int_{r_1}^{r_2} \mathcal{L} dr \right| = 0 \qquad (13.72)$$

The integral is over *position*, not time. In chapter 11, we called this idea the *Principle of Least Action*. Reference [136, p. 52] calls this idea in this context the *Second Hohenberg-Kohn Theorem*. To find the path, we set up and solve the Euler-Lagrange equation. The Euler-Lagrange equation in the case where the independent variable is a vector of the form $\vec{r} = r\hat{a}_r$ instead of a scalar (with no θ or ϕ dependence anywhere) is given by

$$\frac{\partial \mathcal{L}}{\partial (\text{path})} - \vec{\nabla} \cdot \left(\frac{\partial \mathcal{L}}{\partial \left(\frac{d(\text{path})}{dr} \right)} \right) \hat{a}_r = 0 \qquad (13.73)$$

[176, p. 13].

As described above, generalized path is voltage $V = V(r)$, and generalized potential is charge density $\rho_{ch} = \rho_{ch}(r)$. As discussed in Chapter 12, we could have made the opposite choice. In fact, the opposite choice may seem more logical because the words voltage and potential are often used synonymously. The same result is obtained regardless of the choice. However, the algebra is less messy with this choice, and this choice is more consistent with the literature.

Next, evaluate the Euler-Lagrange equation, Eq. 13.73, using the Lagrangian of Eq. 13.67. The resulting equation is the equation of motion. Consider some of the pieces needed. The derivative of the Lagrangian with respect to the path is

$$\frac{\partial \mathcal{L}}{\partial V} = \frac{5}{2} c_0 V^{3/2}. \tag{13.74}$$

In Chapter 11, this quantity was defined as the generalized potential. Above, we defined ρ_{ch} as the generalized potential. Both $\frac{\partial \mathcal{L}}{\partial V}$ and ρ_{ch} have units $\frac{C}{m^3}$. According to Eq. 13.60, $\frac{\partial \mathcal{L}}{\partial V}$ is ρ_{ch} multiplied by a constant, and that constant is close to one. Since $\frac{\partial \mathcal{L}}{\partial V}$ isn't equal to ρ_{ch}, our equations are not completely consistent. However, the difference is small given the extreme assumptions made elsewhere. We also need the generalized momentum.

$$\mathbb{M} = \frac{\partial \mathcal{L}}{\partial \left(\frac{dV}{dr}\right)} = \epsilon \frac{dV}{dr}. \tag{13.75}$$

$$\frac{\partial \mathcal{L}}{\partial \left(\frac{dV}{dr}\right)} \hat{a}_r = \epsilon \vec{\nabla} V \tag{13.76}$$

Next, put these pieces into the Euler-Lagrange equation.

$$\frac{5}{2} c_0 V^{3/2} - \vec{\nabla} \cdot \left(\epsilon \vec{\nabla} V\right) = 0 \tag{13.77}$$

Use Eq. 13.13.

$$\frac{5}{2} c_0 V^{3/2} - \epsilon \nabla^2 V = 0 \tag{13.78}$$

$$\nabla^2 V = \frac{5}{2\epsilon} c_0 V^{3/2} \tag{13.79}$$

Next, following Fermi's original work [177], change variables

$$V = \frac{-y}{r} \tag{13.80}$$

where y has the units V·m. The Laplacian term on the left can be simplified using Eq. 13.12.

$$\nabla^2 V = \nabla^2 \left(\frac{-y}{r}\right) \tag{13.81}$$

$$\nabla^2 V = \frac{1}{r^2} \frac{\partial}{\partial r} \left[r^2 \frac{\partial}{\partial r} \left(\frac{-y}{r}\right)\right] \tag{13.82}$$

$$\nabla^2 V = \frac{1}{r^2} \frac{\partial}{\partial r} \left[r^2 \left(\frac{y}{r^2} - \frac{1}{r} \frac{\partial y}{\partial r}\right)\right] \tag{13.83}$$

13 THOMAS FERMI ANALYSIS

$$\nabla^2 V = \frac{1}{r^2}\frac{\partial}{\partial r}\left(y - r\frac{\partial y}{\partial r}\right) \qquad (13.84)$$

$$\nabla^2 V = \frac{1}{r^2}\left(\frac{\partial y}{\partial r} - \frac{\partial y}{\partial r} - r^2\frac{\partial^2 y}{\partial r^2}\right) \qquad (13.85)$$

$$\nabla^2 V = -\frac{1}{r}\frac{\partial^2 y}{\partial r^2} \qquad (13.86)$$

Eq. 13.79 now simplifies.

$$-\frac{1}{r}\frac{\partial^2 y}{\partial r^2} = \frac{-5}{2\epsilon}c_0\left(\frac{-y}{r}\right)^{3/2} \qquad (13.87)$$

$$\frac{-1}{r}\frac{d^2 y}{dr^2} = \frac{5}{2\epsilon}c_0(-1)^{1/2}\left(\frac{y}{r}\right)^{3/2} \qquad (13.88)$$

$$\frac{d^2 y}{dr^2} = c_1 r^{-1/2} y^{3/2} \qquad (13.89)$$

In the equation above, the constant is

$$c_1 = -\frac{5}{2\epsilon}c_0(-1)^{1/2}.$$

$$c_1 = \frac{-5}{2\epsilon}\left[\left(\frac{-5mq}{3\hbar^2}\right)^{3/2}\left(\frac{-q}{3\pi^2}\right)\right](-1)^{1/2} \qquad (13.90)$$

$$c_1 = \frac{5}{2\epsilon}\left[\left(\frac{5mq}{3\hbar^2}\right)^{3/2}\frac{q}{3\pi^2}\right] \qquad (13.91)$$

To clean Eq. 13.89 up further, choose

$$t = c_1^{-2/3} r. \qquad (13.92)$$

The variable t here is the name of the independent variable, and it does *not* represent time. It is a scaled version of the position r.

$$\frac{d^2 y}{dt^2} = t^{-1/2} y^{3/2} \qquad (13.93)$$

Equation 13.93 is called the *Thomas Fermi equation*. We have finished the derivation. The Thomas Fermi equation along with appropriate boundary conditions can be solved for $y(t)$. Since the equation is nonlinear, numerical techniques are likely used to solve it. Once $y(t)$ is found, Eqs. 13.56 and 13.80 can be used to find $V(r)$ and $\rho_{ch}(r)$. From this equation of motion, we can find $\rho_{ch}(r)$, where, on average, the electrons are likely to be found as a function of distance from the nucleus in spherical coordinates.

13.5 From Thomas Fermi Theory to Density Functional Theory

The analysis considered in this chapter is based on works from 1926 to 1928 [173] [174]. They were early attempts at calculating the location of electrons around an atom, and they were developed when the idea of an atom itself was still quite new. Half of Fermi's work is in Italian, and half is in German. However, it is clearer than most technical papers written in English.

A calculation is called *ab initio* if it is from first principles while a calculation is called *semi-empirical* if some experimental data is used to find parameters of the solution [136, p. 13]. The Thomas Fermi method is the simplest ab initio solution of the calculation of the charge density and energy of electrons in an atom [136]. Since no experimental data is used, the results of the calculation can be compared to experimental data from spectroscopic experiments to verify the results.

We already know that the results are not very accurate because we made a lot of rather extreme assumptions to make this problem manageable. Assumptions include:

- There is no angular dependence to energy, charge density, voltage, or other quantities.

- Temperature is near absolute zero, $T \approx 0$ K, so that all electrons occupy the lowest allowed energy states.

- There is only one isolated atom with no other charged particles around it.

- The atom is not ionized and is not part of a molecule.

- The atom has many electrons, and one electron feels effects of a uniform cloud due to other electrons.

- The electrons of the atom do not have any spin or internal angular momentum.

Refined versions of this calculation are known as *density functional theory*. A function is a quantity that takes in a scalar value and returns a scalar value. A functional takes in a function and returns a scalar value. The name density functional theory comes from the fact that the Lagrangian and Hamiltonian are written as functionals of the charge density. Density functional theory calculations do not make as many or as severe of assumptions as were made above, especially for the $E_{e\ e\ interact}$ term. These calculations have been used to calculate the angular dependence of the charge density, the allowed energy states of electrons that are part of molecules, the voltage felt by electrons at temperatures above absolute zero [136], and many other microscopic properties of atoms. Density functional theory is an active area of research. Often charge density is chosen as the generalized path instead of voltage [136].

Both Thomas [173] and Fermi [174] included numerical simulations. Amazingly, these calculations were performed way before computers were available! More recently, researchers have developed software packages for applying density functional theory to calculate the allowed energy levels, charge density, and so on of electrons around atoms and molecules [178] [179]. Because of the complexity of the calculations, parallel processing is used. Computers with multiple processors, supercomputers, and graphics cards with dozens of processors have all been used.

13.6 Problems

13.1. Generalized momentum is defined as
$$\mathbb{M} = \frac{\partial \mathcal{L}}{\partial \left(\frac{dV}{dr}\right)}.$$

(a) Find the generalized momentum for the system described by the Lagrangian of Eq. 13.67.

(b) The generalized momentum does not have the units of momentum. Identify the units of this generalized momentum.

(c) Write the Hamiltonian of Eq. 13.66 as a function of r, V, and $\mathbb{M}$ but not as a function of $\frac{dV}{dr}$.

(d) Write the Lagrangian of Eq. 13.67 as a function of r, V, and $\mathbb{M}$ but not as a function of $\frac{dV}{dr}$.

(e) Show that the Hamiltonian and Lagrangian found above satisfy the equation $H = \mathbb{M}\frac{dV}{dr} - \mathcal{L}$.

13.2. In the analysis of this chapter, the generalized path was chosen as V and the generalized potential was chosen as ρ_{ch}. The opposite choice is also possible where the generalized path is ρ_{ch} and the generalized potential is V.

(a) Write the Hamiltonian of Eq. 13.66 as functions of ρ_{ch} instead of V, so it has the form $H\left(r, \rho_{ch}, \frac{d\rho_{ch}}{dr}\right)$.

(b) Repeat the above for the Lagrangian of Eq. 13.67.

(c) Find the Euler-Lagrange equation using ρ_{ch} as the generalized path.

13.3. Verify that
$$y = \frac{144}{t^3}$$
is a solution of the Thomas Fermi equation [46].

(While this solution satisfies the Thomas Fermi equation, it is not useful in describing the energy of an atom. In the $t \to 0$ limit, this solution approaches infinity, $y(0) \to \infty$. However, in the $t \to 0$ limit, the solution should approach a constant, $y(0) \to 1$, to correctly describe the physical behavior of an atom [180].)

13.4. The previous problem discussed that
$$y = \frac{144}{t^3}$$
is a solution of the Thomas Fermi equation. Show that
$$y = \frac{72}{t^3}$$
is *not* a solution.

13.5. Prove that the Thomas Fermi equation is nonlinear.

14 Lie Analysis

14.1 Introduction

In Chapter 11, the ideas of calculus of variations were applied to energy conversion processes. We began with two forms of energy and studied how those forms of energy varied with variation in some generalized path and some generalized potential. The result was an equation of motion that described the variation of the generalized path. The equation of motion had the form of conservation of generalized potential. In Chapter 12, conservation laws were listed in the last row of the tables. Knowing how forms of energy vary with path and with potential provide significant information about energy conversion processes. The purpose of this chapter is to show that we can find symmetries, invariants, and other information about the energy conversion process by applying Lie analysis techniques to this equation of motion. If continuous symmetries of an equation can be identified, it is often possible to extract quite a bit of information by starting only with the equation.

The equations of motion that result from calculus of variations are not always linear. It may or may not be possible to solve a nonlinear equation of motion for the path. Even in the cases where it is possible, it is often quite difficult because techniques for solving nonlinear differential equations are much less developed than techniques for linear equations. Furthermore, many nonlinear differential equations do not have closed form solutions. In this chapter, we will see a systematic technique for getting information out of nonlinear differential equations that comes from calculus of variation. The technique is known as *Lie analysis* based on the work of Sophus Lie in the late part of the nineteenth century. Additionally, this chapter introduces Noether's theorem. Using this theorem and an equation of motion, we may be able to derive conserved quantities. The techniques discussed in this chapter apply even for *nonlinear* equations.

Lie analysis is a systematic procedure for identifying *continuous symmetries* of an equation. If the equation possesses continuous symmetries, we may be able to find related conservation laws. Some equations possess multiple symmetries and conservation laws while other equations do not contain any symmetries or conservation laws. Using this procedure with a known generalized path, we may be able to derive conserved quantities even if we do not know how to choose the generalized potential at first. Some systems might even contain multiple conserved quantities, and this procedure will give us a complete set of conserved quantities.

Lie analysis has been used to find continuous symmetries of many fundamental equations of physics, and it has been applied to both classical and quantum mechanical equations. References [164, p. 117] and [181] apply the procedure to the heat equation

$$\frac{dy}{dt} = \frac{d^2y}{dx^2} \tag{14.1}$$

describing the function $y(t,x)$. It has been applied to both the two dimensional wave equation [164, p. 123] and the three dimensional wave equation [181]. Other equations analyzed by this procedure include Schrödinger's equation [182] [183], Maxwell's equations [184] [185], and equations of nonlinear optics [186].

A tremendous amount of information can be gained by looking at the symmetries of equations. Knowledge of continuous symmetries may allow us to solve equations or at least reduce the order of differential equations [164]. If we can identify symmetries, we may be able to simplify or speed up numerical calculations by using known repetition in the form of the solution. If multiple equations contain the same symmetry elements, we can draw comparisons between the equations [164]. We may be able to find invariant quantities of the system from known continuous symmetries of equations. Hopefully this chapter will provide an appreciation for the amount of information that can be gained from applying symmetry analysis to equations of motion describing energy conversion processes.

14.1.1 Assumptions and Notation

The techniques of this chapter are applied to equations of motion that results from describing an energy conversion processes by calculus of variations. All starting equations of motion are assumed to have only one independent and one dependent variable. These equations may or may not be linear. Furthermore, all independent and dependent variables are assumed to be purely real. We made the same assumptions in Chapter 11. Most of the examples in this chapter involve second order differential equations because many of the energy conversion processes studied in Chapter 11 led to equations of motion which were second order differential equations. However, these techniques apply to algebraic equations and to differential equations of other orders.

In this chapter, total derivatives will be denoted as either $\frac{dy}{dt}$ or $\dot{y}$. Partial derivatives will be denoted as either $\frac{\partial y}{\partial t}$ or $\partial_t y$ for shorthand. If the quantity y is just a function of a single independent variable, there is no reason to distinguish between total and partial derivatives, $\frac{dy}{dt} = \frac{\partial y}{\partial t}$. Equations of motion in this chapter will involve one independent and one dependent variable, $y(t)$. However, we will encounter functionals of multiple independent variables such as the Lagrangian $\mathcal{L} = \mathcal{L}(t, y, \frac{dy}{dt})$. For such quantities, we will have to distinguish between total and partial derivatives carefully.

The analysis here is in no way mathematically rigorous. Furthermore, the examples in this chapter are not original. References to the literature are included below.

These techniques generalize to more complicated equations. They apply to equations with multiple independent and multiple dependent variables, and they apply when these variables are complex [164]. Also, these techniques apply to partial differential equations as well as ordinary differential equations, and they even apply to systems of equations [164]. See references [164] for how to generalize the methods introduced in this chapter to the other situations.

14.2 Types of Symmetries

14.2.1 Discrete versus Continuous

This chapter is concerned with identifying symmetries of equations. We say that an equation contains a symmetry if the solution to the equation is the same both before and after a symmetry transformation is applied. The wave equation is given by

$$\frac{d^2y}{dt^2} + \omega_0^2 y = 0 \tag{14.2}$$

where ω_0 is a constant. When t represents time, ω_0 has units of frequency. The wave equation is invariant upon the discrete symmetry

$$y \to \tilde{y} = -y. \tag{14.3}$$

This transformation is a symmetry because when all y's in the equation are transformed, the resulting equation contains the same solutions as the original equation.

$$\frac{d^2\tilde{y}}{dt^2} + \omega_0^2 \tilde{y} = 0 \tag{14.4}$$

$$\frac{d^2(-y)}{dt^2} + \omega_0^2(-y) = 0 \tag{14.5}$$

$$\frac{d^2y}{dt^2} + \omega_0^2 y = 0 \tag{14.6}$$

Symmetries can be classified as either continuous or discrete. *Continuous symmetries* can be expressed as a sum of infinitesimally small symmetries related by a continuous parameter. A *discrete symmetry* cannot be written as a sum of infinitesimal transformations in this way. Three commonly discussed discrete symmetry transformations [187] are:

- Time reversal $t \to \tilde{t} = (-1)^n t$, for integer $\mathfrak{n}$

- Parity $y \to \tilde{y} = (-1)^n y$, for integer $\mathfrak{n}$

- Charge conjugation $y \to \tilde{y} = y^*$, where $*$ denotes complex conjugate.

For example, the wave equation is invariant upon each of these three discrete symmetries because solutions of the equation remain the same before and after these symmetry transformations are performed. The transformation $t \to \tilde{t} = t + \varepsilon$, where ε is the continuous parameter which can be infinitesimally small, is an example of a continuous transformation because it can be separated into a sum of infinitesimal symmetries. Both discrete and continuous symmetries may involve transformations of the independent variable, the dependent variable, or both variables. In this chapter, we will study a systematic procedure for identifying continuous symmetries of an equation, and we will not consider discrete symmetries further.

14.2.2 Regular versus Dynamical

Continuous symmetries can be classified as regular or dynamical. *Regular* continuous symmetries involve transformations of the independent variables and dependent variables. *Dynamical* symmetries involve transformations of the independent variables, dependent variables, and the derivatives of the dependent variables [188]. (Some authors use the term generalized symmetries instead of dynamical symmetries [164, p. 289].) Only regular symmetries will be considered. The techniques discussed here generalize to dynamical symmetries [164], but they are beyond the scope of this text.

14.2.3 Geometrical versus Nongeometrical

Symmetries may also be classified as geometrical or nongeometrical [184] [185]. *Nongeometrical* symmetry transformations involve taking a Fourier transform, performing some transformation of the variables, then taking an inverse Fourier transform. The resulting transformations are symmetries if the solution of the equation under consideration are the same before and after the transformations occur. Nongeometrical symmetries can be written as functions of an infinitesimal parameter but are not continuous. Nongeometrical symmetries will not be discussed here and are also beyond the scope of this text.

14.3 Continuous Symmetries and Infinitesimal Generators

14.3.1 Definition of Infinitesimal Generator

Symmetry transformations can be described as transformations of the independent and dependent variables. Continuous symmetry transformations can be described as transformations of these variables which depend on a, possibly infinitesimal, parameter ε.

$$t \to \tilde{t} = F(\varepsilon)t \tag{14.7}$$

$$y \to \tilde{y} = F(\varepsilon)y \tag{14.8}$$

The operator $F(\varepsilon)$ describes the transformation. It is a function of the infinitesimal parameter ε, and it may also depend on t and y. Furthermore, it is an operator meaning that it may involve derivative operations.

We are considering only continuous symmetries, so we can study the behavior in the limit as $\varepsilon \to 0$. The operator $F(\varepsilon)$ can be written as a Taylor series in the small parameter ε.

$$F(\varepsilon) = 1 + \varepsilon U + \frac{1}{2!}\varepsilon^2 U^2 + \ldots \tag{14.9}$$

The term U in the expansion above is called the *infinitesimal generator*. It may be separated into two components.

$$U = \xi \partial_t + \eta \partial_y \tag{14.10}$$

The function ξ describes infinitesimal variation in the independent variable. The function η describes infinitesimal variation in the dependent variable, and it was introduced in Sec. 11.4. Both ξ and η may depend on both the independent variable and the dependent variable.

$$\xi = \xi(t, y) \tag{14.11}$$

$$\eta = \eta(t, y) \tag{14.12}$$

In the limit of $\varepsilon \to 0$, we can ignore terms of order ε^2 or higher.

$$F(\varepsilon) \approx 1 + \varepsilon U \tag{14.13}$$

An infinitesimal generator describes a continuous symmetry transformation. If we know an infinitesimal generator for some continuous symmetry, we can find the corresponding transformation

$$t \to e^{\varepsilon U} t \quad \text{and} \quad y \to e^{\varepsilon U} y. \tag{14.14}$$

To understand where this relationship between infinitesimal generators and finite transformations come from, consider the Taylor expansion of $e^{\varepsilon U}$ [14, p. 33].

$$e^{\varepsilon U} = 1 + \epsilon U + \frac{1}{2!}(\varepsilon U)^2 + \frac{1}{3!}(\varepsilon U)^3 + \dots \quad (14.15)$$

In the limit as $\varepsilon \to 0$,

$$e^{\varepsilon U} \approx 1 + \varepsilon U. \quad (14.16)$$

Therefore, the corresponding infinitesimal transformation for $\varepsilon \to 0$ is given by

$$t \to t(1 + \varepsilon \xi) \quad (14.17)$$

$$y \to y(1 + \varepsilon \eta). \quad (14.18)$$

14.3.2 Infinitesimal Generators of the Wave Equation

As an example, consider infinitesimal generators of the wave equation

$$\frac{d^2 y}{dt^2} + \omega_0^2 y = 0. \quad (14.19)$$

As mentioned above, the wave equation contains a continuous symmetry of the form $t \to t + \varepsilon$. This continuous symmetry transformation has the form

$$t \to t(1 + \varepsilon \xi) \quad \text{and} \quad y \to y(1 + \varepsilon \eta) \quad (14.20)$$

with $\xi = 1$ and $\eta = 0$. It can be described by the infinitesimal generator

$$U = \xi \partial_t + \eta \partial_y = \partial_t. \quad (14.21)$$

More generally, infinitesimal generators and finite transformations are related by Eq. 14.14, so finite transformations can be derived from infinitesimal generators.

$$t \to \left(e^{\varepsilon \partial_t}\right) t = \left(1 + \varepsilon \partial_t + \frac{1}{2!}(\varepsilon \partial_t)^2 + \dots\right) t = t + \varepsilon \quad (14.22)$$

$$y \to \left(e^{\varepsilon \partial_t}\right) y = \left(1 + \varepsilon \partial_t + \frac{1}{2!}(\varepsilon \partial_t)^2 + \dots\right) y = y. \quad (14.23)$$

While the symmetry transformation was given in this example, below we will see a procedure to derive infinitesimal generators for an equation.

In general, if we know a solution to an equation and we know that a symmetry is present, we can derive a whole family of related solutions to

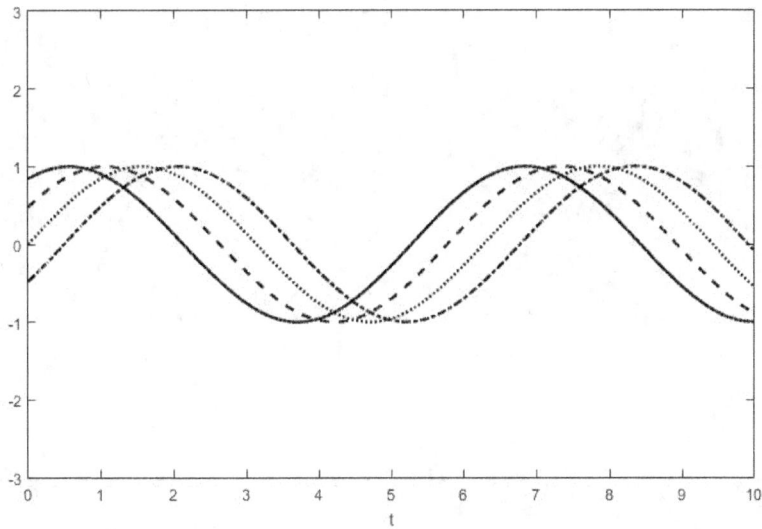

Figure 14.1: The solid line shows a solution to the wave equation. The dotted and dashed lines show solutions found using the symmetry transformation $t \to t + \varepsilon$ and $y \to y$ which has infinitesimal generator $U = \partial_t$.

the equation without having to go through the work of solving the equation again. The wave equation, Eq. 14.19, has solutions of the form

$$y(t) = c_0 \cos(\omega_0 t) + c_1 \sin(\omega_0 t) \tag{14.24}$$

where boundary conditions determine the constants c_0 and c_1. The symmetry described by the infinitesimal generator $U = \partial_t$ tells us that

$$y(t) = c_0 \cos(\omega_0 (t + \varepsilon)) + c_1 \sin(\omega_0 (t + \varepsilon)) \tag{14.25}$$

must also be a solution. Using Eq. 14.24, we have found a family of related solutions because Eq. 14.25 is a solution for all finite or infinitesimal constants ε. Figure 14.1 illustrates this idea. The known solution is shown as a solid line. The dotted and dashed lines illustrate related solutions, for different constant ε values. We encountered the wave equation in the mass spring example of Section 11.5 and the capacitor inductor example of Section 11.6, so symmetry analysis provides information about both of these energy conversion processes. It tells us that if we run the energy conversion process and find one physical path $y(t)$, then for appropriate boundary conditions, $y(t + \varepsilon)$ is also a physical path. This symmetry is present in all time invariant systems. Qualitatively for the mass spring example, it tells us that if we know the path taken by the mass when we

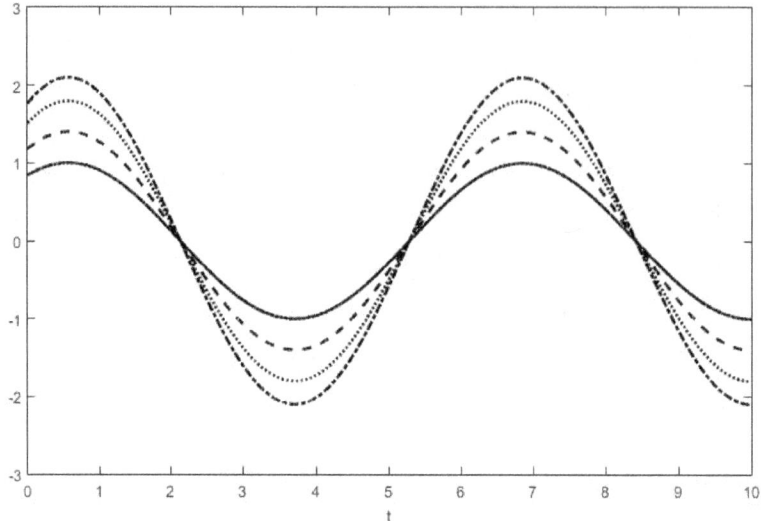

Figure 14.2: The solid line shows a solution to the wave equation. The dotted and dashed lines show solutions found using the symmetry transformation $t \to t$ and $y \to y(1+\varepsilon)$ which has infinitesimal generator $U = y\partial_y$.

remove the restraint today, then we know the path taken by the mass when we repeat the experiment tomorrow, and we know this idea from symmetry analysis without having to re-analyze the system.

All linear equations, including the wave equation, contain a continuous symmetry transformation described by the infinitesimal generator

$$U = y\partial_y \qquad (14.26)$$

which corresponds to $\xi = 0$ and $\eta = y$. Again, we can find the corresponding finite transformation using Eq. 14.14.

$$t \to \left(e^{\varepsilon y \partial_y}\right) t = \left(1 + \varepsilon y \partial_y + \frac{1}{2!}\left(\varepsilon y \partial_y\right)^2 + ...\right) t = t \qquad (14.27)$$

and

$$y \to \left(e^{\varepsilon y \partial_y}\right) y = \left(1 + \varepsilon y \partial_y + \frac{1}{2!}\left(\varepsilon y \partial_y\right)^2 + ...\right) y = y(1+\varepsilon). \qquad (14.28)$$

To summarize this transformation,

$$t \to t \qquad (14.29)$$

and
$$y \to y(1+\varepsilon) \tag{14.30}$$

The above transformation says that if we scale any solution of a linear equation, $y(t)$, by a constant $(1+\varepsilon)$, the result will also be a solution of the equation. By definition, a linear equation obeys exactly this property. By knowing a solution of the wave equation and this symmetry, we can find a whole family of related solutions, and this family of solutions is illustrated in Figure 14.2.

The wave equation also contains the symmetry transformation described by the infinitesimal generator

$$U = \sin(\omega_0 t)\partial_y. \tag{14.31}$$

The operators ξ and η can be identified directly from the infinitesimal generator.

$$\xi = 0 \tag{14.32}$$
$$\eta = \sin(\omega_0 t) \tag{14.33}$$

Again we can find the corresponding finite transformations using Eq. 14.14.

$$t \to e^{\varepsilon U} t = t \tag{14.34}$$

$$y \to e^{\varepsilon U} y = \left(1 + \varepsilon \sin(\omega_0 t)\, \partial_y + \frac{1}{2!}\left(\varepsilon \sin(\omega_0 t)\, \partial_y\right)^2 + ...\right) y \tag{14.35}$$

$$y \to y + \varepsilon \sin(\omega_0 t) \tag{14.36}$$

If we know a solution $y(t)$ to the wave equation, this transformation tells us that $y(t) + \varepsilon \sin(\omega_0 t)$ is also a solution. Since ε can be any infinitesimal or finite constant, we have found another family of solutions using symmetry concepts, and these solutions are illustrated in Figure 14.3.

In this section, we have discussed three of the symmetries of the wave equation. The wave equation actually contains eight continuous symmetry transformations. Deriving these transformations is left as a homework problem.

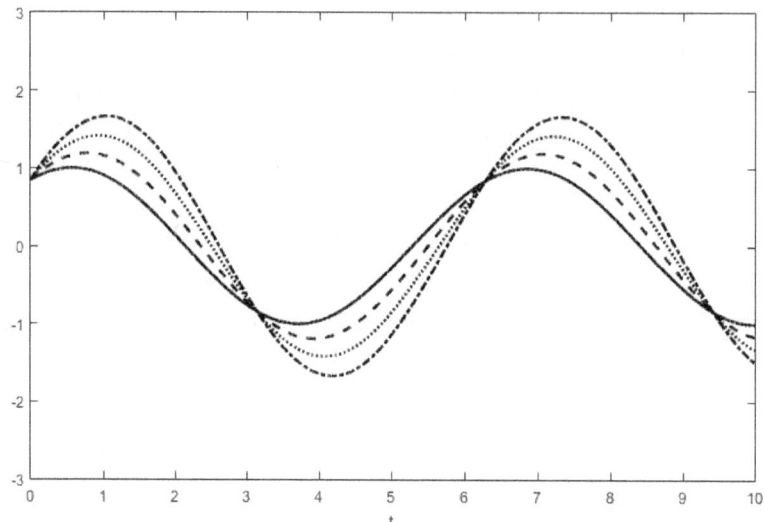

Figure 14.3: The solid line shows a solution to the wave equation. The dotted and dashed lines show solutions found using the symmetry transformation $t \to t$ and $y \to y + \varepsilon \sin(\omega_0 t)$ which has infinitesimal generator $U = \sin(\omega_0 t)\, \partial_y$.

14.3.3 Concepts of Group Theory

The study of symmetries of equations falls under a branch of mathematics called *group theory*. When mathematicians use the word *group*, they have something specific in mind. A *group* is a set of elements along with an operation that combines two elements. The operation is called *group multiplication*, but it may or may not be the familiar multiplication operation from arithmetic. To be a group, the elements and operation must obey four additional properties: identity, inverse, associativity, and closure [14, p. 7] [164, p. 14]. A group is called a *Lie group* if all elements are continuously differentiable [164, p. 14].

The first property of group elements is the identity property which says that every group must have an identity element, X_{id}. When the identity element is multiplied by any other group element, the result is that other element. The inverse property says that each element in the group must have a corresponding inverse element which is also in the group. When the original element is multiplied by its inverse, the result must be the identity element. The associative property says that the product of group elements $(X_1 \cdot X_2) \cdot X_3$ where the first two elements are multiplied first and the product of group elements $X_1 \cdot (X_2 \cdot X_3)$ where the last two elements

Group property name	Summary of property
Identity	$X_1 \cdot X_{id} = X_1$
Inverse	$X_1 \cdot X_1^{-1} = X_{id}$
Associativity	$(X_1 \cdot X_2) \cdot X_3 = X_1 \cdot (X_2 \cdot X_3)$
Closure	$X_1 \cdot X_2$ is an element of the group

Table 14.1: Group properties.

are multiplied first must be equal.

$$(X_1 \cdot X_2) \cdot X_3 = X_1 \cdot (X_2 \cdot X_3) \tag{14.37}$$

The closure property says that when two elements of the group are multiplied together, the result is another element of the group. Table 14.1 summarizes these properties where X_1, X_2, and X_3 are elements of the group, and X_{id} is the identity element which is also a member of the group. However, groups may have more or less than three elements.

In general, the order in which group elements are multiplied matters.

$$X_1 \cdot X_2 \neq X_2 \cdot X_1. \tag{14.38}$$

The quantity $X_1 \cdot X_2 \cdot X_1^{-1} \cdot X_2^{-1}$ is sometimes called the *commutator*, and it is denoted $[X_1, X_2]$. Due to the closure property, the result of the commutator is guaranteed to be another element of the group [14, p. 21,32] [164, p. 39,50].

$$[X_1, X_2] = X_1 \cdot X_2 \cdot X_1^{-1} \cdot X_2^{-1} \tag{14.39}$$

Continuous symmetries of equations are described by infinitesimal generators that form a Lie group. The elements of the group are the infinitesimal generators scaled by a constant [164, p. 52]. The group multiplication operation is regular multiplication also possibly scaled by a constant. According to this definition, $U = \partial_t$, $U = 2\partial_t$ and $U = -10.2\partial_t$ are all the same element of the group because the constant does not affect the element. If we find a few infinitesimal generators of a group, we may be able to use Eq. 14.39 to find more generators. A *complete set* of infinitesimal generators describe all possible continuous (regular geometrical) symmetries of the equation. All continuous (regular geometrical) symmetry operations of the equation can be described as linear combinations of the infinitesimal generators.

14.4 Derivation of the Infinitesimal Generators

14.4.1 Procedure to Find Infinitesimal Generators

We are studying differential equations, which can be written as

$$F(t, y, \dot{y}, ...) = 0 \qquad (14.40)$$

for some function F. We are looking for continuous symmetries that can be applied to this equation such that the original equation and the transformed equation have the same solutions. The symmetries are denoted by infinitesimal generators

$$U = \xi \partial_t + \eta \partial_y \qquad (14.41)$$

that describe how the independent variable t and dependent variable y transform. Upon a symmetry transformation, the independent variable and dependent variable transform, but so do the derivatives of the dependent variable, $\dot{y}$, $\ddot{y}$, ... The *prolongation* of an infinitesimal generator is a generalization of the infinitesimal generator that describes the transformation of the independent variable, the dependent variable, and derivatives of the dependent variable [164, p. 94].

The nth prolongation of a generator U is defined as

$$\mathrm{pr}^{(n)} U = \xi \partial_t + \eta \partial_y + \eta^t \partial_{\dot{y}} + \eta^{tt} \partial_{\ddot{y}} + \eta^{ttt} \partial_{\dddot{y}} + ..., \qquad (14.42)$$

and it has terms involving η^{t^n}. The functions η^t and η^{tt} are defined [164],

$$\eta^t = \eta^t(t, y, \dot{y}) = \frac{d}{dt}(\eta - \xi \dot{y}) + \xi \ddot{y} \qquad (14.43)$$

$$\eta^{tt} = \eta^{tt}(t, y, \dot{y}) = \frac{d^2}{dt^2}(\eta - \xi \dot{y}) + \xi \dddot{y} \qquad (14.44)$$

The quantities η^{ttt}, η^{tttt}, and so on can be defined similarly, but they will not be needed for the examples below. The prolongation of the infinitesimal generator is an operator that describes the transformation of t, y, $\dot{y}$, $\ddot{y}$, and so on up to the nth derivative. Some authors [189] use the term *tangential mapping* instead of prolongation.

The procedure to find all possible continuous symmetries of an equation is based on the idea that the solutions of an equation remain unchanged upon a symmetry operation. For a given transformation to be a symmetry operation, not only must all the solutions remain unchanged, but so must all derivatives of the solutions. Thus, for a differential equation of the form $F(t, y, \dot{y}, ...) = 0$, all symmetries U obey the *symmetry condition*

$$\text{pr}^{(n)} UF = 0. \tag{14.45}$$

We solve this symmetry condition to find all allowed infinitesimal generators that describe continuous symmetries of the original equation.

We can use Eqs. 14.43 and 14.44 to write the symmetry condition in terms of the components of the infinitesimal generators, ξ and η. Then, we solve the symmetry condition for ξ and η. This step involves some algebra, but it can be accomplished with some patience and an adequate supply of ink and paper.

We can solve the symmetry condition for the allowed infinitesimal generators. By careful solution, we find *all* infinitesimal generators of the form $U = \xi \partial_t + \eta \partial_y$. This procedure gives us a systematic way to find *all* continuous symmetries of the equation.

This technique applies to any differential equation. We are most interested in applying it to equations of motion that describe energy conversion processes. From this technique, we get information about solutions of the equation even when the equation of motion is nonlinear. Furthermore, in the Sec. 14.5 we see that we may be able to use the symmetries to find invariants of the equation, and invariants often have physical meaning. All symmetries of calculus of variations problems of the form $\delta \int \mathcal{L} dt = 0$ are necessarily symmetries of the Euler-Lagrange equation. However, the converse is not necessarily true, so not all symmetries of the Euler-Lagrange equation are symmetries of the integral equation [164, p. 255].

14.4.2 Thomas Fermi Equation Example

As an example, we apply this procedure to the Thomas Fermi equation

$$\ddot{y} = y^{3/2} t^{-1/2}. \tag{14.46}$$

This equation was derived in Chapter 13. From the solution of this equation $y(t)$, the charge density $\rho_{ch}(r)$ of electrons around an isolated atom and the voltage $V(r)$ felt by the electrons can be calculated within the, rather severe, assumptions specified in that chapter. The independent variable of the equation is a scaled version of radial position, not time. However, t will be used as the independent variable here because the procedure applies to equations regardless of the name of the variable. Reference [190] applies this procedure to a family of equations known as the Emden-Fowler equations. The Thomas Fermi equation is a special case of an Emden-Fowler equation, so the result of this example can be found in reference [190].

We would like to identify continuous symmetries of Eq. 14.46. These symmetries will be specified by infinitesimal generators of the form

$$U = \xi \partial_t + \eta \partial_y \qquad (14.47)$$

where ξ and η have the form $\xi(t,y)$ and $\eta(t,y)$. Solutions of the equation satisfy

$$\left(\ddot{y} - y^{3/2}t^{-1/2}\right) = 0. \qquad (14.48)$$

For infinitesimal generators that describe symmetries of this equation, the prolongation is also zero.

$$\text{pr}^{(n)}U\left(\ddot{y} - y^{3/2}t^{-1/2}\right) = 0. \qquad (14.49)$$

Eq. 14.49 can be solved for all generators U corresponding to continuous symmetries of the Thomas Fermi equation. Eqs. 14.42 and 14.49 can be combined.

$$\eta^{tt} + \frac{1}{2}\xi y^{3/2}t^{-3/2} - \frac{3}{2}\eta y^{1/2}t^{-1/2} = 0 \qquad (14.50)$$

Next, Eq. 14.44 is used.

$$\partial_{tt}\eta + 2\dot{y}\partial_{yt}\eta + \ddot{y}\partial_y\eta + \dot{y}^2\partial_{yy}\eta - 2\ddot{y}\partial_t\xi - \dot{y}\partial_{tt}\xi - 2\dot{y}^2\partial_{yt}\xi \\ -\dot{y}^3\partial_{yy}\xi - 3\dot{y}\ddot{y}\partial_y\xi + \frac{1}{2}\xi y^{3/2}t^{-3/2} - \frac{3}{2}\eta y^{1/2}t^{-1/2} = 0 \qquad (14.51)$$

Substitute the original equation for $\ddot{y}$.

$$\partial_{tt}\eta + 2\dot{y}\partial_{yt}\eta + y^{3/2}t^{-1/2}\partial_y\eta + \dot{y}^2\partial_{yy}\eta - 2y^{3/2}t^{-1/2}\partial_t\xi - \dot{y}\partial_{tt}\xi - 2\dot{y}^2\partial_{yt}\xi \\ -\dot{y}^3\partial_{yy}\xi - 3\dot{y}y^{3/2}t^{-1/2}\partial_y\xi + \frac{1}{2}\xi y^{3/2}t^{-3/2} - \frac{3}{2}\eta y^{1/2}t^{-1/2} = 0 \qquad (14.52)$$

Regroup terms.

$$\left(\partial_{tt}\eta + y^{3/2}t^{-1/2}\partial_y\eta - 2y^{3/2}t^{-1/2}\partial_t\xi + \frac{1}{2}\xi y^{3/2}t^{-3/2} - \frac{3}{2}\eta y^{1/2}t^{-1/2}\right) \\ +\dot{y}\left(2\partial_{yt}\eta - \partial_{tt}\xi - 3y^{3/2}t^{-1/2}\partial_y\xi\right) + \dot{y}^2\left(\partial_{yy}\eta - 2\partial_{yt}\xi\right) - \dot{y}^3\left(\partial_{yy}\xi\right) = 0 \qquad (14.53)$$

Each of the terms in parentheses in Eq. 14.53 must be zero.

$$\partial_{tt}\eta + y^{3/2}t^{-1/2}\partial_y\eta - 2y^{3/2}t^{-1/2}\partial_t\xi + \frac{1}{2}\xi y^{3/2}t^{-3/2} - \frac{3}{2}\eta y^{1/2}t^{-1/2} = 0 \qquad (14.54)$$

$$2\partial_{yt}\eta - \partial_{tt}\xi - 3y^{3/2}t^{-1/2}\partial_y\xi = 0 \qquad (14.55)$$

$$\partial_{yy}\eta - 2\partial_{yt}\xi = 0 \qquad (14.56)$$

$$\partial_{yy}\xi = 0 \qquad (14.57)$$

Eqs. 14.54, 14.55, 14.56, and 14.57 can be solved for ξ and η. From Eq. 14.57, $\partial_{yy}\xi = 0$, so ξ must have form

$$\xi = (c_1 + c_2 y)\, b(t). \tag{14.58}$$

The quantities denoted c_n are constants. From Eq. 14.56, η must have the form

$$\eta = \left(c_3 + c_4 y + c_5 y^2\right) g(t). \tag{14.59}$$

Functions $b(t)$ and $g(t)$ only depend on t, not y. The condition of Eq. 14.55 can be rewritten.

$$(2c_4 \partial_t g - c_1 \partial_{tt} b) + y(4c_5 \partial_t g - 2c_2 \partial_{tt} b) - 3y^{3/2} t^{-1/2} c_2 b = 0 \tag{14.60}$$

To satisfy Eq. 14.60, c_2 must be zero, and either $c_5 = 0$ or $g(t) = 0$. From Eqs. 14.55 and 14.56, $\partial_y \eta$ and $\partial_t \xi$ must be constant. Therefore, the form of ξ must be

$$\xi = c_6 + c_7 t. \tag{14.61}$$

This form can be substituted into Eq. 14.54.

$$\begin{aligned}&y^{3/2} t^{-1/2}(c_4 + 2c_5 y) - 2 y^{3/2} t^{-1/2} c_7 + \tfrac{1}{2}(c_6 + c_7 t) y^{3/2} t^{-3/2} \\ &- \tfrac{3}{2}(c_3 + c_4 y + c_5 y^2) y^{1/2} t^{-1/2} = 0\end{aligned} \tag{14.62}$$

$$\begin{aligned}&y^{3/2} t^{-1/2}\left(c_4 - 2c_7 + \tfrac{1}{2} c_7 - \tfrac{3}{2} c_4\right) + \tfrac{1}{2} c_6 y^{3/2} t^{-1/2} \\ &- \tfrac{3}{2} c_3 y^{1/2} t^{-1/2} + y^{5/2} t^{-1/2}\left(2c_5 - \tfrac{3}{2} c_5\right) = 0\end{aligned} \tag{14.63}$$

The coefficients c_3, c_5, and c_6 must be zero. Also, $c_4 = -3 c_7$. No other solutions here are possible. Thus, the symmetry condition of Eq. 14.49 can be satisfied by $\xi = t$ and $\eta = -3y$.

This procedure finds one regular continuous infinitesimal symmetry of the Thomas Fermi equation, with infinitesimal symmetry generator

$$U = t\partial_t - 3y\partial_y. \tag{14.64}$$

No other solutions can satisfy the constraints given by Eq. 14.49. Therefore, this equation has only one continuous symmetry.

Finite transformations are related to infinitesimal transformations by Eq. 14.14. In this case, the independent variable transforms as

$$t \to \tilde{t} = e^{\varepsilon(t\partial_t - 3y\partial_y)} t. \tag{14.65}$$

$$t \to \tilde{t} = \left[1 + \varepsilon\left(t\partial_t - 3y\partial_y\right) + \frac{1}{2!}\varepsilon^2 \left(t\partial_t - 3y\partial_y\right)^2 + ...\right] t \tag{14.66}$$

$$t \to \tilde{t} = \left[t + \varepsilon t \left(\partial_t t \right) + \frac{1}{2!} \epsilon^2 t \left(\partial_t t \right) \left(\partial_t t \right) + \ldots \right] \qquad (14.67)$$

$$t \to \tilde{t} = t e^{\varepsilon} \qquad (14.68)$$

The dependent variable transforms as

$$y \to \tilde{y} = e^{\varepsilon(t\partial_t - 3y\partial_y)} y. \qquad (14.69)$$

$$y \to \tilde{y} = \left[1 + \varepsilon \left(t\partial_t - 3y\partial_y \right) + \frac{1}{2!} \varepsilon^2 \left(t\partial_t - 3y\partial_y \right)^2 + \ldots \right] y \qquad (14.70)$$

$$y \to \tilde{y} = y e^{-3\varepsilon} \qquad (14.71)$$

Defining the constant $c_6 = e^{\varepsilon}$, the transformation can be written as

$$t \to c_6 t \qquad \text{and} \qquad y \to (c_6)^{-3} y. \qquad (14.72)$$

The analysis above shows that the original Thomas Fermi equation of Eq. 14.46 and the transformed equation

$$\frac{d^2 \left(y c_6^{-3} \right)}{d \left(t c_6 \right)^2} = \left(y c_6^{-3} \right)^{3/2} \left(t c_6 \right)^{-1/2} \qquad (14.73)$$

have the same solutions. From it, we can conclude that if $y(t)$ is a solution to the Thomas Fermi equation, we know that $c_6^{-3} y(\tau)$ for $\tau = c_6 t$ is also a solution.

14.4.3 Line Equation Example

Consider another example of this procedure applied to the equation $\ddot{y} = 0$. The solution of this equation can be found by inspection

$$y(t) = c_0 t + c_1 \qquad (14.74)$$

because this is the equation of a straight line. The coefficients c_n are constants, and they are different from the previous example. In this example, we will identify the infinitesimal generators for continuous symmetries of this equation, and we will find eight infinitesimal generators. The result of this problem appears in [191], and it is a modified version of problem 2.26 of reference [164, p. 180].

Solutions of the original equation must be the same as solutions of an equation transformed by a continuous symmetry, and this idea is contained in the symmetry condition of Eq. 14.45. In this case, the original equation is $\ddot{y} = 0$, so the prolongation of an infinitesimal generator acting on this equation must also be zero for an infinitesimal generator U that describes a continuous symmetry.

$$\operatorname{pr}^{(n)} U(\ddot{y}) = 0. \tag{14.75}$$

Using Eqs. 14.42, 14.43, and 14.44, we can write this symmetry condition in terms of ξ and η.

$$\eta^{tt} = 0 \tag{14.76}$$

$$\eta^{tt} = 0 = \partial_{tt}\eta + 2\dot{y}\partial_{yt}\eta + \ddot{y}\partial_y\eta + \dot{y}^2\partial_{yy}\eta - 2\ddot{y}\partial_t\xi \\ -\dot{y}\partial_{tt}\xi - 2\dot{y}^2\partial_{yt}\xi - \dot{y}^3\partial_{yy}\xi - 3\dot{y}\ddot{y}\partial_y\xi \tag{14.77}$$

Use $\ddot{y} = 0$, and regroup the terms.

$$(\partial_{tt}\eta) + \dot{y}(2\partial_{yt}\eta - \partial_{tt}\xi) + \dot{y}^2(\partial_{yy}\eta - 2\partial_{yt}\xi) - \dot{y}^3(\partial_{yy}\xi) = 0 \tag{14.78}$$

The above equation is true for all y only if all of the quantities in parentheses are zero.

$$\partial_{tt}\eta = 0 \tag{14.79}$$

$$2\partial_{yt}\eta - \partial_{tt}\xi = 0 \tag{14.80}$$

$$\partial_{yy}\eta - 2\partial_{yt}\xi = 0 \tag{14.81}$$

$$\partial_{yy}\xi = 0 \tag{14.82}$$

The next step is to solve the above set of equations for all possible solutions of ξ and η which will determine the infinitesimal generators of all possible continuous symmetry transformations.

We will consider three cases: case 1 with $\eta = 0$, case 2 with $\xi = 0$, and case 3 with both ξ and η nonzero.

Case 1 with $\eta = 0$: Assume $\eta = 0$. What solutions can be found for ξ? Equation 14.79 to Eq. 14.82 can be reduced.

$$\partial_{tt}\xi = 0 \tag{14.83}$$

$$\partial_{yy}\xi = 0 \tag{14.84}$$

$$\partial_{yt}\xi = 0 \tag{14.85}$$

There are three possible independent solutions for ξ. They are $\xi = 1$, $\xi = t$, and $\xi = y$. So, we found three infinitesimal generators.

$$U_1 = \partial_t \tag{14.86}$$

$$U_2 = t\partial_t \tag{14.87}$$

$$U_3 = y\partial_t \tag{14.88}$$

Case 2 with $\xi = 0$: Suppose $\xi = 0$. What solutions can be found for η? Equation 14.79 to Eq. 14.82 simplify.

$$\partial_{tt}\eta = 0 \tag{14.89}$$

$$\partial_{yt}\eta = 0 \tag{14.90}$$

$$\partial_{yy}\eta = 0 \tag{14.91}$$

There are three possible independent solutions for η. They are $\eta = 1$, $\eta = y$, and $\eta = t$. So, we found three more infinitesimal generators.

$$U_4 = \partial_y \tag{14.92}$$

$$U_5 = y\partial_y \tag{14.93}$$

$$U_6 = t\partial_y \tag{14.94}$$

Case 3 where both ξ and η are nonzero: From Eq. 14.79, we can write

$$\eta = (c_1 + c_2 t)\, b(y). \tag{14.95}$$

Here, b is a function of only y, not t. Therefore,

$$\partial_{yt}\eta = c_2 \partial_y b(y) \tag{14.96}$$

which is not a function of t.

From Eq. 14.82, we can write

$$\xi = (c_3 + c_4 y)\, g(t). \tag{14.97}$$

Here, g is a function of only t, not y. Therefore,

$$\partial_{yt}\xi = c_4 \partial_t g(t) \tag{14.98}$$

which is not a function of y. Now use Eq. 14.80.

$$2c_2 \partial_y b(y) - (c_3 + c_4 y)\, \partial_{tt} g = 0 \tag{14.99}$$

14 LIE ANALYSIS

The first term is not a function of t. Therefore, ξ is at most quadratic in t. So, ξ has the form

$$\xi = (c_3 + c_4 y)(c_5 + c_6 t + c_7 t^2). \tag{14.100}$$

Distribute out the multiplication.

$$\xi = c_3 c_5 + c_3 c_6 t + c_3 c_7 t^2 + c_4 c_5 y + c_4 c_6 yt + c_4 c_7 yt^2 \tag{14.101}$$

$$\partial_{yt}\xi = c_4 c_6 + 2 c_4 c_7 t \tag{14.102}$$

Next, use Eq. 14.81.

$$\partial_{yy}\eta - 2 c_4 \partial_t g = 0 \tag{14.103}$$

The second term is not a function of y. Therefore, η is at most quadratic in y. So, η has the form

$$\eta = (c_1 + c_2 t)(c_8 + c_9 y + c_{10} y^2). \tag{14.104}$$

Distribute out the multiplication.

$$\eta = c_1 c_8 + c_1 c_9 y + c_1 c_{10} y^2 + c_2 c_8 t + c_2 c_9 yt + c_2 c_{10} ty^2 \tag{14.105}$$

$$\partial_{yt}\eta = c_2 c_9 + 2 c_2 c_{10} y \tag{14.106}$$

Now use Eqs. 14.80 and 14.106.

$$2(c_2 c_9 + 2 c_2 c_{10} y) - 2(2 c_3 c_7 + 2 y c_4 c_7) = 0 \tag{14.107}$$

$$(2 c_2 c_9 - 4 c_3 c_7) + y(4 c_2 c_{10} - 4 c_4 c_7) = 0 \tag{14.108}$$

We end up with the pair of equations

$$c_2 c_9 = 2 c_3 c_7 \tag{14.109}$$

$$c_2 c_{10} = c_4 c_7 \tag{14.110}$$

Next use Eqs. 14.81 and 14.102.

$$(2 c_1 c_{10} + 2 t c_2 c_{10}) - 2(c_4 c_6 + 2 c_4 c_7 t) = 0 \tag{14.111}$$

$$(2 c_1 c_{10} - 2 c_4 c_6) + t(2 c_2 c_{10} - 4 c_4 c_7) = 0 \tag{14.112}$$

and we end up with a pair of equations.

$$c_1 c_{10} = c_4 c_6 \qquad (14.113)$$

$$c_2 c_{10} = 2 c_4 c_7 \qquad (14.114)$$

These are the only possible solution of Eqs. 14.110 and 14.114.

Finally, there are two possible solutions which are independent from the previously found solutions. We can set the coefficients of Eq. 14.113 to 1. The first solution is $\eta = y^2$ and $\xi = yt$ corresponding to

$$U_7 = yt\partial_t + y^2 \partial_y. \qquad (14.115)$$

For the second solution, we can set the coefficients of Eq. 14.109 to 1. The second solution is $\eta = yt$ and $\xi = t^2$ corresponding to

$$U_8 = t^2 \partial_t + yt \partial_y. \qquad (14.116)$$

At this point, we have found eight infinitesimal generators. These are *all* possible generators of continuous regular nongeometrical symmetries.

14.5 Invariants

14.5.1 Importance of Invariants

Noether's theorem describes the relationship between continuous symmetries of an equation describing an energy conversion process and invariants of the system. The theorem was originally discovered by Noether around 1918 [165] [166]. The importance of this theorem is described in the introduction to the English translation of the original paper [165]. "The well known theorem of Emmy Noether plays a role of fundamental importance in many branches of theoretical physics. Because it provides a straightforward connection between the conservation laws of a physical theory and the invariances of the variational integral whose Euler-Lagrange equations are the equations of that theory, it may be said that Noether's theorem has placed the Lagrangian formulation in a position of primacy."

14.5.2 Noether's Theorem

Consider an energy conversion process with a known Lagrangian that satisfies an Euler-Lagrange equation. Assume that we have identified continuous symmetries described by infinitesimal generators. Noether's theorem says that there is a relationship between these continuous symmetries and conservation laws which say that some quantity is invariant. We would like to find the corresponding conservation laws and invariants. If we can find a quantity G that satisfies,

$$\frac{dG}{dt} = \text{pr}^{(n)} U \mathcal{L} + \mathcal{L} \frac{d\xi}{dt}, \qquad (14.117)$$

then the quantity

$$\Upsilon = \eta \frac{d\mathcal{L}}{d\dot{y}} + \xi \mathcal{L} - \xi \dot{y} \frac{\partial \mathcal{L}}{\partial \dot{y}} - G \qquad (14.118)$$

is an invariant. For Lagrangians with units of joules, the quantity G also has units joules. In Eq. 14.117, $\text{pr}^{(n)} U \mathcal{L}$ is the prolongation of the infinitesimal generator acting on the Lagrangian where prolongation was defined in Eq. 14.42.

14.5.3 Derivation of Noether's Theorem

We can derive this form of Noether's theorem, and this derivation closely follows the clear and simplified derivation in reference [192]. This theorem is detailed and derived more rigorously in multiple other references [163, p. 208] [164]. For the purpose of this derivation, assume that we begin with an equation of motion that is at most a second order differential equation. However, the ideas generalize to higher order equations too. Also, assume we know the corresponding Lagrangian of the form $\mathcal{L} = \mathcal{L}(t, y, \dot{y})$. The general approach is to assume that we can find a value of G defined by Eq. 14.117. We will perform some algebra on Eq. 14.117 to show that choice of G necessarily implies that Υ is invariant.

Use the definition of the prolongation to write Eq. 14.117 in terms of ξ and η.

$$\text{pr}^{(n)} U = \xi \partial_t + \eta \partial_y + \eta^t \partial_{\dot{y}} \qquad (14.119)$$

For a second order differential equation, no more terms are needed because the Lagrangian depends on, at most, the first derivative $\dot{y}$. Substitute the prolongation acting on the Lagrangian into Eq. 14.117.

$$\frac{dG}{dt} = [\xi \partial_t \mathcal{L} + \eta \partial_y \mathcal{L} + \eta^t \partial_{\dot{y}} \mathcal{L}] + \frac{d\xi}{dt} \mathcal{L} \qquad (14.120)$$

Consider the continuous transformation described by

$$t \to \tilde{t} = (1 + \varepsilon \xi + \varepsilon^2 ...) t \qquad (14.121)$$

and

$$y \to \tilde{y} = (1 + \varepsilon \eta + \varepsilon^2 ...) y \qquad (14.122)$$

in the limit $\varepsilon \to 0$. The Lagrangian $\mathcal{L}(t, y, \dot{y})$ of an energy conversion process represents the difference between two forms of energy. The Lagrangian $\mathcal{L}(\tilde{t}, \tilde{y}, \dot{\tilde{y}})$ represents the difference between two forms of energy upon the continuous symmetry transformation described by infinitesimal generator U. Qualitatively, the quantity $\frac{dG}{dt}$ represents the change in $\mathcal{L}\frac{d\tilde{t}}{dt}$ with respect to ε in this limit [192].

$$\frac{dG}{dt} = \frac{\partial}{\partial \varepsilon}\left[\mathcal{L}\left(\tilde{t}, \tilde{y}, \dot{\tilde{y}}\right) \frac{d\tilde{t}}{dt}\right] \qquad (14.123)$$

Use Eq. 14.43 to substitute for η^t in Eq. 14.120.

$$\eta^t = \frac{d}{dt}(\eta - \xi \dot{y}) + \xi \ddot{y} \qquad (14.124)$$

$$\eta^t = \dot{\eta} - \dot{\xi}\dot{y} - \xi\ddot{y} + \ddot{y} = \dot{\eta} - \dot{y}\dot{\xi} \qquad (14.125)$$

$$\frac{dG}{dt} = \left[\xi \partial_t \mathcal{L} + \eta \partial_y \mathcal{L} + \left(\dot{\eta} - \dot{y}\dot{\xi}\right) \partial_{\dot{y}} \mathcal{L}\right] + \dot{\xi}\mathcal{L} \qquad (14.126)$$

We want to express the right side as the total derivative of some quantity, which we call G. With some algebra, we can write this as a total derivative. We will use the definition of the total derivative.

$$\frac{d\mathcal{L}}{dt} = \partial_t \mathcal{L} + \dot{y}\partial_y \mathcal{L} + \ddot{y}\partial_{\dot{y}} \mathcal{L} \qquad (14.127)$$

$$\partial_t \mathcal{L} = \dot{\mathcal{L}} - \dot{y}\partial_y \mathcal{L} - \ddot{y}\partial_{\dot{y}} \mathcal{L} \qquad (14.128)$$

$$\frac{d}{dt}(\eta \partial_{\dot{y}} \mathcal{L}) = \dot{\eta}\partial_{\dot{y}} \mathcal{L} + \eta \frac{d}{dt}(\partial_{\dot{y}} \mathcal{L}) \qquad (14.129)$$

$$\dot{\eta}\partial_{\dot{y}} \mathcal{L} = \frac{d}{dt}(\eta \partial_{\dot{y}} \mathcal{L}) - \eta \frac{d}{dt}(\partial_{\dot{y}} \mathcal{L}) \qquad (14.130)$$

$$\frac{d}{dt}(\xi \dot{y}\partial_{\dot{y}} \mathcal{L}) = \dot{\xi}\dot{y}\partial_{\dot{y}} \mathcal{L} + \xi \ddot{y}\partial_{\dot{y}} \mathcal{L} + \xi \dot{y}\frac{d}{dt}\partial_{\dot{y}} \mathcal{L} \qquad (14.131)$$

$$\dot{\xi}\dot{y}\partial_{\dot{y}} \mathcal{L} = \frac{d}{dt}(\xi \dot{y}\partial_{\dot{y}} \mathcal{L}) - \xi \ddot{y}\partial_{\dot{y}} \mathcal{L} - \xi \dot{y}\frac{d}{dt}\partial_{\dot{y}} \mathcal{L} \qquad (14.132)$$

Use these pieces to replace the terms of $\frac{dG}{dt}$ in brackets.

$$\frac{dG}{dt} = [\xi \partial_t \mathcal{L}] + \eta \partial_y \mathcal{L} + [\dot{\eta}\partial_{\dot{y}} \mathcal{L}] - [\dot{y}\dot{\xi}\partial_{\dot{y}} \mathcal{L}] + \dot{\xi}\mathcal{L} \qquad (14.133)$$

$$\begin{aligned}\frac{dG}{dt} =\ & \xi\left[\dot{\mathcal{L}} - \dot{y}\partial_y \mathcal{L} - \ddot{y}\partial_{\dot{y}} \mathcal{L}\right] + \eta \partial_y \mathcal{L} + \left[\frac{d}{dt}(\eta \partial_{\dot{y}} \mathcal{L}) - \eta \frac{d}{dt}(\partial_{\dot{y}} \mathcal{L})\right] \\ & - \left[\dot{y}\dot{\xi}\partial_{\dot{y}} \mathcal{L}\right] + \dot{\xi}\mathcal{L}\end{aligned} \qquad (14.134)$$

$$\frac{dG}{dt} = \xi\left[\dot{\mathcal{L}} - \dot{y}\partial_y\mathcal{L} - \ddot{y}\partial_{\dot{y}}\mathcal{L}\right] + \eta\partial_y\mathcal{L} + \left[\tfrac{d}{dt}(\eta\partial_{\dot{y}}\mathcal{L}) - \eta\tfrac{d}{dt}(\partial_{\dot{y}}\mathcal{L})\right] \quad (14.135)$$
$$- \left[\tfrac{d}{dt}(\xi\dot{y}\partial_{\dot{y}}\mathcal{L}) - \xi\ddot{y}\partial_{\dot{y}}\mathcal{L} - \xi\dot{y}\tfrac{d}{dt}\partial_{\dot{y}}\mathcal{L}\right] + \dot{\xi}\mathcal{L}$$

Two terms cancel.

$$\frac{dG}{dt} = \xi\dot{\mathcal{L}} - \xi\dot{y}\partial_y\mathcal{L} + \eta\partial_y\mathcal{L} + \tfrac{d}{dt}(\eta\partial_{\dot{y}}\mathcal{L}) - \eta\tfrac{d}{dt}(\partial_{\dot{y}}\mathcal{L}) \quad (14.136)$$
$$- \tfrac{d}{dt}(\xi\dot{y}\partial_{\dot{y}}\mathcal{L}) + \xi\dot{y}\tfrac{d}{dt}\partial_{\dot{y}}\mathcal{L} + \dot{\xi}\mathcal{L}$$

Regroup terms.

$$\frac{dG}{dt} = \left(\partial_y\mathcal{L} - \tfrac{d}{dt}\partial_{\dot{y}}\mathcal{L}\right)(\eta - \dot{y}\xi) \quad (14.137)$$
$$+ \left[\left(\xi\dot{\mathcal{L}} + \mathcal{L}\dot{\xi}\right) + \tfrac{d}{dt}(\eta\partial_{\dot{y}}\mathcal{L}) - \tfrac{d}{dt}(\xi\dot{y}\partial_{\dot{y}}\mathcal{L})\right]$$

The first term in parentheses is zero because the Lagrangian $\mathcal{L}$ satisfies the Euler-Lagrange equation.

$$\frac{dG}{dt} = \frac{d}{dt}\left(\xi\mathcal{L} + (\eta\partial_{\dot{y}}\mathcal{L}) - (\xi\dot{y}\partial_{\dot{y}}\mathcal{L})\right) \quad (14.138)$$

$$\frac{d}{dt}\left[\xi\mathcal{L} + (\eta\partial_{\dot{y}}\mathcal{L}) - (\xi\dot{y}\partial_{\dot{y}}\mathcal{L}) - G\right] = 0 \quad (14.139)$$

Therefore, if we can find G, then the quantity in brackets Υ must be invariant.

$$\Upsilon = \xi\mathcal{L} + (\eta\partial_{\dot{y}}\mathcal{L}) - (\xi\dot{y}\partial_{\dot{y}}\mathcal{L}) - G = \text{invariant} \quad (14.140)$$

14.5.4 Line Equation Invariants Example

Let us apply Noether's theorem to some examples. First, consider the line equation $\ddot{y} = 0$ which results from application of calculus of variations with Lagrangian

$$\mathcal{L} = \frac{1}{2}\dot{y}^2. \quad (14.141)$$

A continuous symmetry of this equation is described by the infinitesimal generator $U = \partial_y$ with $\xi = 0$ and $\eta = 1$. The prolongation of the generator acting on the Lagrangian is zero.

$$\text{pr}^{(n)}U\mathcal{L} = \eta^t\dot{y} = \dot{y}\left(\frac{d\eta}{dt}\right) = 0 \quad (14.142)$$

Using Eq.14.117, we see that $G = 0$.

$$\frac{dG}{dt} = 0 + \mathcal{L}\cdot 0 = 0 \quad (14.143)$$

Next use Eq. 14.118 to find the invariant.

$$\Upsilon = \eta \frac{\partial \mathcal{L}}{\partial \dot{y}} = \dot{y} \qquad (14.144)$$

Qualitatively, $\dot{y}$ represents the slope of the line, so this invariant tells us that the slope of the solutions to the line equation must be constant.

Another continuous symmetry of this equation is described by the infinitesimal generator $U = t\partial_y$ with $\xi = 0$ and $\eta = t$. We can solve for the prolongation of the generator acting on the Lagrangian.

$$\text{pr}^{(n)} U \mathcal{L} = \eta^t \dot{y} = \dot{y} \left(\frac{d}{dt}(t-0) + 0 \right) = \dot{y}$$

We can find G using Eq.14.117, and we can find the invariant using Eq.14.118.

$$\frac{dG}{dt} = \dot{y} + \frac{1}{2}\dot{y}^2 \cdot 0 = \dot{y} \qquad (14.145)$$

$$G = y \qquad (14.146)$$

$$\Upsilon = y - t\dot{y} \qquad (14.147)$$

Qualitatively, this invariant represents the y-intercept of the line, so this invariant tells us that the y-intercept of the solution to the line equation must be constant.

14.5.5 Pendulum Equation Invariants Example

Consider the equation describing a pendulum, studied in Problem 11.8. The energy conversion process is described by the Lagrangian

$$\mathcal{L} = \frac{1}{2}m\dot{y}^2 - mg\cos y \qquad (14.148)$$

which corresponds to the equation of motion

$$\ddot{y} = g \sin y. \qquad (14.149)$$

In these equations m represents the mass, and g represents the gravitational constants. Both m and g are assumed constant here. This equation of motion has only one continuous symmetry described by the infinitesimal generator $U = \partial_t$ with $\xi = 1$ and $\eta = 0$. We can use Noether's theorem to find the corresponding invariant.

14 LIE ANALYSIS

Use Eq. 14.117 to find G.

$$\frac{dG}{dt} = \mathrm{pr}^{(n)} U \mathcal{L} + \mathcal{L}\frac{d\xi}{dt} \tag{14.150}$$

$$\frac{dG}{dt} = \eta^t m\dot{y} + \eta m g \sin y + \mathcal{L}\frac{d\xi}{dt} \tag{14.151}$$

$$\frac{dG}{dt} = \eta^t m\dot{y} = \dot{y}m\left(\frac{d}{dt}(\eta - \xi\dot{y}) + \xi\ddot{y}\right) \tag{14.152}$$

$$\frac{dG}{dt} = \dot{y}m\left(-\frac{d\xi}{dt}\dot{y} - \xi\ddot{y} + \xi\ddot{y}\right) = 0 \tag{14.153}$$

$$G = 0 \tag{14.154}$$

Use Eq. 14.118 to find the invariant.

$$\Upsilon = \eta\dot{y} + \xi\mathcal{L} - \xi m\dot{y}\dot{y} - 0 \tag{14.155}$$

$$\Upsilon = \frac{1}{2}m\dot{y}^2 - mg\cos y - m\dot{y}^2 \tag{14.156}$$

$$\Upsilon = \frac{-1}{2}m\dot{y}^2 - gm\cos y \tag{14.157}$$

The quantity Υ is conserved, and it is the Hamiltonian which represents total energy.

Whenever the Lagrangian does not explicitly depend on t, the system contains the continuous symmetry described by the infinitesimal generator $U = \partial_t$. This infinitesimal generator has $\xi = 1$ and $\eta = 0$. From Eq. 14.117, G must be zero. From Eq. 14.118, the corresponding invariant has the form

$$\Upsilon = \mathcal{L} - m\dot{y}\dot{y} \tag{14.158}$$

which has the magnitude of the total energy (assuming t is time). This equation is equal to the Hamiltonian of Eq. 11.24. Therefore, if an equation of motion contains the symmetry ∂_t, energy is conserved.

14.6 Summary

In this chapter, a procedure to find continuous symmetries of equations was presented. Also, the relationship between continuous symmetries and invariants, known as Noether's theorem, was discussed. If we can describe an energy conversion process by a Lagrangian, we can use the techniques of calculus of variations detailed in Chapter 11 to find the equation of motion for the path. We can use the procedure discussed in this chapter to identify continuous symmetries of the equation of motion. These symmetry transformations are denoted by infinitesimal generators which describe how the independent and dependent variables transform. We also may be able to use Noether's theorem to find invariants of the system. We can apply this analysis even in cases where the equation of motion is nonlinear or has no closed form solution. The invariants often correspond to physical quantities, such as energy, momentum, or angular momentum, which are conserved in the system. Knowledge of invariants can help us gain insights into what quantities change and what quantities do not change during the energy conversion process under study.

14.7 Problems

14.1. Three commonly discussed discrete symmetry transformations are:
Time reversal $t \to \tilde{t} = (-1)^n t$ for integer n
Parity $y \to \tilde{y} = (-1)^n y$ for integer n
Charge conjugation $y \to \tilde{y} = y^*$

Verify that the wave equation, $\ddot{y} + \omega_0^2 y = 0$, is invariant upon each of these discrete transformations.

14.2. Repeat the problem above for the equation $\ddot{y} + y^{-3} = 0$.

14.3. The Thomas Fermi equation is given by $\ddot{y} = y^{3/2} t^{-1/2}$.

(a) Verify that it is not invariant upon the discrete symmetry transformation of time reversal,
$t \to \tilde{t} = (-1)^n t$ for integer n.

(b) Verify that it is not invariant upon the discrete symmetry transformation of parity,
$y \to \tilde{y} = (-1)^n y$ for integer n.

(c) Verify that it is invariant upon the discrete symmetry transformation
$t \to \tilde{t} = (-1)^n t$ and $y \to \tilde{y} = (-1)^n y$.

14.4. Find the prolongation of the infinitesimal generator
$$U = \xi \partial_t + \eta \partial_y$$
acting on the Lagrangian
$$\mathcal{L} = \frac{1}{2}\dot{y}^2 + \frac{1}{3}ty^2.$$
Write your answer in terms of ξ and η but not η^t or η^{tt}.

14.5. Find the infinitesimal generators for the equation, $\ddot{y} + y^{-3} = 0$. (This problem is discussed in [190].)

Answer:
$$U_1 = \partial_t$$
$$U_2 = 2t\partial_t + y\partial_y$$
$$U_3 = t^2 \partial_t + ty\partial_y$$

14.6. The equation $\ddot{y}+y^{-3}=0$ has the three infinitesimal generators listed in the problem above. These infinitesimal generators form a group. The commutator was defined in Section 14.3.3, and the commutator of any pair of these infinitesimal generators can be calculated by

$$[U_a, U_b] = U_a U_b - U_b U_a.$$

Using the equation above, show that the commutator for each of the three pairs of infinitesimal generators results in another element of the group.

14.7. Derive the infinitesimal generators for the wave equation, $\ddot{y}+\omega_0^2 y = 0$. (This problem is discussed in [191].)

Answer:

$$U_1 = \partial_t$$

$$U_2 = y\partial_y$$

$$U_3 = \sin(\omega_0 t)\,\partial_y$$

$$U_4 = \cos(\omega_0 t)\,\partial_y$$

$$U_5 = \sin(2\omega_0 t)\partial_t + \omega_0 y \cos(2\omega_0 t)\partial_y$$

$$U_6 = \cos(2\omega_0 t)\partial_t - \omega_0 y \sin(2\omega_0 t)\partial_y$$

$$U_7 = y\cos(\omega_0 t)\,\partial_t - \omega_0 y^2 \sin(\omega_0 t)\,\partial_y$$

$$U_8 = y\sin(\omega_0 t)\,\partial_t + \omega_0 y^2 \cos(\omega_0 t)\,\partial_y$$

14.8. The wave equation $\ddot{y}+\omega_0^2 y = 0$ has the eight infinitesimal generators listed in the problem above. The corresponding Lagrangian is

$$\mathcal{L} = \frac{1}{2}\dot{y}^2 - \frac{1}{2}\omega_0^2 y^2.$$

Find the invariants corresponding to the following infinitesimal generators.

(a) $U_1 = \partial_t$

(b) $U_3 = \sin(\omega_0 t)\,\partial_y$

(c) $U_5 = \sin(2\omega_0 t)\partial_t + \omega_0 y \cos(2\omega_0 t)\partial_y$

14.9. In Problem 11.8, we encountered the equation given by $\ddot{y} = g\sin y$ for constant g.

(a) Show that $U = \partial_t$ is an infinitesimal generator of this equation.

(b) Show that $U = y\partial_y$ is *not* an infinitesimal generator of this equation.

14.10. The Lagrangian
$$\mathcal{L} = \frac{1}{2}\dot{y}^2 + \frac{1}{2}y^{-2}$$
corresponds to the equation of motion $\ddot{y} + y^{-3} = 0$. This equation of motion has three infinitesimal generators:
$$U_1 = \partial_t$$
$$U_2 = 2t\partial_t + y\partial_y$$
$$U_3 = t^2\partial_t + ty\partial_y$$

Use Noether's theorem to find the invariants that correspond to each of these infinitesimal generators. (We encountered this Lagrangian in problem 11.3.)

Appendices

Appendix A: Variable List

Vectors are denoted $\vec{v}$, and unit vectors are denoted $\hat{v}$. In the third column, an asterisk * indicates that the units vary depending on the context. In the fourth column, S = scalar, V = vector, C =constant, F = functional, and O = operator. Constants are specified to four significant figures.

Symbol	Quantity	SI Units	Scalar?	Comments
$\hat{a}_x, \hat{a}_y, \hat{a}_z$	Cartesian coordinate unit vectors	unitless	V	
$\hat{a}_r, \hat{a}_\theta, \hat{a}_\phi$	Spherical coordinate unit vectors	unitless	V	
$\vec{a}$	Acceleration	$\frac{m}{s^2}$	V	
A	Cross sectional area	m^2	S	
A_{12}	Einstein A coefficient	s^{-1}	S	
A_{aff}	Electron affinity	$\frac{J}{atom}$	S	
$\vec{b}$	Pyroelectric coefficient	$\frac{C}{m^2 \cdot K}$	V	
B_{12}, B_{21}	Einstein B coefficient	$\frac{m^3}{J \cdot s^2}$	S	
$\vec{B}$	Magnetic flux density	$\frac{Wb}{m^2}$	V	
$\mathbb{B}$	Bulk modulus	$\frac{N}{m^2}$	S	
c	Speed of light in free space	$\frac{m}{s}$	C	$= 2.998 \cdot 10^8$

Symbol	Quantity	Units	Scalar?	Comments
c_n	Coefficient	unitless	S	For integer n
C	Capacitance	F	S	
C_v	Specific heat	$\frac{J}{kg \cdot K}$	S	
d	Piezoelectric strain constant	$\frac{m}{V}$	S	May be a scalar or matrix
d_{thick}	Thickness	m	S	
D	Directivity	unitless	S	For antennas
$\vec{D}$	Displacement flux density	$\frac{C}{m^2}$	V	
e	e	unitless	C	≈ 2.718
e^-	Electron			Used in chem. reactions
E	Energy	J	S	
$\vec{E}$	Electric field intensity	$\frac{V}{m}$	V	
E_f	Fermi energy level	J	S	Also called chemical potential
E_g	Energy gap	J	S	Also called bandgap
f	Frequency	Hz	S	
$\vec{F}$	Force	N	V	
g_n	Degeneracy of energy level n	unitless	S	
G	Component of an invariant	J	S	

Appendices

Symbol	Quantity	Units	Scalar?	Comments
h	Planck constant	J·s	C	$= 6.626 \cdot 10^{-34}$
H	Hamiltonian	J	F	
H_{QM}	Quantum Mechanical Hamiltonian	J	O	
$\vec{H}$	Magnetic field intensity	$\frac{A}{m}$	V	
$[H^+]$	Amount concentration hydrogen ions	$\frac{mol}{L}$	S	
i	Current (AC or time varying)	A	S	
I	Current (DC)	A	S	
$\mathbb{I}$	Moment of inertia	kg·m²	S	
I_{ioniz}	Ionization energy	$\frac{J}{atom}$	S	
j	Imaginary number	unitless	C	$\sqrt{-1}$
$\vec{J}$	Volume current density	$\frac{A}{m^2}$	V	
$\vec{k}$	Wave vector	m^{-1}	V	
k_B	Boltzmann constant	$\frac{J}{K}$	C	$= 1.381 \cdot 10^{-23}$
k_f	Fermi wave vector	m^{-1}	S	
K	Spring constant	$\frac{J}{m^2}$	S	
$\mathbb{K}$	Torsion spring constant	$\frac{J}{rad^2}$	S	
l	Length	m	S	
L	Inductance	H	S	

Symbol	Quantity	Units	Scalar?	Comments
$\vec{L}_{am}$	Angular momentum	J·s	V	
$\mathcal{L}$	Lagrangian	J	F	
m	Mass	kg	S	
$\mathbb{M}$	Generalized momentum	*	S	Many authors use p
$\vec{M}$	Momentum	$\frac{\text{kg·m}}{s}$	V	Many authors use $\vec{p}$
M_{QM}	Quantum mechanical momentum	$\frac{\text{kg·m}}{s}$	V,O	
n	Concentration of electrons	m^{-3}	S	
n	Index of refraction	unitless	S	
$\mathfrak{n}$	Integer	unitless	S	
N	Total number of e^- per atom	$\frac{\text{electrons}}{\text{atom}}$	S	
N_a	Avogadro constant	$\frac{\text{molecule}}{\text{mol}}$	C	$= 6.022 \cdot 10^{23}$
N_v	Number of valence e^-	electrons	S	
$\mathbb{N}$	Number of moles	mol	S	
p	Concentration of holes	m^{-3}	S	
P	Power	W	S	
$\vec{P}$	Material polarization	$\frac{\text{C}}{\text{m}^2}$	V	
$\mathbb{P}$	Pressure	Pa	S	

Symbol	Quantity	Units	Scalar?	Comments
q	Magnitude of electron charge	C	C	$= 1.602 \cdot 10^{-19}$
Q	Charge	C	S	
$\mathbb{Q}$	Heat	J	S	
r	Distance in spherical coordinates	m	S	
$\vec{r}$	Position in spherical coordinates	m	V	
$\tilde{r}$	Distance in reciprocal space	m^{-1}	S	
R	Resistance	Ω	S	
R_H	Hall resistance	Ω	S	
$\mathbb{R}$	Molar gas constant	$\frac{J}{mol \cdot K}$	C	$= 8.314$
R	Mirror reflectivity	unitless	S	
s	Kerr coefficient	$\frac{m^2}{V^2}$	S	
$\$$	Seebeck coefficient	$\frac{V}{K}$	S	
S	Entropy	$\frac{J}{K}$	S	
$\mathbb{S}$	Action	J	F	
t	Time	s	S	
T	Temperature	K	S	
u	Energy density per unit bandwidth	$\frac{J \cdot s}{m^3}$	S	
U	Infinitesimal generator	unitless	O	
$\mathbb{U}$	Internal energy	J	S	

Symbol	Quantity	Units	Scalar?	Comments
$\vec{v}$	Velocity	$\frac{m}{s}$	V	
v	Voltage (AC or time varying)	V	S	
V	Voltage (DC)	V	S	
$\mathbb{V}$	Volume	m^3	S	
V_0	Contact potential of pn junction	V	S	
V_{rp}	Redox potential	V	S	
V_{cell}	Cell potential	V	s	Many authors use Ξ^0 or E^0
W	Mechanical work	J	S	
$\vec{x}$	Positional displacement	m	V	
y	Dependent variable of equation	*	S	
$\dot{y}$	Shorthand for total derivative $\frac{dy}{dt}$	*	S	
Z	Figure of merit	K^{-1}	S	
Z_0	Characteristic impedance	Ω	S	
α	Absorption coefficient	m^{-1}	S	
γ	Pockels coefficient	$\frac{m}{V}$	S	
Δ	Delta (change in)	unitless	O	
ϵ	Permittivity	$\frac{F}{m}$	S	
ϵ_0	Permittivity of free space	$\frac{F}{m}$	C	$= 8.854 \cdot 10^{-12}$

Symbol	Quantity	Units	Scalar?	Comments
ϵ_r	Relative permittivity	unitless	S	
ε	Infinitesimal parameter	unitless	S	
η	Transformation of dependent variable	*	S	
η_{eff}	Efficiency	unitless	S	
θ	Angle (Elevation)	rad	S	
$\vec{\theta}$	Angular displacement vector	rad	V	
κ	Thermal conductivity	$\frac{W}{m \cdot K}$	S	
λ	Wavelength	m	S	
μ	Permeability	$\frac{H}{m}$	S	
μ_0	Permeability of free space	$\frac{H}{m}$	C	$= 4\pi \cdot 10^{-7}$ $= 1.257 \cdot 10^{-6}$
μ_r	Relative permeability	unitless	C	
μ_{chem}	Chemical potential	$\frac{J}{atom}$	S	Also known as Fermi energy level
μ_n	Mobility of electrons	$\frac{m^2}{V \cdot s}$	S	
μ_p	Mobility of holes	$\frac{m^2}{V \cdot s}$	S	
ξ	Transformation of independent variable	*	S	
Π	Peltier coefficient	V	S	

Symbol	Quantity	Units	Scalar?	Comments
ρ	Resistivity	Ωm	S	
ρ_{ch}	Charge density	$\frac{C}{m^3}$	S	
ρ_{dens}	Mass density	$\frac{kg}{m^3}$	S	
σ	Electrical conductivity	$\frac{1}{\Omega \cdot m}$	S	
$\vec{\varsigma}$	Stress	Pa	V	
τ	Thomson coefficient	$\frac{V}{K}$	S	
$\vec{\tau}$	Torque	N·m	V	
ϕ	Angle (Azimuth)	rad	S	
Υ	Invariant	*	S	
χ_e	Electric susceptibility	unitless	S	
χ	Electronegativity	$\frac{J}{atom}$	S	
ψ	Wave function	unitless	S	
Ψ	Magnetic flux	Wb	S	
ω	Frequency	$\frac{rad}{s}$	S	
$\vec{\omega}_{ang}$	Angular velocity	$\frac{rad}{s}$	V	
$\hbar$	Planck constant divided by 2π, also called hbar	J·s	C	$= 1.055 \cdot 10^{-34}$
$\vec{\nabla}$	Gradient operator		O	Also called del
∂_t	Shorthand for partial derivative $\frac{\partial}{\partial t}$		O	

Appendix B: Abbreviations of Units of Measure

Common abbreviations for units of measure are listed in Table 14.2. This table does not cover all units used in this text. Further measures of energy and power are discussed in Section 1.4. For further information, see [68]. The fourth column indicates whether the unit is an SI base unit, an SI derived unit, or not an SI unit. Table 14.3 lists prefixes used with SI units [193].

Abbreviation	unit	Measure	SI unit?
A	ampere	Current	Base
cd	candela	Luminous intensity	Base
C	coulomb	Charge	Derived
°C	degree Celsius	Temperature	Derived
d	day	Time	No
eV	electronvolt	Energy	No
F	farad	Capacitance	Derived
°F	degree Fahrenheit	Temperature	No
H	henry	Inductance	Derived
Hz	hertz	Frequency	Derived
h	hour	Time	No
J	joule	Energy	Derived
K	kelvin	Temperature	Base
kg	kilogram	Mass	Base
L	liter	Volume	No
m	meter	Length	Base
mol	mole	Amount of substance	Base
N	newton	Force	Derived
Pa	pascal	Pressure	Derived
rad	radian	Angle	Derived
s	second	Time	Base
V	volt	Voltage	Derived
W	watt	Power	Derived
Wb	weber	Magnetic flux	Derived
Ω	ohm	Resistance	Derived

Table 14.2: Units and their abbreviations.

Prefix name	Symbol	Value
yotta	Y	10^{24}
zetta	Z	10^{21}
exa	E	10^{18}
peta	P	10^{15}
tera	T	10^{12}
giga	G	10^{9}
mega	M	10^{6}
kilo	k	10^{3}
milli	m	10^{-3}
micro	μ	10^{-6}
nano	n	10^{-9}
pico	p	10^{-12}
femto	f	10^{-15}
atto	a	10^{-18}
zepto	z	10^{-21}
yocto	y	10^{-24}

Table 14.3: Prefixes used with SI units.

Appendix C: Overloaded Terminology

Physicists, chemists, electrical engineers, and other scientists develop their own notation to describe physical phenomenon. However, a single word may be adopted with different meanings by scientists studying different disciplines. In this section, some of these overloaded terms are discussed.

In general, the term *polarization* means splitting into distinct opposite parts. In this text, two types of polarization are discussed: *material polarization* and *electromagnetic polarization*. If an external electric field, a voltage, is placed across a piece of material it will affect the material. If the material is at a temperature other than absolute zero, the electrons are in constant motion. However, the overall electron location will shift when the external electric field is applied. The term *material polarization* refers to the fact that when an external voltage is applied across an insulator, the electrons slightly displace from the nucleus, so the atom is more negatively charged on one end and positively charged on the other. Material polarization is discussed beginning in Section 2.2.1. The other use of the term polarization describes how electromagnetic waves vary with time as they propagate through space. *Electromagnetic polarization* specifies the direction of the electric field with respect to the direction of propagation of an electromagnetic plane wave. It is discussed in Section 4.4.4. A propagating electromagnetic field may be classified as linearly polarized, circularly polarized, or elliptically polarized. To determine the electromagnetic polarization of a plane traveling wave, project the electric field $\vec{E}(t)$ onto a plane perpendicular to the direction that the wave is traveling. If the resulting projection is a straight line, the wave is said to be linearly polarized. If the projection is a circle, the wave is said to be circularly polarized, and if the projection is an ellipse, the wave is said to be elliptically polarized.

Another overused term is *inversion*. *Inversion symmetry* is discussed in Section 2.3.2, and *population inversion* is discussed in Section 7.2.4. If a crystal structure looks the same upon rotation or reflection, the structure is said to have a symmetry. If the crystal structure looks the same after 180° rotation and inversion through the origin, the structure is said to have *inversion symmetry*. This idea is illustrated in Fig. 2.8. The term *population inversion* is defined in the context of lasers. In a laser, LED, lamp, or other device that converts electricity to optical energy, a pump excites electrons from a lower to a higher energy level. A *population inversion* occurs when more electrons of the active material are in the upper, rather than lower, energy level. Lasing requires a population inversion.

The word *potential* is also quite overloaded. Electrical engineers sometimes use *potential*, as well as the term *electromotive force*, as a synonym for voltage. (As an aside, reference [6] carefully distinguishes between these three terms.) In this context, potential, like voltage, has the units volts. The term *chemical potential* μ_{chem} has units of joules per atom, and it represents energy where the probability of finding an electron is one half. For a pure semiconductor, the chemical potential is in the middle of the energy gap. Semiconductor scientists typically use the term *Fermi energy level* E_f instead. These terms are discussed in Sec. 6.3.3 and 9.2.3. Voltage times charge is energy, so chemical potential can be thought of as a voltage times the charge of an electron. The term *redox potential* V_{rp} is equivalent to the term voltage used by electrical engineers, and it has the units volts. It was introduced in Sec. 9.2.5. It is used, typically by experimentalists, in discussing the voltage that develops across electrodes due to oxidation reduction chemical reactions. In the discussion of calculus of variations in Chapter 11, the idea of *generalized potential* was introduced. It is a parameter used to describe the evolution of an energy conversion process. Voltage and chemical potential can both be examples of generalized potentials. Generalized potential has units of joules over the units of the generalized path. The choice of the word generalized potential in calculus of variations follows reference [194, p.II-19]. Another related term is *potential energy*. Potential energy is a form of energy, and it is measured in joules. If we raise an object against gravity, we say that the object gains potential energy, and if we compress a spring, we say the spring gains spring potential energy.

The related words *capacitor, capacitance, theoretical capacity,* and *generalized capacity* are used in this text. A *capacitor* is one of the most common circuit components, and capacitors are discussed in Sec. 1.6.3. A *capacitor* is a device constructed from conductors separated by a dielectric layer. It is specified by a *capacitance* C, in farads, which is a measure of the ability of the device to store a built up charge, hence store energy. The permittivity ϵ describes the distributed capacitance, in $\frac{F}{m}$, of an insulating material. As discussed in Sec. 9.4.1, chemists use the related term *theoretical capacity* in a different way, as a measure of the charge stored in an battery or fuel cell. It is measured in coulombs or ampere hours. The adjective *theoretical* refers to the total amount of charge stored, not the charge that can be practically extracted. The idea of *generalized capacity* was introduced in Sec. 12.2 as the general ability to store energy. As with other concepts of calculus of variations, the units of generalized capacity depend on the choice of generalized path and generalized potential.

Conductivity describes ability of some particles to flow. *Electrical conductivity* σ describes the ability of charges to flow. It was introduced in Sec. 1.6.3 and discussed further in Sec. 8.6.2 and 9.2.1. *Thermal conductivity*

κ describes the ability of heat to flow, and it was introduced in Sec. 8.6.3. Both of these ideas were discussed in the context of thermoelectric devices because understanding these devices requires the understanding of the flow of both electrons and heat. Example values for electrical conductivity and thermal conductivity are given in Table 8.4.

While not identical, the terms *wave vector* and *wave function* are worth distinguishing. The term wave vector was introduced in Sec. 6.4 with the idea of reciprocal space. Functions such as energy or charge density can be described as varying with respect to position specified by the vector $\vec{r}$ in units of meters. These functions can also be described as varying with respect to spatial frequency specified by the *wave vector* $\vec{k}$ in units $\frac{1}{m}$. Wave function ψ was introduced in Section 11.7, and it is a fundamental idea of quantum mechanics. The *wave function* is a measure of the probability of finding an electron or other quantity in a particular state.

Appendix D: Specific Energies

Table 14.4 lists the specific energy of various materials and devices. These are representative values, not values for specific devices. Batteries by different manufacturers, for example, will have a range of specific energy values, and these values are often detailed in a datasheet. See the listed references for additional information on the assumptions made. Two types of values are listed for batteries. Both theoretical specific energy values for the chemical reactions and specific energy values for practical devices. The notation (th) indicates theoretical values while (pr) indicates practical values. NMH is an abbreviation for Nickel Metal Hydride batteries.

Material or Device	Specific Energy in $\frac{J}{g}$	Specific Energy in $\frac{W \cdot h}{kg}$	Ref.
Uranium	$6.77 \cdot 10^{10}$	$1.88 \cdot 10^{10}$	[3]
Hydrogen	$1.18 \cdot 10^5$	$3.28 \cdot 10^4$	[195]
Gasoline	$4.64 \cdot 10^4$	$1.29 \cdot 10^4$	[195]
Petroleum (crude)	$4.4 \cdot 10^4$	$1.2 \cdot 10^4$	[1]
Coal (high quality)	$3.4 \cdot 10^4$	$9.4 \cdot 10^3$	[1]
Methanol	$2.19 \cdot 10^4$	$6.08 \cdot 10^3$	[195]
Ammonia	$2.00 \cdot 10^4$	$5.56 \cdot 10^3$	[195]
Coal (low quality)	$1.6 \cdot 10^4$	$4.5 \cdot 10^3$	[1]
Sugar	$1.57 \cdot 10^4$	$4.36 \cdot 10^3$	[196]
Hydrogen oxygen fuel cell (th)	$1.32 \cdot 10^4$	$3.66 \cdot 10^3$	[128]
Lithium ion battery (th)	$1.61 \cdot 10^3$	448	[128]
Alkaline battery (th)	$1.29 \cdot 10^3$	358	[128]
Lead acid battery (th)	$9.1 \cdot 10^2$	252	[128]
NMH battery (th)	$8.6 \cdot 10^2$	240	[128]
Lithium ion battery (pr)	$7.2 \cdot 10^2$	200	[128]
Alkaline battery (pr)	$5.54 \cdot 10^2$	154	[128]
NMH battery (pr)	$3.6 \cdot 10^2$	100	[128]
Lead acid battery (pr)	$1.3 \cdot 10^2$	35	[128]
Rubber band	7.9	2.2	[195]

Table 14.4: Specific energy of various materials and devices. For batteries, (th) indicates theoretical values, and (pr) indicates practical values.

References

[1] W. H. Wiser, *Energy Resources: Occurence, Production, Conversion, Use.* New York: Springer, 2000.

[2] M. J. Moran, *Availability Analysis.* American Society of Mechanical, 1990.

[3] M. A. Kettani, *Direct Energy Conversion.* Addison Wesley, 1970.

[4] R. Decher, *Direct Energy Conversion.* Oxford, 1996.

[5] S. L. Soo, *Direct Energy Conversion.* Prentice Hall, 1968.

[6] A. R. von Hippel, *Dielectrics and Waves.* Wiley, 1954.

[7] A. Thompson and B. N. Taylor, *Guide for the Internation System of Units.* NIST, 2008. http://physics.nist.gov/Pubs/SP811/appenB9.html Date accessed 6-1-18.

[8] "Energy calculators." http://www.eia.gov/kids/energy.php?page=about_energy_conversion_calculator-basics. Date accessed 6-1-18.

[9] B. Streetman, *Solid State Electronic Devices*, 4 ed. New York: Prentice Hall, 1999.

[10] B. E. A. Saleh and M. C. Teich, *Fundamentals of Photonics.* NY: John Wiley and Sons, 1991.

[11] N. N. Rao, *Elements of Engineering Electromagnetics*, 6 ed. New Jersey: Prentice Hall, 2004.

[12] J. W. Hill and R. H. Petrucci, *General Chemistry.* Pearson, 1999.

[13] M. G. Mayer, "Rare-earth and transuranic elements," *Physical Review*, vol. 60, 1941.

[14] B. G. Wybourne, *Classical Groups for Physicists.* New York, NY: Wiley, 1974.

[15] E. C. Jordan and K. G. Balmain, *Electromagnetic Waves and Radiating Systems*. New York: Prentice Hall, 1968.

[16] "Ford focus review." http://www.edmunds.com/ford/focus/review.html. Date accessed 6-1-18.

[17] http://www.oreo.co.uk/products/original-oreo. Date accessed 6-1-18.

[18] T. K. Liu, "Gate dielectric scaling-integrating alternative high k gate dielectrics." http://www.cs.berkeley.edu/~tking/high.html. Date accessed 6-1-18.

[19] D. Wolfe, K. Flock, R. Therrien, R. Johnson, B. Rayner, L. Gunther, N. Brown, B. Claflin, and G. Lucovsky, "Remote plasma-enhanced-metal organic chemical vapor deposition of zirconium oxide/silicon oxide alloy thin films for advanced high-k gate dielectrics," in *Materials Research Society Symposium Proceedings*, vol. 567 (Warrendale, PA), Materials Research Society, 1999, pp. 343–348.

[20] Y. Nishioka, "Ultrathin tantalum pent-oxide films for ulsi gate dielectrics," in *Materials Research Society Symposium Proceedings*, vol. 567 (Warrendale, PA), Materials Research Society, 1999, pp. 361–370.

[21] http://www.digikey.com.

[22] P. Horowitz and W. Hill, *Art of Electronics*. Cambridge, 1989.

[23] B. Dobkin and J. Williams, *Analog Circuit Design*. Newnes, 2011.

[24] C. Klein, *Mineral Science*, 22 ed. New York: John Wiley, 2002.

[25] C. Kittel, *Introduction to Solid State Physics*, 7 ed. New York: John Wiley and Sons, 1996.

[26] N. W. Ashcroft and N. D. Mermin, *Solid State Physics*. Fort Worth: Saunders, 1976.

[27] R. J. Pressley, *CRC Handbook of Lasers*. Chemical Rubber Co, 1971.

[28] M. Tinkham, *Group Theory and Quantum Mechanics*. Dover, 1964.

[29] A. Schöenflies, *Kristallsysteme und Kristallstruktur*. 1891.

[30] "Crystal systems." https://en.wikipedia.org/wiki/Crystal_system. Date accessed 6-1-18.

[31] A. Yariv, *Quantum Electronics*, 3 ed. Wiley, 1989.

[32] "Mindat website." http://www.mindat.org. Date accessed 6-1-18.

[33] R. Goldman, *Ultrasonic Technology*. Reinhold, 1962.

[34] P. K. Panda, "Review: environmentally friendly lead-free piezoelectric materials," *Journal of Materials Science*, vol. 44, pp. 5049–5062, 2009.

[35] A. V. Carazo, "Micromechatronics, inc. website." http://www.mmech.com/transformers/dc-dc-piezo-converter. Date accessed 6-1-18.

[36] S. R. Anton and H. A. Sodano, "A review of power harvesting using piezoelectric materials 2003-2006," *Smart Materials and Structures*, vol. 16, pp. R1–R21, 2007.

[37] H. E. Soisson, *Instrumentation in Industry*. Wiley, 1975.

[38] W. G. Cady, "Electroelastic and pyroelectric phenomena," *International Critical Tables*, pp. 207–212, 1929.

[39] B. Ertug, "The overview of the electrical properties of barium titanate," *American Journal of Engineering Research*, vol. 2, no. 8, pp. 1–7, 2013.

[40] W. C. Röntgen, "Pyro und piezoelektrische untersuchungen," *Annalen der Physik*, vol. 350, pp. 737–800, 1914.

[41] https://scientech-inc.com/categories/laser-power-measurement.html. Date accessed 6-1-18.

[42] R. W. Boyd, *Nonlinear Optics*. Academic press, 2003.

[43] S. R. Hoh, "Conversion of thermal to electrical energy with ferroelectric materials," *Proceedings of the IEEE*, vol. 51, pp. 838–845, 1963.

[44] H. Fritzsche, "Toward understanding the photoinduced changes in chalcogenide glasses," *Semiconductors*, vol. 32, no. 8, pp. 850–856, 1998.

[45] V. K. Tikhomirov, "Photoinduced effects in undoped and rare-earth doped chalcogenide glasses, review," *Journal of Non-crystalline Solids*, vol. 256, pp. 328–336, 1999.

[46] G. Baym, *Lectures on Quantum Mechanics*. Addison Wesley, 1990.

[47] N. Cohen, "Fractal antenna applications in wireless telecommunication," *Electronics Industries Forum of New England*, pp. 43–49, 1997.

[48] D. H. Werner and S. Ganguly, "An overview of fractal antenna engineering research," *IEEE Antennas and Propagation Magazine*, vol. 45, pp. 38–58, Feb. 2003.

[49] E. A. Wolff, *Antenna Analysis*. Wiley, 1966.

[50] S. H. Ward, *ARRL Antenna Book*, 22 ed. ARRL, 2012.

[51] P. S. Carney and J. C. Schotland, "Near-field tomography," *Inside Out: Inverse Problems and Their Applications*, vol. 47, pp. 133–168, 2003.

[52] T. O'Laughlin, "The ELF is here," *Popular Communications*, pp. 10–13, April 1988.

[53] B. Villeneuve, "ELF Station Republic, MI." http://ss.sites.mtu.edu/mhugl/2015/10/10/elf-sta-republic-mi/, 2015.

[54] R. S. Carson, *Radio Communication Concepts, Analog*. Wiley, 1990.

[55] R. Wallace, "Antenna selection guide," *TI Application Note AN058*, 2010.

[56] R. Lewallen. http://www.eznec.com. Date accessed 6-1-18.

[57] "Hall effect sensing and application." http://sensing.honeywell.com/index.php?ci_id=47847. Date accessed 6-1-18.

[58] D. J. Epstein, "Permeability," *Dielectric Materials and Applications, ed. A. R. von Hippel*, pp. 122–134, 1954.

[59] B. Jeckelmann and B. Jeanneret, "The quantum Hall effect as an electrical resistance standard," *Reports on Progress in Physics*, vol. 64, pp. 1603–1655, 2001.

[60] J. O. Bockris and S. Srinivasan, *Fuel Cells Their Electrochemistry*. McGraw Hill, 1969.

[61] B. D. Iverson and S. Garimella, "Recent advances in microscale pumping technologies: a review and evaluation," *Birck and NCN Publications*, vol. 81, 2008. http://docs.lib.purdue.edu/nanopub/81.

[62] J. de Vicente, D. J. Klingenberg, and R. Hidalgo-Alvarez, "Magnetorheological fluids a review," *Soft Matter*, vol. 7, pp. 3701–3711, 2011.

[63] K. von Klitzing, G. Dorda, and M. Pepper, "New method for high-accuracy determination of the fine-structure constant based on quantized Hall resistance," *Physical Review Letters*, vol. 45, no. 6, pp. 494–498, 1980.

[64] D. C. Tsui, H. L. Störmer, and A. C. Gossard, "Two-dimensional magnetotransport in the extreme quantum limit," *Physical Review Letters*, vol. 48, no. 22, pp. 1559–1562, 1982.

[65] T. Chakaraborty and K. von Klitzing, "Taking stock of the quantum Hall effects: Thirty years on," *arXiv preprint:1102.5250*, 2011. https://arxiv.org/pdf/1102.5250.pdf.

[66] "A turning point for humanity: redefining the world's measurement system." https://www.nist.gov/si-redefinition/turning-point-humanity-redefining-worlds-measurement-system. Date accessed 6-25-18.

[67] E. Bellini, "Global installed PV capaicty exceeds 300GW, IEA PVPS," *PV magazine*, 2017.

[68] A. Thompson and B. N. Taylor, *Guide for the Use of the International System of Units*. 2008.

[69] D. M. Chapin, "How to make solar cells," *Radio Electronics*, pp. 89–94, Mar. 1960.

[70] A. Kramida, Y. Ralchenko, J. Reader, and NIST ASD Team. NIST Atomic Spectra Database (ver. 5.1), [Online]. Available: http://physics.nist.gov/asd [2014, May 14]. National Institute of Standards and Technology, Gaithersburg, MD.

[71] C. Downs and T. E. Vandervelde, "Progress in infrared photodetectors since 2000," *Sensors*, vol. 13, no. 4, 2013.

[72] S. Graham, "Remote sensing," 1999. https://earthobservatory.nasa.gov/Features/RemoteSensing/remote.php.

[73] M. G. Thomas, H. N. Post, and R. DeBlasio, "Photovoltaic systems an end-of-millennium review," *Progress in Photovoltaics Research and Applications*, vol. 7, pp. 1–19, 1999.

[74] http://www.nrel.gov/learning/re_photovoltaics.html. Date accessed 9-6-12.

[75] http://www.mit.edu/~6.777/matprops/ito.htm. Date accessed 6-1-18.

[76] J. J. Wysocki and P. Rappaport, "Effect of temperature on photovoltaic solar energy conversion," *Journal of Applied Physics*, vol. 31, p. 571, Mar. 1960.

[77] S. Kurtz, D. Levi, and K. Emery. https://www.nrel.gov/pv/assets/images/efficiency-chart.png, 2017.

[78] L. E. Chaar, L. A. Lamont, and N. E. Zein, "Review of photovoltaic technologies," *Renewable and Sustainable Energy Reviews*, vol. 15, pp. 2165–2175, 2011.

[79] B. Kippelen and J. L. Bredas, "Organic photovoltaics," *Energy and Environmental Science*, vol. 2, 2009.

[80] "The 2009 Nobel Prize in Physics." https://www.nobelprize.org/nobel_prizes/physics/laureates/2009/press.html, 2009.

[81] http://hyperphysics.phy-astr.gsu.edu/hbase/vision/rodcone.html. Date accessed 6-1-18.

[82] "Hamamatsu infrared detectors, selection guide." https://www.hamamatsu.com/resources/pdf/ssd/infrared_kird0001e.pdf, 2018. Date accessed 6-15-18.

[83] S. M. Sze, *Physics of Semiconductor Devices*. Wiley, 1969.

[84] W. T. Silfvast, *Laser Fundamentals*. Cambridge University press, 1996.

[85] D. Kule. https://en.wikipedia.org/wiki/File:Black_body.svg. Date accessed 6-10-18.

[86] J. T. Verdeyen, *Laser Electronics*. Prentice Hall, 1995.

[87] J. D. Cobine, *Gaseous Conductors*. Dover, 1958.

[88] M. F. Gendre, "Two centuries of electric light source innovations." http://www.einlightred.tue.nl/lightsources/history/light_history.pdf. Date accessed 6-1-18.

[89] J. B. Calvert. http://www.physics.csbsju.edu/370/jcalvert/dischg.htm.html, 2005.

[90] S. Nakamura, "GaN-based blue green semiconductor lasers," *IEEE Journal of Selected Topics in Quantum Electronics*, vol. 3, pp. 435–442, 1997.

[91] C. P. B. Geffroy, P. Roy, "Organic light-emitting diode technology: Materials, devices, and display technologies," *Polymer International*, vol. 55, 2006.

[92] M. G. Bernard and G. Duraffourg, "Laser conditions in semiconductors," *Physica Status Solidi*, vol. 1, pp. 699–703, 1961.

[93] R. N. Hall, G. E. Fenner, J. D. Kingsley, T. J. Soltys, and R. O. Carlson, "Coherent light emission form GaAs junctions," *Physical Review Letters*, vol. 9, no. 9, pp. 366–368, 1962.

[94] H. Nelson, "Epitaxial growth from the liquid state and its application to the fabrication of tunnel and laser diodes," *RCA Review*, vol. 24, pp. 603–615, 1963.

[95] A. Y. Cho and J. R. Arthur, "Molecular beam epitaxy," *Progress in Solid-State Chemistry*, vol. 10, pp. 157–191, 1975.

[96] R. D. Dupuis, P. D. Dapkus, N. Holonyak, E. A. Rezek, and R. Chin, "Room-temperature laser operation of quantum well $Ga_{1-x}Al_xAs$-GaAs laser diodes grown by metalorganic chemical vapor deposition," *Applied Physics Letters*, vol. 32, pp. 295–297, 1978.

[97] M. N. Polyanskiy, "Refractive index database." http://refractiveindex.info/?shelf=main&book=GaAs&page=Skauli. Date accessed 6-1-18.

[98] H. Kressel and H. Nelson, "Close-confinement gallium arsenide pn junction lasers with reduced optical loss at room temperature," *RCA Review*, vol. 30, pp. 106–113, 1969.

[99] V. A. Donchenko, Y. E. Geints, V. A. Kharenkov, and A. A. Zemlyanov, "Nanostructured metal aggregate-assisted lasing in rhodamine 6G solutions," *Optics and Photonics Journal*, vol. 3, no. 8, 2013. http://file.scirp.org/Html/2-1190302_40925.htm.

[100] A. Ishibashi, "II-VI blue-green laser diodes," *IEEE Journal of Selected Topics in Quantum Electronics*, vol. 1, pp. 741–748, 1995.

[101] J. L. Jewell, J. P. Harbison, A. Scherer, Y. H. Lee, and L. T. Florez, "Vertical-cavity surface-emitting lasers: design, growth, fabrication, and characterization," *IEEE Journal of Quantum Electronics*, vol. 27, pp. 1332–1346, 1991.

[102] M. J. Moran, H. N. Shapiro, D. D. Boettner, and M. B. Bailey, *Fundamentals of Engineering Thermodynamics*. Wiley, 2014.

[103] B. R. Munson, A. P. Rothmayer, T. H. Okiishi, and W. W. Huebsch, *Fundamentals of Fluid Mechanics*. Wiley, 2012.

[104] "Azo Materials website, diamond properties, applications." http://www.azom.com/properties.aspx?ArticleID=262, 2001.

[105] "Azo Materials website, stainless steel grade 304 (uns s30400)." http://www.azom.com/properties.aspx?ArticleID=965, 2001.

[106] "Azo Materials website, graphite." http://www.azom.com/properties.aspx?ArticleID=1630, 2002.

[107] "Azo Materials website, silicone rubber." http://www.azom.com/properties.aspx?ArticleID=920, 2001.

[108] J. M. Smith, H. C. V. Ness, and M. Abbott, *Introduction to Chemical Engineering Thermodynamics*. McGraw Hill, 2000.

[109] M. J. Moran and H. N. Shapiro, *Fundamentals of Engineering Thermodynamics*. Wiley, 2004.

[110] P. H. Egli, *Thermoelectricity*. New York: John Wiley, 1958.

[111] S. G. Carr, *Essential Linear Circuit Analysis*. 2019. Preprint.

[112] G. S. Nolas, J. Sharp, and H. J. Goldsmid, *Thermoelectrics, Basic Principles and New Materials Developments*. Germany: Springer, 2001.

[113] N. F. Mott and E. A. Davis, *Electronic Processes in Non-Crystalline Materials*, 2 ed. Oxford: Clarendon Press, 1979.

[114] W. M. Haynes, *CRC Handbook of Chemistry and Physics*, 93 ed. CRC press, 2013.

[115] R. Venkatasubramanian, E. Silvola, T. Colpitts, and B. O'Quinn, "Thin-film thermoelectric devices with high room-temperature figure of merit," *Nature*, vol. 43, pp. 597–602, Oct. 2001.

[116] H. Beyer, J. Numus, H. Bottner, A. Lambrecht, T. Roch, and G. Bauer, "PbTe based superlattice structures with high thermoelectric efficiency," *Applied Physics Letters*, vol. 80, pp. 1215–1217, Feb. 2002.

[117] C. A. Gould, N. Y. A. Shammas, S. Grainger, and I. Taylor, "A comprehensive review of thermoelectric technology, micro-electrical and power generation properties," *Proceedings of the 26th International Conference on Microelectronics*, pp. 978–982, May 2008.

[118] S. B. Riffat and X. Ma, "Thermoelectrics a review of present and potential applications," *Applied Thermal Engineering*, vol. 23, pp. 913–935, 2003.

[119] G. Brumfiel, "Curiosity's dirty little secret," *Slate*, 2012. http://www.slate.com/articles/health_and_science/science/2012/08/mars_rover_curiosity_its_plutonium_power_comes_courtesy_of_soviet_nukes_.html.

[120] "Radioisotope power systems, power and thermal systems." https://rps.nasa.gov/power-and-thermal-systems/power-systems/current/. Date accessed 6-10-18.

[121] B. C. Sales, B. C. Chakoumakos, and D. Mandrus, "Thermoelectric properties of thallium-filled skutterudites," *Physical Review B*, vol. 61, pp. 2475–2481, Jan. 2000.

[122] A. Watcharapsorn, R. S. Feigelson, T. Caillat, A. Borshchevsky, G. Snyder, and J.-P. Fleurial, "Preparation and thermoelectric properties of $CeFe_4As_{12}$," *Journal of Applied Physics*, vol. 91, no. 3, pp. 1344–1348, 2002.

[123] "Apple product information sheet." https://images.apple.com/legal/more-resources/docs/apple-product-information-sheet.pdf, 2018. Date accessed 6-10-18.

[124] https://www.apple.com/iphone-x/specs/. Date accessed 6-10-18.

[125] D. Wright, "Request for issuance of certificate of conformity." https://iaspub.epa.gov/otaqpub/display_file.jsp?docid=39828&flag=1, 2016. The curb mass of the 2017 Tesla Model S AWD 90D is 2172 kg, and the weight of its battery is 580 kg.

[126] E. F. Kaye, "Chairman's hoverboard press statement." https://www.cpsc.gov/about-cpsc/chairman/elliot-f-kaye/statements/chairmans-hoverboard-press-statement/, 2016.

[127] J. McCurry, "Samsung blames two separate battery faults for Galaxy Note 7 fires," *The Guardian*, Jan. 2017.

[128] T. Reddy, *Linden's Handbook of Batteries*, 4 ed. McGraw Hill, 2010.

[129] https://www.tesla.com/gigafactory. Date accessed 6-10-18.

[130] R. S. Mulliken, "A new electroaffinity scale, together with data on valence states and on valence ionization potentials and electron affinities," *Journal of Chemical Physics*, vol. 2, pp. 782–793, Nov. 1934.

[131] R. G. Pearson, "Absolute electronegativity and hardness: application to inorganic chemistry," *Inorganic Chemistry*, vol. 27, pp. 734–740, 1988.

[132] H. O. Pritchard and F. H. Sumner, "The application of electronic digital computers to molecular orbital problems II. A new approximation for hetero-atom systems," *Proceedings of the Royal Society of London Series A Mathematical and Physical Sciences*, vol. 235, pp. 136–143, Apr. 1956.

[133] R. P. Iczkowski and J. L. Margrave, "Electronegativity," *Journal of the American Chemical Society*, vol. 83, pp. 3547–3553, Sept. 1961.

[134] L. Pauling, "The nature of the chemical bond IV. The energy of single bonds and the relative electronegativity of atoms," *Journal of the American Chemical Society*, vol. 54, pp. 3570–3583, 1932.

[135] W. Gordy, "A new method of determining electronegativity from other atomic properties," *Physical Review*, vol. 69, pp. 604–607, June 1946.

[136] R. G. Parr and W. Yang, *Density Functional Theory of Atoms and Molecules*. New York: Oxford University Press, 1989.

REFERENCES

[137] S. G. Bratsch, "Standard electrode potentials and temperature coefficients in water at 298.15 K," *Journal of Physical Chemical Reference Data*, vol. 18, pp. 1–21, 1989. http://www.nist.gov/srd/upload/jpcrd355.pdf.

[138] R. Shapiro, "Oxidation-reduction potential," *Water and Sewage Works*, vol. 101, pp. 185–188, Apr. 1954.

[139] C. E. Wallace. https://www.youtube.com/watch?v=RAFcZo8dTcU. Date accessed 6-1-18.

[140] C. K. Morehouse, R. Glicksman, and G. S. Lozier, "Batteries," *Proceedings of the IRE*, pp. 1462–1483, 1958.

[141] https://www.energy.gov/eere/fuelcells/fuel-cell-animation. Date accessed 6-1-18.

[142] "Overview of battery technologies for MEMS applications," *MEMS Journal*, 2011. http://www.memsjournal.com/2011/02/overview-of-battery-technologies-for-mems-applications-.html.

[143] http://www.frontedgetechnology.com/gen.htm. Date accessed 6-1-18.

[144] "Polaroid p100 polapulse/powerburst battery." http://users.rcn.com/fcohen/P100.htm. Date accessed 6-1-18.

[145] http://en.wikipedia.org/wiki/Instant_film. Date accessed 6-1-18.

[146] "Energizer nickel metal hydride handbook and application manual," 2001.

[147] "Energizer nickel metal hydride handbook and application manual." http://data.energizer.com/PDFs/nickelmetalhydride_appman.pdf, 2010.

[148] P. J. Dalton, "International space station nickel hydrogen batteries approach 3 year on orbit mark." https://ntrs.nasa.gov/archive/nasa/casi.ntrs.nasa.gov/20050215412.pdf. Date accessed 6-10-18.

[149] http://www.eaglepicher.com/ips-2/medical-power-technology. Date accessed 6-1-18.

[150] J. P. Owejan, T. A. Trabold, D. L. Jacobson, M. Arif, and S. G. Kandlikar, "Effects of flow field diffusion layer properties on water accumulation in a pem fuel cell," *International Journal of Hydrogen Energy*, vol. 32, pp. 4489–4502, 2007.

[151] S. O. Farwell, D. R. Gage, and R. A. Kagel, "Current status of prominent selective gas chromatographic detectors: a critical assessment," *Journal of Chromatographic Science*, vol. 19, 1981.

[152] "Background on smoke detectors." https://www.nrc.gov/reading-rm/doc-collections/fact-sheets/smoke-detectors.html. Date accessed 6-1-18.

[153] http://www.landauer.com/Industry/Products/Dosimeters/Dosimeters.aspx. Date accessed 6-1-18.

[154] http://www.perkinelmer.co.uk/product/tri-carb-4910tr-110-v-a491000. Date accessed 6-1-18.

[155] C. D. F. Massaad, T. M. Lejeune, "The up and down bobbing of human walking," *Journal of Physiology*, vol. 582, pp. 789–799, 2007.

[156] A. J. Bur, "Measurements of the dynamic piezoelectric properties of bone as a function of temperature and humidity," *Journal of Biomechanics*, vol. 9, pp. 495–507, 1976.

[157] T. J. Anastasio, *Tutorial on Neural System Modeling*. Sinauer, 2009.

[158] "The principles of nerve cell communication," *Alcohol Health and Research World*, vol. 21, pp. 107–108, 1997. https://pubs.niaaa.nih.gov/publications/arh21-2/107.pdf.

[159] H. A. Stone, A. D. Stroock, and A. Ajdari, "Engineering flows in small devices: Microfluidics towards a lab-on-a-chip," *Annual Review of Fluid Mechanics*, vol. 36, pp. 381–411, 2004.

[160] P. Gravesen, J. Branebjerg, and O. S. Jensen, "Microfluidics - a review," *Journal of Micromechanical Microengineering*, vol. 3, pp. 168–182, 2004.

[161] P. K. Wong, T. Wang, J. H. Deval, and C. Ho, "Electrokinetics in micro devices for biotechnology applications," *IEEE ASME Transactions on Mechatronics*, vol. 9, pp. 366–377, June 2004.

[162] A. Tipler, *Physics*, 3 ed., Vol. 1. Worth Publishing, 1991.

[163] B. V. Brunt, *The Calculus of Variations*. Springer, 2002.

[164] P. J. Olver, *Applications of Lie Groups To Differential Equations*. New York: Springer, 1986.

[165] E. Noether, "Invariant variation problems," *Transport Theory and Statistical Physics*, vol. 1, no. 3, pp. 183–207, 1971. English Translation.

[166] E. Noether, "Invariant variation problems," *Nachrichten von der Gesellschaft der Wissenschaften zu Gottingen, Mathematisch-Physikalische Klasse*, vol. 235, 1918.

[167] V. I. Arnol'd, *Mathematical Methods of Classical Mechanics*, 2 ed. Springer, 2010.

[168] C. L. Nachtigal and M. D. Martin, *Instrumentation and Control*. Wiley, 1990.

[169] http://hyperphysics.phy-astr.gsu.edu/hbase/electric/watcir.html. Date accessed 6-1-18.

[170] P. Salamon, B. Andresen, and R. S. Berry, "Thermodynamics in finite time. II potentials for finite-time processes," *Physical Review A*, vol. 15, no. 5, 1977.

[171] P. Salamon, B. Andresen, and R. S. Berry, "Minimum entropy production and the optimization of heat engines," *Physical Review A*, vol. 21, no. 6, 1980.

[172] R. G. Parr, R. A. Donnelly, M. Levy, and W. E. Palke, "Electronegativity: the density functional viewpoint," *Journal of Chemical Physics*, vol. 68, pp. 3801–3808, Apr. 1978.

[173] Thomas, "The calculation of atomic fields," *Proceedings of the Cambridge Philosophical Society*, vol. 23, pp. 542–548, 1927.

[174] E. Fermi, *Collected Papers*, Vol. 1. Chicago: University of Chicago Press, 1962. See the article "Über die Anwendung der statistischen Methode auf die probleme des Atombaues, " 1928, the article "Un metodo statistico per la determinazione di alcune proprieta dell'atome," 1927, and the article "Zur Quantelung des idealen einatomigen gases," 1926.

[175] G. K. Woodgate, *Elementary Atomic Structure, 2nd ed.* Oxford, 1980.

[176] S. J. A. Malham, *An Introduction to Lagrangian Mechanics, Lecture Notes.* 2016. http://www.macs.hw.ac.uk/~simonm/mechanics.pdf.

[177] E. Fermi, "Un metodo statistico per la determinazione di alcune prioreta dell'atome," *Rendicondi Accademia Nazionale de Lincei*, vol. 6, no. 32, pp. 602–607, 1927.

[178] http://www.abinit.org/. Date accessed 6-1-18.

[179] http://departments.icmab.es/leem/siesta/. Date accessed 6-1-18.

[180] H. Krutter, "Numerical integration of the Thomas-Fermi equation from zero to infinity," *Journal of Computational Physics*, vol. 47, pp. 308–312, 1982.

[181] N. H. Ibragimov, *Lie Group Analysis of Differential Equations*, Vol. 3. CRC press, 1996.

[182] L. Gagnon and P. Winternitz, "Lie symmetries of a generalized nonlinear Schrödinger equation: I. the symmetry group and its subgroups," *Journal of Physics A*, vol. 21, pp. 1493–1511, 1988.

[183] E. G. Kalnins and W. Miller, "Lie theory and separation of variables 5," *Journal of Mathematical Physics*, vol. 15, pp. 1728–1738, 1974.

[184] V. I. Fushchich and A. G. Nikitin, "New and old symmetries of the Maxwell and Dirac equations," *Soviet Journal of Particles and Nuclei*, vol. 14, pp. 1–22, Jan. 1983.

[185] W. I. Fushchich and A. G. Nikitin, *Symmetries of Maxwell's Equations.* Boston, MA: D. Reidel Publishing, 1987.

[186] A. R. Chowdhury and P. K. Chanda, "On the Lie symmetry approach to small's equation of nonlinear optics," *Journal of Physics A*, vol. 18, pp. 117–121, 1985.

[187] J. J. Sakurai, *Modern Quantum Mechanics.* Massachusetts: Addison-Wesley Publishing Company, 1994.

[188] M. Lutzky, "Dynamical symmetries and conserved quantities," *Journal of Physics A*, vol. 12, no. 7, pp. 973–981, 1979.

[189] L. V. Ovsiannikov, *Group Analysis of Differential Equations*. New York, NY: Academic Press, 1982.

[190] P. G. L. Leach, R. Maartens, and S. D. Maharaj, "Self similar solutions of the generalized Emden Fowler equation," *International Journal of Nonlinear Mechanics*, vol. 27, pp. 575–582, 1992.

[191] R. L. Anderson and S. M. Davison, "A generalization of Lie's counting theorem for second order ordinary differential equations," *Journal of Mathematical Analysis and Applications*, vol. 48, pp. 301–315, 1974.

[192] E. A. Desloge and R. I. Karch, "Noether's theorem in classical mechanics," *American Journal of Physics*, vol. 45, pp. 336–339, Apr. 1977.

[193] http://www.nist.gov/pml/wmd/metric/prefixes.cfm. Date accessed 6-1-18.

[194] R. P. Feynman, *Feynman Lectures on Physics*. MA: Addison Wesley Publishing Company, 1963.

[195] F. R. Whitt and D. G. Wilson, *Bicycling Science*. MIT, 1974.

[196] R. S. Shallenberger, *Sugar Chemistry*. Connecticut: Avi Publishing Company, 1975.

Index

Absorption, 139
Action, 247
Anisotropic, 28
Anode, 201
Antenna, 67
Arc discharge, 150
Azimuth plot, 79

Battery, 201, 286
Battery model, 219
Bernoulli's equation, 241, 284
Bioluminescence, 141
Blackbody radiation, 141
Boltzmann constant, 118
Boltzmann statistics, 146
Bravais lattice, 34
Brillouin zone, 120
Bulk modulus, 176

Calculus of variations, 247
Capacitance, 271, 352
Capacitor, 23
Capacity, 217
Carnot efficiency, 188, 190
Cathode, 201
Chalcogenide, 63
Charge density, 217
Chemical hardness, 207
Chemical Potential, 118
Chemical potential, 205, 301, 352
Chemiluminescence, 140
Commutator, 321
Compressibility, 176
Conduction band, 113
Conductor, 13, 114
Conservation of angular momentum, 281
Conservation of momentum, 281
Constitutive relationship, 177, 256, 271, 279

Continuous symmetry, 313
Coulomb's law, 16, 297
Crystal basis, 34
Crystal momentum, 120, 300
Crystal point group, 35, 39
Crystal structure, 34

Density functional theory, 308
Detectivity, 134
Dielectric, 13, 114
Dielectrophoresis, 243
Dipole, 68
Directivity, 78
Discrete symmetry, 313
Displacement flux density, 273
Doping, 105

Efficiency, 9
Electric field intensity, 273
Electrical conductivity, 18, 183, 203, 352
Electro-optic effect, 56, 274
Electrohydrodynamic, 241
Electrokinetic, 241
Electroluminescence, 141
Electrolyte, 201
Electrolytic capacitor, 31
Electromagnetic polarization, 82, 140, 351
Electron affinity, 204
Electron configuration, 13
Electron-hole pair, 104, 110
Electronegativity, 205
Electroosmosis, 243
Electrophoresis, 242
Electrostriction, 61
Electrowetting, 243
Elevation plot, 79
Emden-Fowler equation, 323
Energy band, 113

INDEX

Energy conservation, 9
Energy density, 217
Energy gap, 113, 203
Entropy, 175, 285
Epitaxial layer, 152
Equation of motion, 247
Erbium doped fiber amplifier, 165
Euler-Lagrange equation, 247
Exciton, 63, 104
Extensive property, 174

Faraday constant, 218
Fermat's principle, 265
Fermi Dirac distribution, 118
Fermi energy, 117, 203, 301, 352
Ferroelectricity, 42, 55, 63
Ferromagnetism, 42
Figure of merit, 186
First law of thermodynamics, 179
Fluorescent lamp, 150
Flywheel, 276
Fuel cell, 201, 229, 286
Functional, 247, 308

Gas discharge, 148
Gauss's law, 273
Geiger counter, 238
Generalized capacity, 247, 271, 352
Generalized momentum, 246
Generalized potential, 246, 352
Glow discharge, 150
Gradient, 17, 293
Group theory, 320

Hall effect, 91, 274
Hall resistance, 95
Hamiltonian, 246, 298
Heat engine, 188
Heisenberg uncertainty principle, 15, 291
Hole, 92, 104
Hooke's law, 279

Hysteresis, 42

Ideal gas law, 178
Impedance, 77
Incandescent lamp, 147
Index of refraction, 26
Infinitesimal generator, 315
Insulator, 13, 114
Intensive property, 174
Invariants, 330
Inversion symmetry, 38, 351
Ionization chamber, 238
Ionization energy, 204
Isotropic antenna, 78

Joule-Thomson coefficient, 177

KCL, 271
Kerr coefficient, 58
Kerr effect, 57
Kinetic energy, 297
Kramers Kronig relationship, 20
KVL, 271

Lagrangian, 246, 298
Laser, 139, 152
Lattice, 33
Lie analysis, 311
Light emitting diode, 150
Line equation, 326, 333
Liquid crystals, 61
Lorentz force equation, 91, 242

Magnetic field intensity, 273
Magnetic flux density, 15, 273
Magnetohydrodynamics, 96, 274
Mars rover, 192
Material polarization, 24, 351
Maxwell's equations, 16
Microfluidic device, 241
Mobility, 184
Moment of inertia, 279
Mulliken electronegativity, 204

Nernst equation, 220
Neuron, 239
Noether's theorem, 248, 330

Operator, 262
Orbital, 14
Oxidation, 201

Peltier coefficient, 182
Peltier effect, 182
Permeability, 18
Permittivity, 19
pH, 208
Phase change, 63
Phonon, 102
Photoconductivity, 133
Photodetector, 132
Photoelectric emission, 133
Photoluminescence, 140
Photomultiplier tube, 133
Photon, 101
Photovoltaic device, 101
Photovoltaic effect, 101
Piezoelectric strain constant, 32
Piezoelectricity, 31, 274
Pirani hot wire gauge, 240
Plasma, 96, 148
Pn junction, 122, 150
Pockels coefficient, 58
Pockels effect, 57
Poling, 42
Population inversion, 147, 351
Potential, 351
Potential energy, 252, 352
Potentiometer, 240
Primary battery, 224
Primitive lattice vectors, 33
Principle of least action, 247
Prolongation, 322
Pyroelectricity, 53, 274

Quantum Hall effect, 97

Quantum mechanics, 262
Quantum number, 13, 68, 109
Quartz, 43

Radiation pattern plot, 70, 79
Reciprocal lattice, 120
Reciprocal space, 296, 301
Reciprocity, 69
Redox potential, 207, 352
Reduction, 201
Resistance temperature detector, 240
Resistivity, 18

Schrödinger's equation, 262
Scintillation counter, 238
Second harmonic generation, 61
Second law of thermodynamics, 285
Secondary battery, 224
Seebeck coefficient, 181
Seebeck effect, 181
Semiconductor, 13, 114
Semimetal, 115, 187
Shell, 13
Simple compressible systems, 178
Snell's law, 266
Solar cell, 101
Sonoluminescence, 141
Specific capacity, 217
Specific energy, 217
Specific heat, 177, 186
Spontaneous emission, 139
Stimulated emission, 139
Strain gauge, 240
Streaming potential, 242
Superposition, 69

Theoretical capacity, 352
Theoretical cell voltage, 216
Thermal conductivity, 184, 352
Thermionic device, 237
Thermocouple, 192
Thermodynamic cycle, 188

INDEX

Thomas Fermi analysis, 292
Thomas Fermi equation, 307, 323
Thomson coefficient, 183
Torsion spring, 264
Townsend discharge, 150
Transmission line, 74

Valence, 14
Valence band, 113
Volume expansivity, 177

Wave equation, 255, 260, 316
Wave function, 68, 262, 296, 353
Wave number, 102
Wave vector, 121, 299, 353
Work function, 205

Young's elastic modulus, 32

About the Book

Direct Energy Conversion discusses both the physics behind energy conversion processes and a wide variety of energy conversion devices. A direct energy conversion process converts one form of energy to another through a single process. The first half of this book surveys multiple devices that convert to or from electricity including piezoelectric devices, antennas, solar cells, light emitting diodes, lasers, thermoelectric devices, and batteries. In these chapters, physical effects are discussed, terminology used by engineers in the discipline is introduced, and insights into material selection is studied. The second part of this book puts concepts of energy conversion in a more abstract framework. These chapters introduce the idea of calculus of variations and illuminate relationships between energy conversion processes.

This peer-reviewed book is used for a junior level electrical engineering class at Trine University. However, it is intended not just for electrical engineers. Direct energy conversion is a fascinating topic because it does not fit neatly into a single discipline. This book also should be of interest to physicists, chemists, mechanical engineers, and other researchers interested in an introduction to the energy conversion devices studied by scientists and engineers in other disciplines.

About the Author

Andrea M. Mitofsky received her B.S., M.S., and Ph.D. degrees in Electrical Engineering from the University of Illinois at Urbana-Champaign. In 2008, she graduated with her Ph.D. degree and began teaching at Trine University in Angola, Indiana. She is currently an Associate Professor in the Electrical and Computer Engineering department at Trine University. She can be contacted at mitofskya@gmail.com.

www.ingramcontent.com/pod-product-compliance
Lightning Source LLC
Chambersburg PA
CBHW062317220526
45469CB00008B/2546